Ecological Studies

Analysis and Synthesis

Edited by

W.D. Billings, Durham (USA) F. Golley, Athens (USA)
O.L. Lange, Würzburg (FRG) J.S. Olson, Oak Ridge (USA)
K. Remmert, Marburg (FRG)

Volume 37

Hans Jenny

The Soil Resource

Origin and Behavior

With 191 Figures

Springer-Verlag
New York Heidelberg Berlin

Hans Jenny
Professor Emeritus
Department of Plant and Soil Biology
College of Natural Resources
University of California
Berkeley, California 94720
USA

Library of Congress Cataloging in Publication Data
Jenny, Hans, 1899—
    The soil resource.
    (Ecological studies, v.37)
    Bibliography: p.
    Includes index.
    1. Soil formation. 2. Soil ecology. I. Title.
II. Series.
S592.2.J46   631.4   80-11785

9 8 7 6 5 4 3 (Third printing, 1986)
ISBN 0-387-90453-X Springer-Verlag New York Heidelberg Berlin
ISBN 3-540-90453-X Springer-Verlag Berlin Heidelberg New York

This book is dedicated to my wife
Jean Jenny
in appreciation of her persistent, successful efforts—with important help from many
supporters—to preserve for research and education the natural landscape areas:

Apricum Hill laterite crust near Ione, California

Mt. Shasta mudflow research area

Pygmy forest ecological staircase (Jug Handle Reserve)

and others still in the slow process of acquisition.

Foreword: Hans Jenny and Fertile Soil

Hans Jenny has an unusual capacity for originating ideas, applying them and communicating them to others. In any field of knowledge, truly original ideas are rare, and particularly those which are pursued to the point of stimulating an entire generation of scholars in the field. Hans Jenny has done precisely that for soil science. Those who are or have been privileged to be his colleagues, students and friends are deeply grateful. (1)

PAUL R. DAY

The soil resource is supporting a growing world. Hans Jenny's fertile imagination has supported the science of origin of soils. Now it supports a broadening and deepening of the foundations of ecology.

Vignettes from Jenny's life story tell something of how these twin contributions came to pass. They illustrate how ideas and traditions can crossbreed with one another for hybrid vigor in the history of science. I hope they show a bit deeper insight into creativity than we can usually learn from leading architects of science. Historians of science can follow details in an *Oral History* of Jenny's life (7). The Editors of *Ecological Studies* and Springer-Verlag feel especially privileged to make available not only the present book as a major historic contribution to pedology and ecology, but also the following personal sidelights about how that contribution was created over a full lifetime of observations, analysis, and synthesis.

The second half of this book is a substantial extension of the natural philosophy that lit up the world in the classic *Factors of Soil Formation* in 1941 (5), a popular textbook, long out of print and pageworn by heavy use. Jenny's *factor–function paradigm*, or framework for organizing thought, is as lively as ever today. It seems likely to me to enjoy a still broader renaissance for the remainder of the twentieth century and for the twenty-first, as the world's people come to realize—and try harder to understand and reverse—our wasting of the rich and varied resources of soils and ecosystems now exploited ever more intensively.

The first half of the present book reaches far back to the foundations of many sciences. It shows concisely, in as self-contained a manner as possible, how these contribute to understanding the many processes that maintain our soil resource as an operating system—part of the larger ecosystems of various sizes.

The fundamental concept of balance or imbalance of forces, of energy budgets, of rates of mass income and loss, here emerge from physics and chemistry to give a view of changing soil (or changing ecosystems) in terms of quantitative rates. Where the change is not zero (the ideal balanced system, rarely found in Nature), the *rate* of change is simply described by the rate of income and rate of loss. Our home's energy budget, our firm's inventory, our nation's debt, and humanity's numbers all have accounts that change at rates that are equal to the inputs minus the outputs. Jenny's "system view" of the soil was carried into the fertile fields of Midwestern American prairies from the laboratories of Switzerland in the late 1920s. Jenny's rate equations provided the other paradigm or world view that, I recall, brought us to the threshold of systems ecology as it later evolved in the second half of the twentieth century.

As if world renown in the specialties of pedology and soil chemistry were not enough for one lifetime, excerpts below remind us that Hans Jenny has also been a perceptive outdoor field ecologist since his early Alpine expeditions with Braun-Blanquet in the mid 1920s. Jenny's ecosystem studies in the pygmy forest, a further classic example of a soil−plant system "run down" over hundreds of thousands of years since its origin, continue to occupy some of the vigorous retirement time near his farm in Mendocino County. But each specific, quantitative case study, and each research area conserved (with additional hard work) for further study by future generations, fits into Jenny's coherent world view. It is that view, and its legacies of discovery and of tangible landscape preserves, which we are privileged to share with their originator in this volume. In his eighties, Hans Jenny has also remained active in relating some of his favorite research areas to our global ecosystem problem of carbon and nitrogen cycling.

Jenny's notes for his *Oral History* (7) and the personal knowledge I have enjoyed with Hans Jenny since my own semester pilgrimage to Berkeley in 1950, may therefore help many kinds of readers to find more of Jenny's own life story between the lines in the remainder of this book. Little Hans (born 1899) grew up mostly in Basel, with a student's sparkling view of the world from the extraordinarily powerful technical institute in Zurich, and later with a series of very productive positions in his adopted America. Typical of Hans's always fresh observations was a teen-age incident while on farming "duty" in the Swiss valleys in World War I. The boy from town was awed by the distant view of the "snow mountain"—but his host family had lived with it for years without even noticing. Hans still points out fresh views of a world we have lived in without fully seeing. He shares with us his special scientific perception, as well as his admiration for each view. The beauty lies not only in landscapes he describes here (reviewed from an artistic standpoint in his famous Vatican lecture on pedology and art) (6). The Platonic beauty of Jenny's ideas emerges from their elegant ways of connecting so many fragments of knowledge that might otherwise become lost in their various pigeonholes—if not forgotten altogether.

Each of Jenny's life sojourns brought new insights, which then hybridized with one another in a kind of cross-fertilizing of ideas. This hybrid vigor of definitions and hypotheses was then subject to severe natural selection, a "survival of the fittest" theory, to make sense of the world. That selection involved the balance in a system of ideas: the fresh mutations of the intellectual world. Whence came this originality? the cross-breeding? the survival of two paradigms?

Georg Wiegner, Jenny's mentor at Zurich (7), was the Professor of Agricultural Chemistry, a physical chemist who motivated and guided Jenny's early work on ion exchange. (That continuing interest is reflected later in the present book.) Jenny's famous syllabus on colloid chemistry (4) not only reflects this individual interest, but the power of the giants of physical science whom Hans could see and hear around Zurich. Yet Wiegner opened Hans's eyes to the fresh view of Hilgard from Berkeley and Dokuchaev from Russia, still unconventional for European agriculture of the 1920s. With more field experience than his mentor had time for, as in his vacation excursions collecting samples for analysis, Hans soon gained the precocious power to make connections that none of his predecessors was equipped for. The early "hybrid" book on soil acidification and successions (3) promptly gained worldwide attention in the Fuller/Conard English edition of Braun-Blanquet's *Plant Sociology* (2).

When Tüxen in the 1950s praised the little book as a classic, I replied laughingly that it was a Jugendsünde (an illegitimate product of youth). In one of the sections I drew a speculative curve relating humus to climate, relying on some of Hilgard's analyses. It was a premature embryo, but nevertheless the concept of a climofunction was clearly stated. . . . . I never dreamed that a decade later I would sit in his chair in Berkeley, California, and teach his course on soil formation.

When I expressed a desire to study plant nutrition with Professor Hoagland in California I was politely told to forget it and select a place nearer the Atlantic Coast [by a visitor named A. R. Mann when he probed Jenny's interest in an International Education Board Fellowship from the Rockefeller Foundation] . . .

Not long after Mann's proposal, Professor S. A. Waksman, the soil microbiologist from the University of New Jersey (Rutgers) visited our group and urged me to come to New Brunswick and work in his lab on colloid chemistry of humus. The prospect of seeing the United States was exciting. . . . I landed one evening in the fall of 1926 in New York. . . . While Wiegner's physiochemical lab was spick and span with ultraclean benches and mirror-like floors, Waksman's biological lab in the basement of an old building was messy and required galoshes whenever it rained. As an interlude, I switched to Professor Shives's greenhouse, working on effects of ions on transpiration by barley plants. Years later I realized that lab lifestyles are not crucial, for it was Waksman who got a Nobel Prize, not Wiegner. . . .

In the summer of 1927 at Washington, D.C., the first International Congress of Soil Science was ably organized by Lipman, Director of the New Jersey Agricultural Experiment Station. Wiegner lectured on my Ph.D. thesis work on ion exchange and I reported on my Alpine soil analyses. I was thrilled to get to know in person the great soil scientists of the era. . . . A few months later the great Marbut reviewed the Congress and remarked that the paper by Wiegner and Jenny was probably the best, but that it wasn't soil science. . . . I had been taught that ion exchange was the heart of soil science. Later, at Missouri, I got to know Marbut better and appreciated his credo that there is more to soil science than colloid chemistry.

For a grand, transcontinental tour to the West Coast and back, an entire Pullman train was chartered. . . . When the train came to an official halt, owner-driven cars were waiting and took us to the sites. . . . Many of us fell asleep during the afternoon tours because we had been arguing too long the night before. . . . The tour was a thrill and opened a new world. The red soils of the South and black soils of Canada were showcases of the climatic theory of soil formation. On the trip, Bradfield of Missouri decided to spend a sabbatical year with Wiegner in Zurich and he offered me his lab at Columbia, Missouri, an opportunity I grabbed with both hands.

Now, I was an instructor in the Department of Soils, a combination that was still a novelty to me. In central Europe soil science was taught by the professor of agricultural chemistry and the scope of the lectures reflected the docent's personal, often one-sided, interests. Here, an entire staff dealt with the extensive domain of soil science and did it on a broader scale than what I had seen at New Brunswick. I attended Professor M. F. Miller's lectures on soil classification which centered on the soon to be abandoned system of soil provinces. Pedologist Professor H. H. Krusekopf took me along on his many trips to the loessial prairies and the forested Ozarks and taught me soil survey which then still relied on the plane table. To satisfy my curiosity about land and people I cross-examined and pumped him relentlessly in field and office. Soon he became my pedological guru and a personal friend. Before long I teamed up with broad-visioned Professor W. A. Albrecht, microbiologist, and grew in the greenhouse soybean plants on mixtures of sand, and Ca, H-clay.

The tumultuous year of 1927 culminated in the presentation of my first nitrogen paper at the meeting of the intellectually vigorous American Soil Survey Association held at Chicago on November 5. . . .

Although Columbia, Missouri, lacked the international reputation of Zurich, it was in turn spared the exclusive cliques of prima donnas of science. Instead, the university's stimulating faculty fostered intercommunication. I was close to Brody (animal physiology), Kirby-Miller (philosophy), Robbins (botany), Schlundt (chemistry), Stadler (genetics), and Stearn (physical chemistry). The favorable environment encouraged my professional growth. It proceeded along the lines of pedology and soil colloid chemistry.

Pedologic Work

Still fascinated by the summer's transcontinental experience I kept pondering about the contrast between the dark-colored Canadian prairie soils and the lighter-colored counterparts in Missouri. From bulletins and unpublished reports I gathered soil nitrogen and carbon analyses and arranged them in tables and graphs. Within a few weeks it became clear that from Canada to Louisiana along a temperature transect having similar moisture conditions soil organic matter was sharply declining in exponential fashion. Well-wishers encouraged me to present the findings at the upcoming meeting in November in Chicago, which I did, as already mentioned. Most everybody seemed impressed, especially Marbut who became a friendly counselor. The response spurred me on to search for additional, supportive N-T-functions.

Russians had long known that in their country humus is linked to climate, a trend they attributed to moisutre stress because cool, wet countries dominate their North and hot deserts their South. The transcontinental tour taught me that in the Great Plains area temperature and moisture trends are not synchronic, as in Russia, but intersect at nearly right angles, which provides an ideal climatic checker board. I made the most of Nature's fortunate American design.

Climatic comparisons of Europe and the United States suggested that the famous Russian Chernozem, a soil rich in humus and having a calcium carbonate horizon in the subsoil, might also have arisen in central Kansas and Nebraska. I was longing to meet this idol of a climatic soil profile, and on a field trip to western Missouri I talked my graduate student companion (Roy Hockensmith) into driving the old model-T Ford westward deep into Kansas, at the 25 miles per hour it was capable of doing. We found the missing link and were elated, having seemingly confirmed the climatic principle of soil genesis. Alas, back in Columbia we were lambasted for having taken a state car across the state line without prior permission.

Eventually, I acquired a substantial collection of soils across the Great Plains along a transect of increasing mean annual precipitation but constant, mean annual temperature. Soil organic matter responded to rainfall logarithmically. The individual moisture and temperature relations enabled me to formulate an "equation of state," a three-dimensional N-climate surface, which was published in 1930 in the Journal of Physical Chemistry. I enjoyed seeing field data aligned by equations and derived aesthetic pleasure from the shapes of the curves. Several pedologists, however, accused me of trying to be erudite.

In the same year I wrote an extensive overview for *Research Bulletin* **152.** During the absence of the department head (M. F. Miller) in Europe, I smuggled in a discourse on the low corn yields in the South, attributing them to the scarcity of nitrogen rather than to summer dry spells, as was customary. Miller thought the thesis too speculative. Two decades later the idea was given credit for having helped foster the spectacular rise in corn yields that followed on the heels of advances in hybrid corn genetics. For my friends abroad I wrote an account in German in the *"Naturwissenschaften"* but Wiegner and his staff kept mum.

By now I was an assistant professor at $2300 a year. In 1931 I was awarded the Nitrogen Research Award of the American Society of Agronomy.

Meanwhile, the stock market crash of 1929 initiated economic upheaval that engulfed the university. Young faculty members were laid off. I was spared in spite of complaints that the College of Agriculture favored a foreigner over native sons. Fortunately, the emerging Soil Conservation Service (SCS) absorbed many of the department's unemployed. The new adventure was organized by soil surveyor H. H. Bennett, and Missouri received special consideration because years before, F. L. Duley and M. F. Miller had established the first, systematic soil erosion experiments. The SCS was to occupy me profoundly at a much later date.

One day Professor Miller asked me to examine the effect of agricultural cultivation on contents of soil organic matter. I don't remember what prompted the inquiry; it put me on the trail leading to American soil history. Along old Highway 40 we selected the grazed but otherwise virgin Tucker prairie and adjacent cultivated land, both on level Putnam clay pan soil. We found that 60 years of farming had induced a loss of 35% of total soil nitrogen and a slightly higher amount of carbon. Additional comparisons discovered in the literature permitted the construction of a declining nitrogen-time curve (*Bull.* **324,** 1933). Years later I developed a simple mathematical model of gains and losses of N and C that envisioned an equilibrium or steady-state condition of soil organic matter.

Further work disclosed the secondary role of vegetation, topography, and parent material, telling me that I had "rediscovered" the five soil-forming factors of Hilgard and of the Russian Dokuchaev school. Still, the nitrogen-climate surface qualifies as an original contribution to the storehouse of pedologic knowledge. The logical extension of the factor scheme to the clan of pedogenic functions did not emerge until the subsequent California position enhanced my self confidence. . . .

Pedologist at Berkeley

When the Department of Plant Nutrition became my abode in 1936, it was already famous for its work on metabolic uptake and for establishing the existence of micronutrients. . . . In 1939 Professor C. F. Shaw unexpectedly died of a heart attack. The Shaw hiatus was resolved by creating at Berkeley a Department of Soils (1940) with Kelley as head, joined by Weir, Bodman, and the entire soil survey staff. In addition, Jenny and Overstreet left Plant Nutrition and became members of the new department, all of them located in Hilgard Hall. I was assigned to teaching a course in pedology, following Shaw, and to bringing Dean Lipman Hilgard's roll-up desk. However, I continued teaching the colloid course.

Struggling with Definitions

I felt an urgent need to formulate a conceptual scheme that would embrace climate and time functions and others as well—in short, a logical theory of soil-forming factors. For over a year I pondered intensely about the meaning of soil formers, and what I read in the literature about them appeared logically inconsistent and often indulging in circular reasoning.

Why did a physical chemist choose temperature and pressure as crucial variables? What puts time and climate into the same category? I was led to conclude that soil-forming factors were neither causes nor forces, as claimed, but factors that define the state and history of a soil. The immense variations that exist among soils on the Earth were variations within a collection or ensemble of properties. Moreover, the factors could vary independently, whereas many properties could not. The independency criterion was widely misunderstood. It earned me such attributes as being deceitful and naive.

The meaning of parental material [the French and Germans speak of "mother rock"] kept me in suspense for quite some time. Finally, I pinpointed it as the initial stage of a soil at soil formation.

The real bugbear was the biotic factor. Like everybody else, I could see that vegetation affects the soil and that soil affects the vegetation, the very *circulus vitiosus* that I was trying to avoid. It took me a while to abandon "vegetation" and substitute "flora," eventually thinking of "flora of influx." Yet, in *Factors of Soil Formation* I was not fully consistent. Years later, prolonged discussions with R. L. Crocker, Jack Major, and N. C. W. Beadle confirmed the central core of thought.

Factors of Soil Formation

As far as I could tell, no pedologist has been digging for definitions more intensely than I. What is of interest to me, I reasoned, might interest other pedologists as well. I decided to write a book about concepts, supported by whatever evidence I could put my hands on.

The book is based on the soil-forming factor equation, dubbed by students as *"clorpt."* On the left side are the soil properties s, on the right side the factors climate (cl), organisms (o), topography (r), parent material (p), and time (t). Soil-forming factor equations were nothing new, as I learned subsequently. They try to fuse properties and factors into a conceptual scheme. But by redefining the factors as possible independent variables, *I could solve the equation. That was the new approach.*

I submitted the manuscript to McGraw-Hill, whose editor, a botanist, turned it down. Again, R. Bradfield rescued it. The reactions were mixed. Overstreet said; "It's the kind of book I like." Thorpe didn't like it at first, then changed his mind, he told me.

Interest in the Ecosystem

The concern with soil nitrogen and soil carbon stemmed from considerations of soil fertility. The two elements enter the soil along biological pathways, hence vegetation is important in soil genesis. In the 1941 book, *I saw soil and vegetation as coupled systems* and designated the two together as the "larger system." Crocker then called my attention to Tansley's term of *ecosystem.* The letter l in the equations goes back to the "larger system." I became curious whether in a forest the vegetation contains more or less C and N than the soil. At Hobergs, California, in collaboration with A. Schultz, H. Biswell, J.R. Sweeney, and R. Glauser, we felled three trees, dug up the roots and collected the soil quantitatively. Analytical work and calculations dragged on for years, and, finally, a brief summary appears in this book.

To appraise the build-up of C and N in an entire ecosystem, my wife (Jean) and I spent three summers (1955, 1956, 1957) in the Swiss Alps collecting over 30 soil and vegetation (shrubs) tesseras on historically dated moraines of the Rhoneglacier. It took one-half to a whole day to collect a tessera in the rocky drift. In spite of cold and rainy weather, it was an enjoyable experience. An account appeared in a German journal in 1965.

The full impact of a holistic ecosystem view on soil genesis manifested itself in collaboration with R. J. Arkley and A. M. Schultz working on the Pygmy Forest Ecological Staircase near Mendocino, California. We had to answer the basic question whether the Spodosol created the pygmy forest, or the pygmy forest the Spodosol (7).

For more, various readers may find quite different paths through this book. After an overview of Contents, chapter summaries, and many of the pictures, a sequential reading straight through could construct the soundest foundations on which Jenny erects his building of ideas and equations. Sections in finer print—like closets that are opened later to reveal a storehouse of additional, useful details—can be skipped while first touring the overall structure and making ourselves more "at home" in the book. But the hallways and staircases connecting different levels of abstraction can be explored and then retraced in several orders, each pathway making more connections than our first guided tour.

The windows from this book looking out to the real world are its pictures. Many are by Jenny's own hand. To carry through the artistic touch inherent in this book and in Jenny's life (6), the editors have deliberately retained many of his original sketches that he expected to be redrawn or relettered to a more standardized, but less spontaneous style.

Like the "snow mountain," spied from afar by the farm youth; like each peak which the Alpinist Jenny continues to see as a challenge, the pinnacles of his career are reflected in the chapters and tables to follow.

Many other volumes of the *Ecological Studies* series contain material closely related to that of the present one (especially Volumes 1, 2, 4, 5, 6, 8, 10, 11, 15, 16, 17, 19, 26, 27, 29, 32, 34). Volume 35 by John Fortescue, on *Environmental Geochemistry: A Holistic Approach,* in particular can be viewed as a companion or complement to the present one. From the view of a geochemist exploring the periodic table as well as the factors controlling the landscape's state, Fortescue is explicit about the interdisciplinary character of geochemistry, ecology, and pedology; also about the paradigms which have been changing the character of each in recent decades. Now we have a fuller working out of the closely related paradigms, which Hans Jenny brings together in the present volume.

References

1. Day, Paul. 1971. *Soil Science* **111**: 87−90.
2. Braun-Blanquet, J., and G. Fuller (Ed). 1929. *Plant Sociology.* McGraw-Hill, New York.
3. Braun-Blanquet, J., and H. Jenny. 1926. *Denkschr. Schweiz. Nat. Ges.* **63:** 183−349.
4. Jenny, H. 1937. *Colloid Chemistry.* Stanford University Press, Palo Alto.
5. Jenny, H. 1941. *Factors of Soil Formation.* McGraw-Hill, New York.
6. Jenny, H. 1968. The image of soil in landscape art, old and new. *Pont. Acad. Sci. Scripta var.* **32:** 948−979 (illustr.). Vatican, Rome. (Obtainable: Wiley Interscience Division, New York.)
7. Jenny, H. *Oral History,* Bancroft Library. University of California, Berkeley, California (in progress, D. Maher, Editor).

Oak Ridge, Tennessee   Jerry Olson
September 1980

Preface

Soil is a natural resource. On human time scales its mass is nonrenewable. Today, people are becoming aware of resource limitations and they want to know more about the soil, its behavior, and its fate.

In this book soils are treated as parts of land ecosystems and as structured bodies made up of biotic and abiotic components.

The book is written in textbook style. It is an expansion and updating of the author's lectures on soil genesis given for many years to students of soil science, ecology, forestry, geology, geography, and related fields. Had the treatise been written by a group of specialists, each chapter would be more authoritative and detailed, but the integration of the subject matter to a broad and long view would have suffered.

The book contains more information than can be covered thoroughly in one semester. It is a many-sided companion to teachers and students of various backgrounds, interests, and accomplishments. A course in high school chemistry is a prerequisite. Simple calculus is used for those readers who are prepared to use it, but equations using calculus may be skipped.

Chapter 1 serves as an introduction for the book. Part A, Processes of Soil Genesis, consists of six chapters on soil developmental factors. Part B, Soil and Ecosystem Sequences, consists of six chapters on state factor analysis and a seventh that offers an overview of the book. The Appendix lists the scientific names of plant species.

Acknowledgments. The author is grateful for encouragement and help from many colleagues, friends, and former and present co-workers and students. Several took the trouble of reading the entire manuscript, others examined parts of it, and all responded graciously to inquiries. I am alone responsible for what appears in print, the more so as I did not heed all the advice given.

Among the Department colleagues I should like to single out are, in alphabetical order, Rodney J. Arkley (soil morphology), Kenneth L. Babcock (soil physical chemistry), Isaac Barshad (pedochemistry), Paul R. Day (soil physics), Harvey E. Doner (soil chemistry), Mary K. Firestone (soil microbiology), Paul Gersper (pedology), Louis Jacobson (plant nutrition), John G. McColl (forest soils), the late A. Douglas McLaren (soil biochemistry), and Larry Waldron (soil physics). During the many years of "retirement" the Department generously provided me with quiet office space, and Elizabeth Little, administrative assistant, saw to it that I got help (besides my wife's) in typing. I enjoyed participating in the seminars and field trips that enabled me to fuse the old with the new.

I am deeply indebted to Jerry S. Olson (ecology, Oak Ridge National Laboratory, Tennessee), and I extend my gratitude for valuable comments to Frank E. Riecken (soil genesis, Ames, Iowa), Arnold M. Schultz (ecology, Berkeley, California), and Max E. Springer (soil survey, Knoxville, Tennessee). I thank Alice Q. Howard (Herbarium, University of California, Berkeley) for the names of the plant species in the appendix, and Egolfs V. Bakuzis (St. Paul, Minnesota) for sending me his informative lectures on forest ecosystems.

At the end, doctoral candidate Jennifer Harden examined the entire manuscript and cross-examined me skillfully, which was a novel experience for a senior citizen. Thank you, Jennifer.

Science Editor Philip Manor, Ph.D., has been a sympathetic counselor and Miss Betty Sun an efficient production editor, and I thank both of them.

Berkeley, California   Hans Jenny
November 1980

Contents

The Soil Resource

1. Ecosystems and Soils

On the 3000 mile flight from New York to San Francisco the plane traverses a landscape of gigantic proportions. The rugged mountains, smooth hills, rolling plains, broad valleys, rivers, and lakes are *ecosystems* with wide variations in climate, bodies of water, soils, vegetation, animal life, and human activities. This book brings into focus the soil resource, its creation by nature, its variation over continents, and its husbandry by man.

A. Concept of Ecosystem

In 1935 the botanist Tansley defined the ecosystem as the aggregate of plants, animals, and microbes *plus* the environment in which they live. The term has gained great popularity. *Eco* reminds one of ecology, the study of the relations of organisms to their environment; and *system* is the magic word of the exact sciences.

The Finite Ecosystem

The entire Earth may be considered one giant ecosystem that extends outward to infinity. It is overwhelmingly complex. A convenient small segment, a piece of landscape with arbitrary boundaries, is sketched in Figure 1.1. The boundaries create a *finite* physical system that has finite volume and mass and permits evaluation of energy and mass flow across the imposed boundaries. System boundaries may be placed anywhere: around a watershed, a forest, a pond, a soil, or an organism. Always, the system is what is inside the boundaries; on the outside is the surroundings or environment. Ecosystem and surroundings together constitute the ecosystem universe. The distinctions are important for system analysis. Whereas the finite ecosystem becomes organized and ordered, the ecosystem universe always gains disorder or entropy.

The Land Ecosystem

When, as in Figure 1.1, soil is inside the chosen boundaries, the entire soil-plant body is termed "larger system" (6), terrestrial ecosystem, or land ecosystem. Terrestrial ecosystems are contrasted with aquatic ecosystems: rivers, ponds, lakes, and oceans. At

Fig. 1.1. Ecosystem with arbitrary boundaries. In assay of biomass the upper boundary is made to coincide with contours of vegetation (Soil Science Soc. Am., Madison, WI)

the shores, transitions may be sharp or gradual. Lindeman (10) designates the substratum of a lake a benthic soil, and, if that idea is accepted, soil also is a component
of aquatic ecosystems.

Soil by itself has still broader connotations. It need not harbor living things because,
as seen on alkali flats and serpentine barrens, spores and seeds that have access to it
may perish. In Figure 1.1 the arbitrarily positioned upper-boundary plane is envisioned
whenever measurements are made of incoming precipitation, flux density of light, and
"ambient temperature." Below the lower-boundary plane, below the soil, lies the subterranean surrounding termed *nonsoil*. The decision where to divide soil from nonsoil
is perplexing. Hilgard (15) placed the boundary at the depth of root penetration, a
notion that has been revived (18). A depth of 1 m is often taken as a limit, but live
shrub and tree roots have been traced to over 50 m deep (12). The lower system
boundary is arbitrary, and its selection is guided by the interests and inquiries of the
investigator (Fig. 1.2).

In deciding on the *side boundaries* of ecosystems the premise that vegetation reflects
soil conditions is of limited help in the present-day, man-modified landscape. On soil
maps separation lines often coincide with changes in topography, ground water table,
parent rock, and climate.

Figure 1.2. Gradual change from rocks to soil (after a drawing by J. B. Bodman)

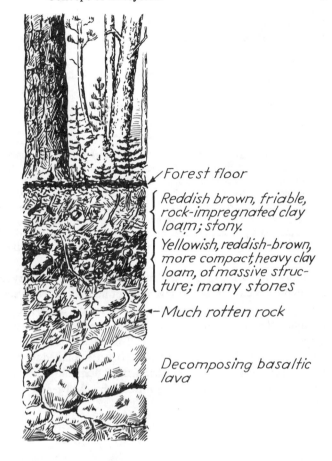

Forest floor

Reddish brown, friable, rock-impregnated clay loam; stony.

Yellowish, reddish-brown, more compact, heavy clay loam, of massive structure; many stones

Much rotten rock

Decomposing basaltic lava

Soil Space and Vert Space

Portions of a chosen ecosystem are treated as subsystems or compartments. In land ecosystems there is "soil space" on one hand and "green space" or "vert space" (vert = green) on the other. Vert space comprises all the above-ground parts of the land ecosystem, the plants, the animals, and the voids between them. Soil space includes the soil horizons and their mineral and humus particles, also the gas- and water-filled pores, and all the organisms that inhabit it. Soil is *not* an *abiotic* environment of plants (13, 15), for it is teaming with billions of minute animals and microorganisms. Soil defies summation of parts. Though not an organism that can multiply, soil on the Earth is a living system.

Tracing the boundary between soil space and vert space may become physically frustrating. Few are inclined to designate the roots of a tree as a property of the soil. Yet, if that is not done the boundary of soil and vegetation would have to follow the contours of the root surfaces to the finest root hairs and epidermal protrusions, producing a phase boundary of unmanageable tortuosity. Conveniently, soil begins with the forest floor and at the base of a tree, as if the ecosystem had been clearcut. The boundary between vert space and soil space is operational.

Advantages of Ecosystem View

The close accord of soil and vegetation is persuasively illustrated in the case of a redwood seed (*Sequoia*) germinating on a fresh sand bank in a river bend.

> As the tree grows the sand beneath is altered by root action, microbe influx, plant debris and its extracts, and percolating water. In turn, the mineral- and nitrogen-enriched soil material influences the growth rate of the tree. Action and reaction are continually associated. In a millenium or two the seed will have grown into a tall redwood tree standing on a humus-rich soil with observable horizon differentiations. Tree and soil have evolved together. This is ecosystem genesis, and cytoplasm, bark, ground litter, humus layers, and associated microbiota mark the redwood specimen a "definer" of a cluster of species and soil attributes.

Along the Pacific Coast the sight of ash-colored soils carrying a dwarf forest of pines and cypresses challenges the ingenuity of naturalists. Plant ecologists seek the cause of this pygmy forest in the strongly acidic and impoverished soils, whereas the pedologists attribute the soils' bleached strata and the hardpan to the acid humus substances derived from the peculiar vegetation. This circular reasoning hopelessly entangles cause and effect because vert space and soil space are coupled systems that interact continually. It is such striking situations that make it desirable to use the concept of the land ecosystem as the broad framework for research in soil genesis and vegetation development.

From Disorder to Order

When the functioning of ecosystems is discussed in broad terms, the expressions *closed* and *open systems* and *energy flow* and *entropy* invariably appear (11, 13). Entropy has achieved ecological status through the physicist Schrödinger's comment (14) that organisms feed on "negative entropy."

A stoppered bottle containing water, clay particles, and salt is a closed system. Heat, but not matter, may enter or leave it. When the reaction between clay and salt comes to rest, the system is at *equilibrium*. Soil and vegetation are open systems; both heat and matter may enter or leave simultaneously. When all entries and exits are balanced, the system is at *steady state*. Portions of organic and inorganic properties come and go yet the system appears stabilized. Energy flow, in the ecological sense, traces the *caloric* changes in food and egesta, as when a sheep eats grass, a beetle feeds on sheep dung, and soil microbes digest the beetle droppings.

> The science of thermodynamics deals with the energetics of processes. It strives to evaluate the free energy change ΔG, which is the maximum amount of useful work that may be obtained from a chemical reaction. It utilizes the heat ΔH gained (considered positive) or lost (considered negative) when the process occurs in a constant pressure calorimeter. The heats of reactions (ΔH) reflect the making and breaking of chemical bonds. A third thermal quantity is $T\Delta S$, which is the heat the

system takes from or gives to the environment in "reversible" fashion, that is, slowly in small steps without friction. T denotes the absolute temperature K (=273 + °C), and ΔS denotes the change in *entropy*, S, of the reacting partners, measured as calories/°K. At atmospheric pressures and ordinary temperatures, the three components are related as

$$\Delta G = \Delta H - T\Delta S$$

which is the "Gibbs function." When a reaction proceeds spontaneously by itself ΔG assumes a negative value; thus, the symbol $-\Delta G$ attached to a chemical equation signals reactivity and amount of work expected, although catalysts may have to be present to speed up the process.

To understand ΔS, consider a large beaker with ice water and ice cubes as "the system" at 0°C. Ice and water are in equilibrium with one another. When heat diffuses slowly (viz. reversibly) from the environment into the beaker, some of the ice cubes will melt, yet the ice-water mixture remains at 0°C. If eventually 1436 cal of heat are conveyed to the beaker, 1 mole (18 g) of ice will have melted. As the conversion of ice into water performs no useful external work, $\Delta G = 0$ and the heat, ΔH (1436 cal), is equal to $T\Delta S$.

Mechanistically, the calories taken in by the ice-water mixture are used up in separating the H_2O molecules in the ice crystals from their orderly fixed positions, causing them to become disordered, mobile molecules in the liquid. This conversion is identified as entropy changes $\Delta S = \Delta H/T$, which is 1436 cal/273°K or 5.3 cal/mole/degree. The entropy of liquid water at 0°C is greater than that of frozen water by 5.3 cal/degree; that is, the randomness of the molecules has been increased by 5.3 entropy units. These are transferred as $T\Delta S$ from the environment, which thereby experiences a corresponding reduction in its own entropy reservoir. Conversely, when liquid water slowly freezes at 0°C and crystallizes, -1436 cal are given up to the environment; entropy in the beaker is reduced by -5.3 cal/degree, which may be looked upon as a gain in negentropy. For systems and environment together, the universe, $\Delta S_u = 0$; only an exchange of entropy takes place. In system nomenclature, with Δ meaning a change,

$$\Delta S_u \quad = \quad -\Delta S_e \quad + \quad \Delta S_{sy} \quad = \quad 0.$$

universe   environ-   system's
ment's loss   gain

It follows that increasing order in a system necessitates disordering the environment. In any real reaction, heat-producing friction cannot be avoided, and the universe continually gains entropy or disorder, $\Delta S_u > 0$.

The connection of heat exchange and degree of order goes back to Boltzmann who set entropy proportional to the logarithm of disorder. The idea is applied to *macroecosystems* in a trend of thought like this: In a large deposit of well-mixed sand many mineralogical and chemical properties are randomized. When the body is exposed to rain and sunshine, weathering and leaching move alteration products into select positions parallel to the land surface. Now, the formerly isotropic arrangement has become anisotropic, that is, ordered to some extent.

In the the air above the bare sand the molecules of carbon dioxide (CO_2) gas also are randomized because of wind and turbulent diffusion. Under the influence of light energy, plants convert water and carbon (C) of CO_2 gas to sugar molecules that are highly organized ensembles of low entropy, compared to gas and water. They sustain organic life processes in respiration, which returns CO_2 and heat (ΔH) to the environment, thereby increasing its entropy. Macroscopically, plant life becomes arranged as vertical trees, as grids of cornstalks and grass, and as horizontal strata (synusias) of crowns, shrub thickets, herbs, and mosses that are paralleled below by the horizons of the soil, which vegetation reinforces. The landscape gains order or negentropy.

We may conclude that sun-driven genesis of ecosystems transforms disordered portions of the Earth's surface to more ordered ones. Photosynthesis creates negentropy and respiration converts it to positive entropy. At steady state the latter always exceeds the former, owing to unavoidable frictional losses of heat. Quantification of ΔG, ΔH, and ΔS of entire ecosystems is still far in the future.

B. Soil as an Object of Nature

As explored here, soil is more than farmer's dirt, or a pile of good topsoil, or engineering material; it is a body of nature that has its own internal organization and history of genesis.

The Soil Profile

Let us follow a road crew that scrapes the road bank and exposes a fresh soil face. In Hawaiian sunlight the cut may glow in Titian reds and yellows of iron oxides and be delicately sculptured; in Manitoba's prairies it may be humus-rich and ebony-black, grading along vertical structure grooves to dark and light grays of the subsoil; in Denmark a snow-white sand may rest on rustbrown hardpan of maroon and tan.

A vertical cut of a soil body is a *soil profile* that resembles an abstract painting on a sculptured surface (8). Its bands and blotches of color, its texturologies of fine silts, coarse sands, and firm nodules, and its structural designs of platelets, crumbs, and columns are key features of genesis.

Soil Horizons

The profile's colorations, cleavages, plasticities, and moisture patterns line up as "horizons" that follow the surface of the land, reflecting the impacts of climate and organic life. The farmer's topsoil and subsoil have long been particularized as A, B, C horizons that have pedogenic (Gr. "pedon," soil or ground) implications. A is the zone of surface depletion of mineral matter by leaching (eluviation); B, which lies beneath A, is the zone of infiltration and accumulation (illuviation) of A substances; and C, which lies below B, is meant to simulate the original rock or parent material from which and in which A and B evolved. While the simple A, B, C scheme has merits for young soils in glaciated regions, it requires expansion to encompass the multitude of soil bodies in warm temperate and tropical lands.

Nowadays, the master A and B horizons are subdivided variously as A1, A2, A3; A11, A22, A23; B1, B2, B3; and B21, B22, B23. Additional layers are recognized as morphologically and chemically different entities, such as plowed (p), carbonate (ca), and concretionary (cn) layers; illuviated humus (h); iron (ir) and clay (t); brown-white color spots of mottling (g); and hardpan (m). The small letters are attached to the symbols of the master horizons and their subdivisions as, for example, Ap, A1ca, A3g, Bh, B2t, Bmir, and Cg. Below the C horizon, which has become merely the stratum below B, unweathered bedrock is marked R, and litter above the mineral soil is O.

Horizon differentiation has developed into a fine art, and the Soil Taxonomy of 1975 (18) must be consulted for details. Two common profiles with horizon designations are shown in Figures 1.3 and 1.4.

To what extent the horizon theme is mirrored in biological responses is still being pursued (16). Roots of larger plants penetrate several horizons and proliferate in the fertile ones that have low water stress. That hardpans hamper root growth is elementary; what biological reactions mottling instigates may have to be answered by microbiologists.

The Tessera

A three-dimensional, vertical soil prism or cylinder is a *soil tessera,* and when the vegetation above it is included it is an *ecotessera.* The term tessera refers to the small stone cubes that make up a mosaic art work. The soil profile is a two-dimensional face of a soil tessera. Vegetation tesseras are familiar to ecologists as "quadrats." Tesseras are landscape elements of arbitrary cross sections, but soil tesseras rarely exceed 1 m^2 of actual excavation. Common are prisms of 20 X 20 cm area to chosen depth. In mellow soil a barrel auger readily retrieves 20-cm-long cylindrical segments. For detailed work a pit is dug, and on one of its sides a shaft conveniently 20 X 20 cm is secured in small increments with knives, trowels, and brushes. The procedure yields stones, gravel, roots, and fine earth on a volume basis in place (see Fig. 9.1). The total nitrogen (N) and organic carbon (C) inventory of a large pine tree ecosystem obtained in collaboration with H. H. Biswell, R. Glauser, and A. M. Schultz in 1958 is shown in Figure 1.5. The two elements are more abundant in soil space than in vert space.

Fig. 1.3. Virgin prairie soil (Mollisol). Gradual fading of black humus with depth, related to distribution of grass roots. Subsoil B horizon is rich in clay, which provides angular aggregates of soil structure (Courtesy Soil Conservation Service)

Figure 1.4. Forested Podzol or Spodosol (6). Surface 0 horizon is underlain by dark, humic A1, followed by whitish, albic A2, which rests on a brown B horizon (spodic horizon) composed of a narrow humus band (Bh) on an iron- and aluminum-rich ortstein layer (Bmir)

A tessera may itself be the "system of interest," or it may serve as an operational sample in a larger piece of landscape.

The Pedon

Soil Survey Staff followed the tessera with the concept of pedon, which is also a three-dimensional soil body. In soils of uniform horizons the minimal area is about 1 m^2; it may extend to 10 m^2 when horizons are intermittent or cyclic. In the latter situation the pedon comprises half a cycle. Unlike the arbitrary tessera the pedon is meant to be akin to the "unit cell" of a crystal, which is the theme of a repetitive pattern like the basic design on a wall paper. The pedon is the natural, elementary unit of a soil in the field and covers areal extent by lateral displacement. In this respect the concept deserves further inquiry. How are its dimensions assessed when profile features vary randomly over an area or when they change continuously along vectors of state factors, as along a slope? Whiteside (20) concludes that sampling, describing, and classifying soils is accomplished with tesseras, not pedons.

Fig. 1.5. Inventory of organic nitrogen (N) and carbon (C) of a cylindrical pine ecotessera of 117 m²
cross-section on stony Ultisol
(Table 5.2). Vert, vert space
(needles, branches, trunk); F.F.,
forest floor; Min. soil, mineral soil
(stones plus fine earth of < 2 mm
particle size)

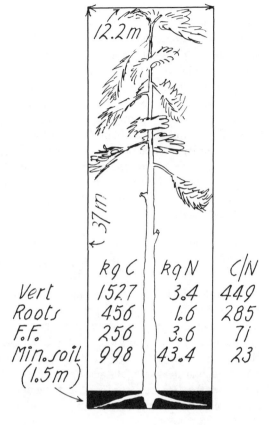

	kg C	kg N	C/N
Vert	1527	3.4	449
Roots	456	1.6	285
F.F.	256	3.6	71
Min. soil	998	43.4	23
(1.5 m)			

Soil as System

Theorists view soils as systems that can be described with the symbols of mathematical
language. Prominent in their vocabulary is the term "a variable," which is anything
that varies.

Soil properties, like pH, stonyness, moisture, humus, and bacteria content, are
variables designated by the letter s as s_1, s_2, s_3, . . . , etc. The number is very large
and many of the properties are interrelated. When one of them changes, many
others also assume different values. Adding water to a soil not only increases its
moisture content but also its bulk density and vapor pressure, and it lowers salt
concentration and heat conduction. Pairs of soil properties that tend to stay together, that are collinear, include pH and titratable acidity, carbon and nitrogen in
humus, and percentage of iron oxide and soil color.

In system language the value of any property, s, may be written as the deterministic
dependency:

$$s = f(s_1, s_2, s_3, . . .)$$

where f means "function of" or "dependent on." For example, in some soils the contents of calcium relate to amounts of clay, humus, and acidity.

Various combinations of s properties and their magnitudes make different kinds of soils, that is, different states of the general soil system. Admitting small differences (ds), the number of soil states is infinitely large.

A soil is defined when its properties are described. Because of the dependencies of s it is not necessary to enumerate and quantify all of the properties. For stabilized soils the kinds of s that together fix the values of all remaining s are known as *state variables.* Their number is still unknown, but soil moisture and temperature are candidates of high priority. Soil taxonomy emphasizes the soil horizons as cardinal characteristics.

Many s properties are shared by other natural bodies, e.g., desert sands, mud flats, and turbid ponds. Discrimination requires assignment of special limits to the ranges of s. Too high a water content might shift "soil" to "swamp." Quantification of limits as to what soil is and is not still awaits consent.

State Factors and Soil Processes

A group of variables, historically known as soil-forming factors, are ecosystem determinants or conditioners. In the symbolic triangle of Figure 1.6 the ecosystem sketched has a beginning, an *initial state.* It might have been a freshly blown sand dune or a glacial deposit of composition p (parent material) and shape r (topography). In time the system acquires the *age* t. During its evolution it is subjected continually to *influxes,* such as heat, rainfall, and light, assembled as cl (climate). Organism specimens, too, enter the area, as seeds, eggs, and migrants, termed *biotic factor* ϕ.* Occasionally fires sweep the ecosystem and dust storms have an impact. Such influx variables are listed as dot factors (. . .).

Total ecosystem properties l, vegetation properties v, animal properties a, and soil properties s become functions of the *state factors:*

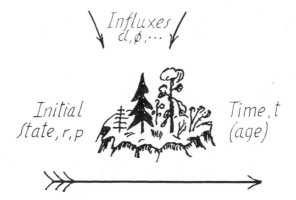

Fig. 1.6. Symbolic triangle of three groups of state factors that determine ecosystems: initial state, age of system, and influx variables. cl, climate; ϕ, biotic factor; r, topography; p, parent material; t, time or age

*Represents letter "o" throughout.

$$l,v,a,s, = f(cl,\phi,r,p,t, \ldots)$$

The state factors are "independent variables" in the sense that in a given area one of them may vary a great deal while the others vary but little; they are "constant." Thus, soils on two different, adjacent rock materials may be compared pedologically when both have similar air climate, flora influx, topography, and age. In favorable settings state factor analysis promises ordering the diversity of soils and vegetation as *sequences* in space and time, to be elaborated later.

Any *process* taking place in soil space is a pedogenic process. It may be physical infiltration of rain water, chemical dissolution of salt, microbiological respiration, or tunneling by invertebrates.

Soil genesis integrates gains, losses, and transformations of energy and matter, hence it embraces build-up as well as break-down of the soil body. Elucidation requires the efforts of botanists, ecologists, foresters, geologists, and soil biochemists, soil chemists, soil fertility specialists, soil microbiologists, soil microscopists, soil mineralogists, soil organic chemists, soil physical chemists, soil physicists, soil surveyors, soil taxonomists, and soil zoologists.

A given process, e.g., erosion or weathering, may be "geologic" in the geologist's vocabulary and "pedologic" in the soil science dictionary. It depends on who studies what, and no conflict need arise. State factors and soil processes are considered broad, fundamental approaches to comprehending genesis and behavior of soils.

C. Soil Colors and Soil Particles

Colors and particle assemblages (textures) are present in all soils. And these colors and textures are impressive enough to cause people to name towns and cities after the soil on which they are built, e.g., Black Earth (in Wisconsin), and Redlands, White Plains, Sand Flat, Clay, Gravel, Adobe Meadows, Salt Marsh (in California) (4).

Color of Soils

The names an artist might give to a color, such as Titian red, are too subjective to be helpful to the surveyor in the field or to achieve international consent. Soil colors are matched with the colored papers of the Munsell color chart (18). Some 175 chips are lined up in a loose-leaf notebook in decimal notation by "hue" (related to wavelength of light), "value" (relative lightness), and "chroma" (relative purity diluted by grayness); the last two are expressed as fractions. Groups of formulas or notations are accorded color names that vary from nation to nation.

Red assumes many notations, among them 7.5R 4/8 and 2.5YR 4/6. It is modified as yellowish red (5YR 4/8), dusky red (7.5R 3/3), and many others. Black has more than eight formulas, and its nearest members include very dark brown (10YR 2/2), dark olive gray (5Y 3/2), and others.

Red soil colors indicate richness in iron oxides, good drainage, and aeration; yellows also denote high iron content but tend to follow sites of higher moisture. Black correlates with high humus content, but not infallibly, and light gray and whitish imply near-absence of iron and humus. Mottling consists of brown, irregular spots and blotches on a gray background and indicates excessive moisture, now or in the past.

Particle Sizes

Individual soil particles comprise stones, cobbles, gravel, and sand at one extreme and molecules, ions and electrons at the other, more than a billion-fold span in diameters.

Immersed in water, soil slackens. Coarse particles settle out quickly, small ones slowly. At 20°C the settling velocity v, in centimeters/second, is given by *Stokes's law* as $v = 36,700\ r^2$, in which r is the effective particle radius in centimeters. It is most accurate for intermediate sizes. As seen in Table 1.1, it takes coarse sand grains about 1 sec to fall 10 cm in water; ultrafine clay may stay in suspension for centuries.

The highly standardized method of physical soil analysis (2) determines the weights of particles of any size group. Prior to sedimentation, carbonates are commonly dissolved in acid and humus is oxidized. Any particle less than 2 μm in effective diameter, but larger than small molecules, is "clay" regardless of chemical or mineralogical composition. In fine earth ($<$ 2 mm diameter) the many mixtures of sand, silt, and clay are known as soil textures, and they are assigned to divisions in the texture triangle of Figure 1.7. Popularly, a loam is a crumbly soil rich in humus, but pedologically it is a mixture of sand, silt, and clay comprising, for example, the proportions 40, 40, and 20%. In the field, texture is determined by feel with fingers, and experienced soil surveyors achieve high accuracy.

The Colloidal Realm

Submicroscopic particles known as colloidal particles are of paramount significance for soils and plants.

While atoms have radii of about one Angstrom unit (Å), which is one hundred-millionth of a centimeter, or 10^{-8} cm, small molecules measure up to 10 Å. A near spherical particle is said to be colloidal when its diameter lies between 10 and 2000 Å, which is 1-200 nm, or 0.001-0.2 μm. Nonspherical corpuscles, like threads and plates, are colloidal when at least one dimension is below 0.2 μm. Depending on features emphasized, colloidal particles qualify as polymers and giant macromolecules.

Table 1.1.   Time Required for a Soil Particle to Fall in Water a Vertical Distance of 10 cm (Stokes's Law)[a]

Diameter (μm)	Particle designation	Settling time for 10 cm
2000 (= 2 mm)	Very coarse sand	0.03; 1.0^{b} sec
200	Fine sand	2.7; 4.8^{b} sec
20	Silt	4.5 min
2 (= 0.002 mm)	Coarse clay	7.7 hr
1	Coarse clay	30 hr
0.2	Fine, or colloidal clay	32 days
0.02 = 200 Å	Fine colloidal clay	8.6 years
0.002 = 20 Å	Ultrafine clay (large "molecule")	860 years

[a]Temperature, 20°C; density difference of particles and water, 1.70; viscosity, 0.1000 units.
[b]More realistic values by the Stokes-Oseen formula (3).

Vert space and soil space abound in colloidal particles of highly divergent behaviors. In organic tissues the colloidal proteins, pectins, starches, and celluloses are held in place by membranes and walls, or they interlace to fibrous frameworks of high stability. A clod of soil, instead, muddies the waters with myriads of colloidal particles of humus and clay that had adhered to each other and to coarser grains with relatively weak chemical bonds.

Sol and Gel

A *sol* is a suspension of colloidal particles in a liquid, like clay and humus in turbid ponds or pore waters. The solid is the dispersed phase, the liquid the continuous one. In a *gel*—represented by a jelly, a sponge, or a mud flat—the solid is the continuous phase, and the pores and capillaries of colloidal diameters are the dispersed phase.

Cube Division. Consider a cube of shiny silver of 1 cm edge length. It weighs about 10 g and has a surface area of 6 cm^2. Divide the cube into 1000 cublets of 1 mm edge length. The combined surface is 60 cm^2. Divide each millimeter cublet into micron cublets, of which there are 1000 million or 10^9; the combined surface of these coarse colloidal particles is 6 m^2. Divide each micron cublet into 10^9 ultrafine nanometer cublets. The total surface is 6000 m^2 or nearly 1.5 acres. For but 1 g of silver the "specific surface area" is 600 m^2. As measured by gas uptake, clay powders have surface areas of 10 to > 500 m^2/g. They attract a multitude of molecules and ions.

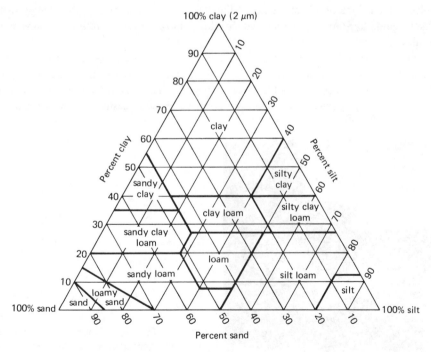

Fig. 1.7. Texture triangle (18) in which sand, silt, and clay percentages of soil add up to 100%

Tyndall Beam. Colloidal silver no longer appears silvery. Suspended in water, the particles form colloidal solutions or sols that are yellow, red, blue, or green because the silver corpuscles being smaller than the wavelength of light, scatter it optically, as do colloidal clays, humus, and iron oxides.

> This can be easily tested in a darkened corner of a room. Using a hand lens, focus a beam of light from a projector. Have it pass horizontally through a test tube held vertically and filled with dilute sol of clay or humus. By looking at the test tube at right angles to the direction from which the projector light is coming, one can see that the suspension displays an illuminated "Tyndall" beam of light scattering that is sensitive to colloidal size. Larger particles reflect light like a mirror, smaller ones (molecules) do not exhibit a beam.

Colloidal Fine Structures in Soils. When suspended colloidal platelets and rods settle out, they lie flat like a thrown deck of cards or matches sorted in a box. The sol becomes an oriented gel. The speed of light rays passing through it differs in vertical and horizontal direction, causing double refraction, observable in a polarizing microscope.

> Kubiena (9) pioneered soil micromorphology by impregnating small soil slabs with liquid resin, cutting thin sections from the hardened material, and observing pores, cavities, and channels microscopically in polarized light. The often vividly colored microscope images unfold a vast domain of microarchitecture illustrated in Figure 1.8.

Colloid Chemical Regimes. The visible interactions of colloidal particles with water molecules and ions of all sorts control the stability of suspensions, influence porosity of gels, guide hydration, swelling, and shrinking, and facilitate or retard the movement of salt constituents.

Colloid science has come forth with two conceptual schemes for comprehending the multitude of chemical reactions. One is the Donnan membrane equilibrium, the other the diffuse electrical double layer theory. In this book the latter is emphasized (see Chapter 3) because its qualitative contents provide easily perceptible models.

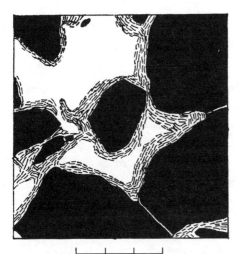

0   0.1   0.2   0.3 mm

Fig. 1.8. Microscopic view of soil matrix. Skins of oriented clay platelets cover sand grains (black). White areas are pore voids (Adapted from photo by Soil Survey Staff)

D. Soil Taxonomy and Classification

Most nations develop their own systems of soil classification. All face a conceptual dilemma: unlike chemical elements, crystals, organic molecules, and organisms that vary discontinuously from one to the other, soils may change continuously over a landscape, yet soil maps and classifications set arbitrary boundaries.

The New Soil Taxonomy

During the last half century the United States Department of Agriculture (USDA) experimented with several national soil classifications, while, simultaneously, individual states produced their own preferences. In 1975 the Soil Conservation Service of the USDA published a new, comprehensive Soil Taxonomy (18) that had been preceded by a series of approximations of which the seventh comes closest to the final work.

As this book utilizes soil studies published under the old and the new systems, and since the names cannot be transferred unequivocally, especially since soil names (series) may have changed, both nomenclatures are used in accordance with the original author's preferences.

The Ten Orders

The new system is a vigorous hybrid that aims to satisfy agricultural demands and pedogenic criteria. Basic is the grouping of all soils into 10 orders that end in the syllable -sol (L. "solum," soil). The sol-ending bears no relation to the sols of colloid science previously mentioned. For quick reference the descriptions that follow (in alphabetical order) are stated as briefly as possible.

Alfisols (aluminum, Al, and iron, Fe). Shallow penetration of humus, translocation of clay (argillic horizon), relatively high base content, rather well-developed horizons. Includes the former Gray-Brown Podzolic soils, some of the Planosols, Noncalcic Browns, and Gray Wooded soils.

Aridisols (L. "aridus," dry). Lack of available water to plants for extended periods; low in humus, high in base content; may have carbonate (calcic), gypsum, and clay horizons. Contains the former Sierozems, Red Desert soils, and Solonchaks.

Entisols (from "recent"). Dominance of mineral soil materials and absence of distinct, pedogenic horizons; very young soils, e.g., on flood plains. Includes former Alluvial soils, Regosol, Lithosol, and Azonal soils.

Histosols (Gr. "histos," tissue). High content of organic matter, as in peats, bogs, mucks.

Inceptisols (L. "inceptum," beginning). Little clay translocation, texture finer than loamy sand, variable chemical properties; often shallow, moderate horizon development, e.g., cambic horizon; relatively young soils (postglacial) of humid climates. Includes Brown Forest soil, Sol Brun Acide, Humic Gley, weak Podzols. The order is heterogeneous and was the last one established.

Mollisols (L. "mollis," soft). Dark brown to black surface horizons of soft consistence, rich in bases; may have argillic or calcic horizon. Soils of semihumid natural grasslands and of some forests with rich understories (e.g., ferns). Includes former Prairie soils, Chernozems, Chestnut soils, Humic Gley, and some Planosols and Rendzinas.

Oxisols (Fr. "oxyde," oxide). Highly weathered soils on old landscapes in tropics and subtropics; red, yellow, or gray, rich in kaolinite, iron oxides, and often in humus. Former Laterites and some of the Latosols.

Spodosols (Gr. "spodos," wood ash). A light-gray, whitish (albic) A2 horizon rests on a black and reddish B horizon high in extractable iron and aluminum, the spodic horizon. Identical with well-developed Podzols.

Ultisols (L. "ultimus," last). Strong clay translocation, intensely leached, low base content; on Pleistocene and older surfaces of humid, usually warm climates. Includes the former Red-Yellow Podzolic soils, Red-Brown Lateritic soils, and some Latosols.

Vertisols (L. "verto," to turn). Dark clay soils that feature wide, deep cracks when soil dries out. Formerly Grumusol, Smolnitza, Regur.

The five geographically dominant orders are traced in Figure 1.9, copied from a more detailed map in Soil Taxonomy (18). As a companion illustration, Carl Sauer's modified (6) vegetation map is appended in Figure 1.10. Causal connections between the two will be conjectured in ensuing chapters.

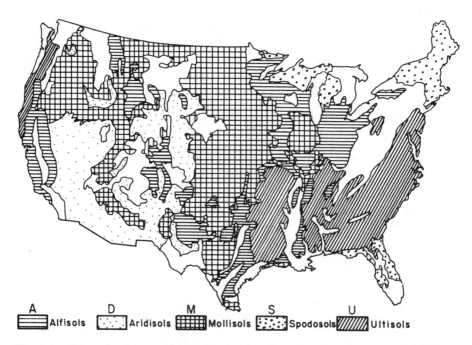

Fig. 1.9. Generalized map of five dominant soil orders. The blank spaces comprise Inceptisols, followed by Entisols and small areas of Histosols and Vertisols. Oxisols are absent

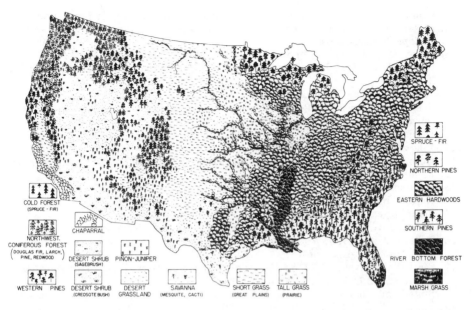

Fig. 1.10. Natural vegetation map of the United States. Adapted from C. Sauer (6)

Orders are subdivided into *suborders*, often according to the soil climate regime. Further divisions lead to *great groups, subgroups,* and *families.* The time-honored *soil series* is the lowest category and is named after the locality (e.g., Fresno, Marshall) near which it was first found. With its phases (slope, erosion, stonyness, etc.) it appears on the standard large-scale soil survey maps of over 3.2 cm to 1 km, that is, 2 or more inches to the mile (1:31,680). So far, over 10,000 series have been identified, described, and classified (17). Progress is being made in numerical taxonomy (1).

E. Review of Chapter

Soil, vegetation, and animals constitute the land ecosystem. This ecological troika and its environment are chosen as the scene for the evaluation of soil genesis in time and space.

The time-honored soil attributes of color and texture (whether sand or clay) are joined to the soil profile and its horizons and to the colloidal state of macromolecules that permeate the entire ecosystem.

A brief description of the 10 orders of the new Soil Taxonomy is presented.

References

1. Arkley, R. J. 1976. *Adv. Agron.* 28: 37-70.
2. Black, C. A. (Ed.) 1965. *Methods of Soil Analysis* (2 vols. Agron. Ser. No. 9). Amer. Soc. Agron., Madison, Wisc.
3. Gessner, H. 1931. *Die Schlämmanalyse.* Akad. Verl., Leipzig.

 4. Gudde, E. G. 1949. *California Place Names.* Univ. California Press, Berkeley.
 5. Hilgard, E. W. 1860. *Report on the Geology and Agriculture of the State of Mississippi.* Jackson, Miss.
 6. Jenny, H. 1941. *Factors of Soil Formation.* McGraw-Hill, New York.
 7. Jenny, H. 1961. *E. W. Hilgard and the Birth of Modern Soil Science.* Agrochimica 3, Pisa.
 8. Jenny, H. 1968. *The Image of Soil in Landscape Art, Old and New.* Pont. Acad. Sci. Scripta var. 32, 948-979 (illustr.). Vatican, Rome. (Obtainable: Wiley Interscience Division, New York.)
 9. Kubiena, W. 1938. *Micropedology.* Collegiate Press, Ames, Iowa.
10. Lindeman, R. L. 1942. *Ecology* 23: 399-417.
11. Odum, E. P. 1971. *Fundamentals of Ecology,* 3d ed. W. B. Saunders, Philadelphia.
12. Phillips, W. S. 1963. *Ecology* 44: 424.
13. Richards, B. N. 1974. *Introduction to the Soil Ecosystem.* Longman Group Lim., Harlow, Essex.
14. Schrödinger, E. 1945. *What is Life?* Macmillan, New York.
15. Sheals, J. G. (Ed.) 1969. *The Soil Ecosystem.* Systematics Assoc. No. 8, London.
16. Soileau, J. M. 1966. *Root Growth in Soils as Related to Pedological Features.* TVA Muscle Shoals, T66-1SF.
17. Staff (Soil Survey). 1972. *Soil Series of the United States, Their Taxonomic Classification.* U.S.D.A., S.C.S., Wash. D. C.
18. Staff (Soil Survey). 1975. *Soil Taxonomy.* U.S.D.A., S.C.S., Agr. Hand. 436.
19. Tansley, A. G. 1935. *Ecology* 16: 284-307.
20. Whiteside, E. P. 1965. *Soil Surv. Horizons* 6(1): 17-20.

Part A.
Processes of Soil Genesis

Soil is a fundamental resource. In order to use it wisely we must understand its inner workings, the processes that keep it dynamic. The strategy is reductionism, the reformation of complex notions and events into simpler, more basic ones. The task makes use of the concepts of forces and potentials, and of atoms, ions, molecules, colloids, enzymes, and organisms, the whole armory of modern science.

In the six chapters of Part A the behavior of water in ecosystems and the interplay of ions with inorganic and organic colloidal particles, including root surfaces, is first examined. In sequence, the genesis of clays from rocks and the creation of humus from plant remains and by microbial synthesis are explored. Also discussed are how the cardinal macromolecules, clay and humus, interact with one another and form aggregates, how bacteria guide the nitrogen cycle, and how invertebrates build structures and facilitate energy flow. The principles arrived at are brought to bear on horizon formation, as in clay and carbonate accumulations, gleysation, humus-depth functions, podzolization, and laterization.

2. Water Regimes of Soils and Vegetation

Growth of land plants is sustained by the water supply in the soil. How the soil recepta-cle is filled up and how it is drained by roots are questions of fateful concern to agri-culture and forestry. Unlike the water in ponds and lakes, stored soil water is "hanging water."

A. Soil, a Climostat

Recording instruments placed in vert space and soil space, as in Figure 2.1, expose the ecosystems as climatic modulators.

Damping

Daily and seasonal oscillations dampen with increasing soil depth. Temperatures be-come stabilized (Fig. 2.2) and, at proper depth, assume the annual mean of the envi-ronment (25). The paths of temperature and the flows of heat into and out of soils have been tied mathematically to temperature variations at the surface (23). Moisture too is less variable inside the soil than it is above it, and subsoil provides water to organisms long after the surface has dried out. Soil is a buffer that tempers climatic transitions from the poles to the equator, but this sheltering impedes organic evo-lution, as Delamare Debouteville (8) avers. He cites the primitive morphology of the tiny soil insects *Collembola* and *Ciplur* and the spiderlike *Ricinulida* that trace back to very ancient geologic periods.

Water Fluxes

Raindrops that pass in random fashion through an imaginary plane above the forest canopy are intercepted by leaves and twigs and channelled into distinctive vert space patterns of through-drip, crown-drip, and stem flow. The soil surface, as receiver, transmits the "rain message" downward, but as the subsoils lack a power source to mold a flow design, the water tends to leave the ecosystem as it entered it, in rando-mized fashion.

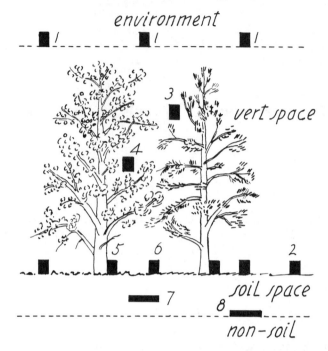

Fig. 2.1. Soil-forest ecosystem with climatic recording instruments placed in environment (1, 2), vert space (3, 4), and soil space (7, 8); 5 records stem flow; 6 records crown drip

With rain gauges placed above and inside the canopy and in the soil space as ceramic suction plates, the amount and fate of influx water are being traced. In plant succession and horizon genesis the readings would shift during centuries and millennia even if the climate above the system should remain invariant.

Water Potentials

A coherent theoretical framework of water continuity in the *soil-plant-atmosphere system* is slowly emerging. In it several types of "water potentials," ψ, play a central role because their variations in space and time are the driving forces of water flow. Potentials relate to the *energy status of water* which is different in roots and in leaves, in small soil pores and in large ones. Conventional records of soil water content, θ, must be supplemented with measurements of potentials, ψ. Total water potential is subdivided into gravitational, sorptive, matric, osmotic, and others.

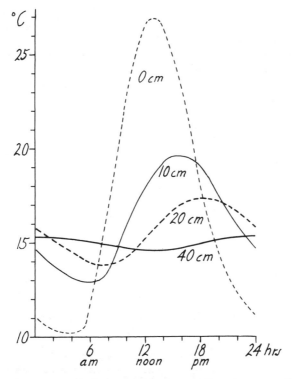

Fig. 2.2. Soil temperatures in a bare sand deposit during August at Pawlowsk (23). Note damping and retardation with increasing soil depth (cm)

B. Gravitational Potential of Water

The gravitational potential alludes to the chore of lifting a bucket of water from the floor onto a table.

As seen in Figure 2.3, to raise the mass of m grams of water—the system—from the lower to the higher reservoir, somebody or something in the "surrounding" must do work to overcome the downward pull of gravity (g). The larger the height (h), the greater the effort. The transported water gains "free energy," the term loosely implying the capacity to perform useful work. Such work, accompanied by frictional heat, is done during downhill flow over a paddle wheel. Water returns to the surrounding the energy it acquired from it. A system's gain in energy is marked as plus, a loss as work done and heat given off is marked as minus, in accord with the acquisitive convention.

The work of lifting, W, measured in calories or in ergs, is given by the equation $W = mgh$, in which g is the Earth's gravitational constant of 981 dynes/g. The mass, m, is a capacity factor whereas gh is an intensity factor that is independent of quantity, designated as potential—specifically, *gravitational potential.*

Viewed differently, division of the work equation by m, as $W/m = gh$, again gives the potential; in other words, the work or energy per *unit mass of water* is equal to the gravitational potential in ergs/gram. The potential is set to zero at the lower

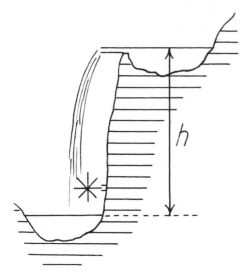

Fig. 2.3. Gravitational potential. One unit of water lifted from the lower to the upper reservoir gains the energy of position gh. In the waterfall the energy is converted to work (paddle wheel) and heat of friction

reservoir, which is taken as a reference point or standard state.

In some field studies volume is preferred to mass. Density, ρ, is defined as mass per unit volume (V), which for water is 0.997 g/cm^3 at 25°C. Accordingly, $m = \rho V$ and $W/V = \rho gh$, which is the gravitational potential per *unit volume of water*, designated ψg. The dimensions are ergs/cubic centimeter and because 1 erg = 1 dyne · centimeter they are dynes/square centimeter which signifies pressure per unit area.

The gravitational potential plays a role whenever water descends in a soil or rises to the top of a tree.

C. Sorptive Potential of Water

A dry sponge, wood chip, earth clod, or glass capillary takes up water, either directly as a liquid or as condensation of molecules from moist air. The event is initiated by chemical, not gravitational, forces.

Sorption Experiment

When a few grams of dry clay or soil used as a sample in a dish is placed above water in a sealed desiccator (Fig. 2.4, inset), the sample gains weight as water. The process is absorption, or adsorption, or simply sorption of water molecules.

The water molecule H_2O is a dipole, an unsymmetrical structure of protons and electrons, and is attracted by the electric force fields of the ions and molecules that make up the solid materials. Over a period of months certain colloidal particles capture so much water from vapor that they swell, separate, and change from a dry powder to a liquid suspension. Sand grains by themselves have negligible sorption,

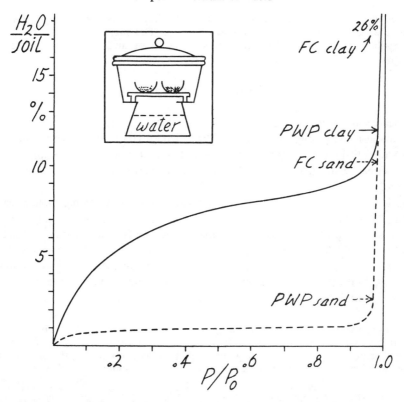

Fig. 2.4. Sorption isotherm for a sand and a clay displaying grams of sorbed water per 100 grams of soil in relation to relative vapor pressure P/P_0. Inset shows desiccator with water or solution in lower part. Adapted from Bodman and Edlefsen (4)

but, in a pile of grains, water wedges form at the points of contact. The presence of salts modifies the sorptive power of the materials because salts themselves attract water molecules.

Sorption Isotherm

An extension of the sorption experiment employs a series of desiccators each carrying different mixtures of water and sulfuric acid that lower the vapor pressures and maintain relatively dry air. At equilibrium the pressure of water vapor of the soil sample in the dish equals the vapor pressure, p, above the solution.

Typical "sorption isotherms," meaning sorption of water vapor (grams H_2O/gram soil) at constant temperature ($25°C$) and under variable vapor pressures, p, are shown for sand and clay in Figure 2.4. The vapor pressures range from very low to atmospheric (p_0), which is that over a level, free water surface of a pool. They are plotted on the horizontal axis as fractional pressures or relative humidities p/p_0.

The initial rising portions of the curves are characteristic of a gas adsorption isotherm portraying the gradual build-up of a monolayer of H_2O molecules on the surfaces of the solids. At higher humidities multiple water layers develop. These films are the "bound water" of earlier investigators. The sharp rise on the right marks "capillary condensation," which is the filling up of small and large crevasses and pores.

Energetics of Soil Water

Water molecules held on a solid surface give up kinetic energy of motion and release heat, the heat of wetting. A spoonful of dry, reddish Hawaiian clay in the palm of a hand with a little water added gets uncomfortably hot (see Fig. 6.6). Compared to free water, sorbed water has a lower escaping tendency, which explains its lower vapor pressure.

While no useful work is done in the desiccator experiment, an ingenious apparatus housing a paddle wheel or a piston might harness the flow of water molecules from pool to soil sample and get work performed. Theoretically, the maximum amount of useful work (W_s) that could be obtained in the environment in the transfer of 18 g or 1 mole (M) of water vapor from p_0 to p is given by the logarithmic expression $RT\ln(p_0/p)$ in which R is the universal gas constant and T the absolute temperature. This work term is also known as free energy change, $-\Delta G$.

When water vapor expands from p_0 to p, its own energy diminishes by an amount that is equal to the work that *could* be performed (regardless whether work is actually performed or not). Counting the work term as positive, W_s, the diminution of energy of water is viewed as negative, $-W_s$. For 1 g of water

$$-\frac{W_s}{M} = -\frac{RT}{M} \ln (p_0/p) = \frac{RT}{M} \ln (p/p_0) = \psi_s \qquad (2.1)$$

sorptive potential
of water

The formula describes the relative energy status of 1 g of sorbed water and as such is an intensity factor, the *sorptive potential,* ψ_s, expressed as ergs/gram.

Multiplication by the density of water (ρ) provides the energy of 1 cm^3 of sorbed water, which has the dimensions of a force or pressure in dynes/square centimeter. In the desiccator experiment $p < p_0$; therefore, ψ_s is negative and the state of sorbed water is said to correspond to a negative pressure, a pressure deficit or tension.

As seen in Figure 2.4 at any chosen relative vapor pressure (p/p_0) two substances may have very different moisture contents, yet, when placed adjacent to each other, water will not move from the wetter to the drier material because the water in both samples has the same potential, the same escaping tendency. But water will flow from high to low potentials, and that condition must be fulfilled for water to be transported from soil to roots.

Actual vapor pressures of leaves and soils *in situ* are measured with sensitive thermocouple psychrometers, and the ψ_s calculated are termed "leaf water" and "soil water" potentials.

D. Matric Potential of Water

Instead of measuring the quantity of sorbed water and its vapor pressure and then calculating its energy status, the latter information may be obtained by observing the changes produced in the surroundings of the sorbing sample.

Tensiometer

In the arrangement of Figure 2.5, based on the ideas of Richards (19), a moist soil sample free of salt is placed on a ceramic plate that has its fine pores filled with water connecting to water in a U-tube. The right-hand upper end carries a stopper and a pressure gauge. As the soil attracts liquid water the vacuum P is created. It is read-off the gauge as "suction" in positive numbers (e.g., atmospheres). Suction P is a manifestation of the work of sorption of 1 cm³ of water,

$$\frac{RT\rho}{M} \ln \frac{p_0}{p} = P \tag{2.2}$$

and its negative value, divided by ρ, is the matrix or *matric potential*, ψ_m, which is equal to the sorptive potential, ψ_s, of a salt-free sample. The drier the soil the larger are ψ_m and P, though of opposite sign.

For field work the tensiometer is a long tube with a porous ceramic cup at the lower end. Inserted to various soil depths the tool calibrates suction up to about 0.8 atm, which is of main concern in irrigation practice.

Example of Calculation of Suction. A certain loam soil exposed to a relative humidity of 75% ($p_0/p = 1.333$ at 25°C) holds 5.5 cm³ of water on the basis of 100 g of oven-dry soil. What is the suction?

According to chemistry handbooks the symbols in Eq. (2.2) have the following values: $R = 8.134 \cdot 10^7$ ergs/mole/degree, $T = 273.2 + 25 = 298.2°$K (absolute), $\rho = 0.997$ g/cm³, $M = 18.02$ g, hence $RT\rho/M = 137.17 \cdot 10^7$ dynes/cm². From tables, $\ln 1.333 = 0.2877$, therefore, $P = 39.346 \cdot 10^7$ dynes/cm², or, since 10^6 dynes/cm² equals 1 bar or 0.987 atm, suction $P = 388.35$ atm.

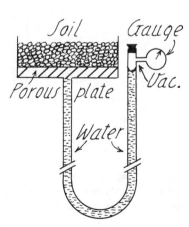

Fig. 2.5. Principle of tensiometer for measuring suction or negative pressure of water in soil. Vac., vacuum created by small transport of water from porous plate into soil

The suction force in a porous body may also be measured by an equivalent counter-force of outside pressure that expels water. For soils in the laboratory, Richards (18) employs a pressure membrane apparatus whereas Scholander (22) uses a pressure chamber for leaves, twigs, and roots.

E. Capillary Phenomena

The study of capillaries, which are tubes with very small bores, is helpful in comprehending diverse aspects of soil moisture such as surface tension, capillary rise, hanging water, repulsion of water, and vapor pressures of drops and of liquids in cavities.

Surface and Adhesion Tension

Water molecules attract each other, which is *cohesion*. To enlarge the surface area of a body of water, molecules must be brought from the interior of the liquid to the surface, and the work required is 73 ergs at 20°C for every square centimeter of new surface created. This surface work or energy σ is numerically equal to the "surface tension," which is a hypothetical surface force in dynes/centimeter. The work of cohesion of water W_c is 2 σ ergs/cm² because the breaking of a water column of 1 cm² cross-sectional area into two pieces would create two surfaces of 1 cm² area each.

Liquid water held by solids is *adhesion* and its degree is manifest in the angle of contact α measured inside the liquid phase (Fig. 2.6). The adhesion tension is $\sigma \cos \alpha$,

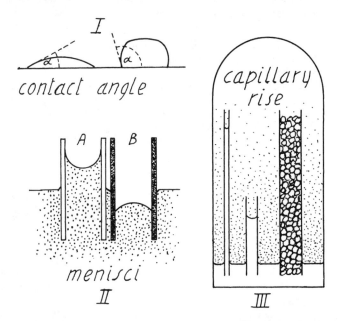

Fig. 2.6. Capillary phenomena. I: angle of contact α of a liquid on a solid; II: shape of meniscus of water in a glass (A) and paraffin (B) capillaries; III: capillary rise in narrow and wide glass capillaries and in soil column, all surrounded by equilibrium water vapor molecules

and the work needed to create 1 cm^2 of new water-solid interface is the work of adhesion W_a which is $\sigma + \sigma \cos \alpha$. When it exceeds the work of cohesion, a drop of water placed on a surface will spread and wet the surface, as is the case for most soil particles. Contrariwise, when $W_c > W_a$, a drop of liquid put on a solid assumes semispherical shape, like a dew drop on a glossy leaf. For complete wetting, $\alpha = 0°$; for no wetting whatsoever, $\alpha = 180°$.

Adhesion tension induces the rising or falling of water in a narrow tube and determines the curvature of its water meniscus, as shown in Figure 2.6. In the glass capillary (A), wetting is complete, water rises, and the meniscus is concave. In the paraffin capillary (B) of the same bore size, $\alpha = 105°$ and water molecules are repelled by the wall. Water descends and the meniscus is convex, yet σ is unchanged in either case. In the wake of forest fires organic compounds may impart repulsion to walls of soil pores and water will not wet the soil (7).

Liquid Curvature and Internal Pressure

In the paraffin capillary of Figure 2.6 (IIB), water under the convex meniscus is compressed by the "contracting skin" of surface tension, as is the water in a raindrop. For a small sphere of radius r the pressure P inside exceeds the outside air pressure P_0 by $P - P_0 = 2\sigma/r$, which is the Laplace boundary jump, formulated in 1805. Since the work of surface energy σ must be done to create 1 cm^2 of droplet surface, drops contain more free energy than water in bulk and they tend to evaporate.

For the rise in Figure 2.6, II, the concave, half-spherical meniscus has a negative surface configuration compared to the drop, a negative r, and the Laplace jump across the liquid-air boundary becomes negative,

$$- (P - P_0) = \frac{2\sigma \cos \alpha}{r} \qquad (2.3)$$

where r is the radius of the capillary and $\sigma \cos \alpha$ the adhesion tension.

The term $-(P - P_0)$ is a *negative pressure* that is viewed as a tension or stress. For example, when a liquid is being sucked out of a bottle with a straw, the liquid in the straw is under tension. Much of the sorbed water in soil and biological cavities is under tension and it is common to speak of plant water stress and soil water stress.

Capillary Rise

It is the negative pressure that drives water up a glass capillary or soil column until $-(P - P_0)$ is balanced by the weight of the water column ($\rho g h$). Since $\alpha = 0°$, Eq. (2.3) becomes $\rho g h = 2\sigma/r$, which is the textbook equation of capillary rise.

Exit of Capillary Water

To appreciate soil as a body of capillaries, lift the two tubes in Figure 2.6 III vertically out of the water basin. They will *not drain*. Equilibrated water in a capillary is hanging and will not flow into an empty wider one unless suction is applied below or pressure

exerted above. Slowly adding water to the upper meniscus bulges the lower one until a
drop descends. The situation corresponds to the difficulty of capillary water passing
from a fine-pore clay layer into a coarse sandy substratum, or into a worm or gopher
channel, or from soil into a drain.

When dry air is blown across the tops of the two upraised capillaries, *water evaporates* from them, the water columns shorten and move upward because the lower
meniscuses bearing the weights of the liquids have less curvature, hence less negative
pressure, than the upper ones. The behavior is similar to the way bare soil dries out. In
Arizona, capillary *water moves upward* on occasion as fast as water evaporates from a
bare surface layer 9 cm thick (11). Below a depth of 30 cm, evaporation is exceedingly
slow.

Curvature and Vapor Pressure

In the air-evacuated, constant-temperature chamber of Figure 2.6C, in which two capillaries and a soil column stand in a pool of water, the little dots connote the equilibrium distribution of water vapor molecules.

Their concentration gradually diminishes with elevation in accordance with Laplace's logarithmic law of atmospheric pressure,

$$RT \ln (p/p_0) = -gh \qquad (2.4)$$

in which p_0 refers to vapor pressures at the water surface and p at height h.

At the concave meniscuses of the capillaries, their equilibrium vapor pressures
must be identical with those of water vapor in the chamber at the same height.
Lord Kelvin in 1871 connected curvature and vapor pressure, written here for unit
volume (1 cm^3) of water as follows:

$$\frac{RT\rho}{M} \ln \frac{p}{p_0} = -\rho gh = -\frac{2 \sigma \cos \alpha}{r} = (P-P_0) \qquad (2.5)$$

For wet soil and plant materials that are free of dissolved solutes, a lowered vapor
pressure implies concave air-water surfaces within the pores and the water under
negative pressure.

The porous matrix in the soil column of Figure 2.6C is near saturation in the lower
parts and relatively dry in the upper. And although the height of the "capillary fringe"
cannot be predicted, when equilibrium is reached the vapor pressure in the soil at any
height is equal to that in the adjacent chamber space, and the potentials vary accordingly.

F. Osmotic Potential

Salts, bases, acids, and many organic substances, called solutes when dissolved in water,
which is then the solvent, lower the vapor pressure of free water. In porous bodies
they contribute to sorption potentials.

Osmotic Model

In Figure 2.7 pure water with vapor pressure p_0 is separated from salt solution with vapor pressure p, with $p < p_0$, by the membrane m, which is permeable to water but not to solute. To some extent plant membranes and clay layers perform this function. When the weightless pistons in both arms of the apparatus are under the same external pressure, e.g., atmospheric, water diffuses into the salt solution from right to left. The process is the time-honored *osmosis*.

Eventually so much water will "osmose" that the weight, $\rho g h/cm^2$, of the rising liquid in the left arm stops further influx. Equilibrium is then achieved. Osmosis can be prevented either by applying excess pressure on the left piston, the increase being the osmotic pressure π—which is equal to $\rho g h$—or by inducing the suction $-\pi$ on the right side arm.

At equilibrium, p, p_0, and π are connected as

$$\frac{RT\rho}{M} \ln \frac{p}{p_0} = -\pi \qquad (2.6)$$

The left-hand term is again the sorption potential, and here it is equal to the negative of the osmotic pressure, which is osmotic suction. It is the *osmotic potential* ψ_o in dynes/square centimeter. For a mole of solute (e.g., 342 g of cane sugar) in 1000 g of water, π is about 24 bars, provided the solute does not dissociate or react chemically with water.

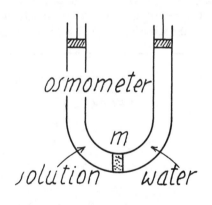

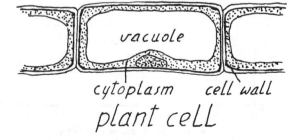

Fig. 2.7. Osmotic potential illustrated in an osmometer and
in a plant cell (simplified)

Applications

When nutrient concentration in the cell sap of roots is higher than in the outside solution, water enters the cell, inflates, and stiffens it, which is turgor having the pressure *TP*. Maximum turgor equals the osmotic pressure and as long as *TP* is less than
π, water keeps entering the cell. Since the osmotic potential is the negative of π, the
cell water potential is *TP* - π, which is a negative quantity.

Water extracts of saline and alkali soils may generate potentials of over 10 atm
which seriously interfere with good growth of crop plants (20). Roots penetrating a
strongly saline soil and leaf tissue becoming impregnated with salts from ocean sprays
may suffer plasmolysis, which is water flowing out of the cells causing plasma disruptions and eventual death.

G. Summation of Potentials

Above a ground water table the major potentials of soil-plant ecosystems are gravitational (ψ_g) and sorptive (ψ_s), the latter being composed of matrix (ψ_m) and osmotic
potential (ψ_o), or $\psi_s = \psi_m + \psi_o$. The two potentials are additive, provided the salt
ions do not alter the soil matrix configuration. Tensiometers measure ψ_m but not ψ_o
because their porous cups are permeable to solute. Likewise, pressure chambers for
detecting water stress in plant materials respond to ψ_m only. The ψ_o is determined in
expressed cell saps and displaced soil solutions by physicochemical methods.

How sharply biomass of soybean plants responds to the sum of $\psi_m + \psi_o$, which is
a composite stress or suction, is shown vividly in Figure 2.8 in which the decimal numbers indicate the salt percentage on a dry soil basis (20).

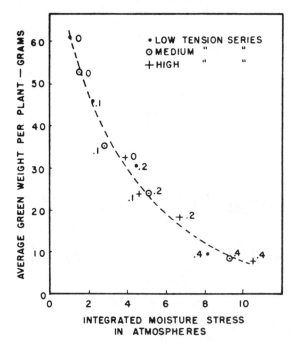

Fig. 2.8. Growth of bean plants
as influenced by total soil
moisture stress. The salinity
level for teach treatment is indicated as percentage on a dry
soil basis. Graph from U.S. Salinity Laboratory (20)

Highest yield (61.0 g) is obtained at low matric tension (ψ_m = 1.1 atm) with no salt present. As chemically harmless salt is added to it, the yields decline along the curve to 9.6 g. The compensation of potentials is marked in the middle of the curve where similar yields (30-35 g) are produced by low ψ_m (1.1 atm) plus medium salt (0.2%), medium ψ_m (1.5 atm) plus low salt (0.1%), and high ψ_m (3.9 atm) without salt.

The most general index of energy of moisture is provided by the "chemical potential" (6). In soils it is still fraught with conceptual difficulties (3). Soil moisture literature is extensive (e.g., 16, 28). Water potentials in plants are detailed in the books of Slatyer (24) and Kramer (12).

On the potential scale of Figure 2.9 a level water surface at atmospheric pressure (1.013 bar) has a potential of zero that denotes "free water." Air dry soils have large negative potentials, that is, high suctions. Positive pressures are reached below ground water tables or in vines and twigs before the leaves emerge. Flow of water is from right to left.

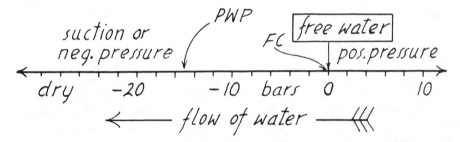

Fig. 2.9. Memory aid for distinguishing positive and negative pressures and suctions. Note approximate positions of field capacity (FC) and permanent wilting percentage (PWP) to be elaborated on later

H. Movement of Water

Moisture flows downward, upward, and sideways along gradients of decreasing water potentials.

Darcy's Law (1856)

In a uniform soil the volume v of water passing every second in the x-direction through a cross-sectional plane of 1 cm^2 area is proportional (K) to the potential gradient ($\Delta\psi/\Delta x$) across the area

$$v = -K \frac{\Delta\psi}{\Delta x} ,$$

the negative sign indicating a "downhill" force (Fig. 2.10). The simplicity of the Darcy equation is deceptive, for its application to specific conditions in plants and soils taxes experimental and mathematical ingenuity.

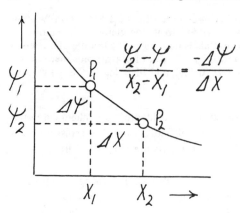

Fig. 2.10. Explaining negative gradient or slope from point P_1 with potential ψ_1 to point P_2 with potential ψ_2

$$\frac{\psi_2 - \psi_1}{X_2 - X_1} = \frac{-\Delta \psi}{\Delta X}$$

Flow in Saturated Soil

The tube in Figure 2.11 has in its lower part a metal sieve to hold the soil column, which is L cm long. A constant water level, that is, a positive water pressure, is maintained at height H above the base where water flows out continually, the system being at steady state. The driving forces are liquid pressures on the water and the pull of gravity. The potential difference is known as hydraulic head difference H. The gradient is H/L, and flux v equals $K \cdot H/L$. For unit gradient, v in a chosen time interval is equal to K, the hydraulic conductivity. It is an important and sensitive soil characteristic.

Water moves rapidly through the network of large pores of sands and sluggishly through the fine pores of loams and clays. In some clay soils K is less than 1 cm per day and the soils stay wet and are "poorly drained."

For three types of soil textures the K values at low suction are plotted on the vertical axis in Figure 2.12, adapted from data from the U.S. Salinity Laboratory in Riverside, California. There is a 10- to 100-fold increase in conductivity from clay to loam to sand.

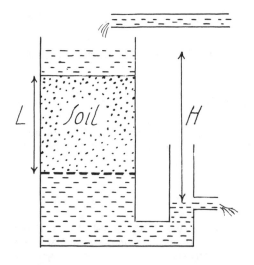

Fig. 2.11. Illustration of saturated flow; H = hydraulic head composed of water pressure and gravity pull, H/L = gradient. All large and small soil pores are filled with water

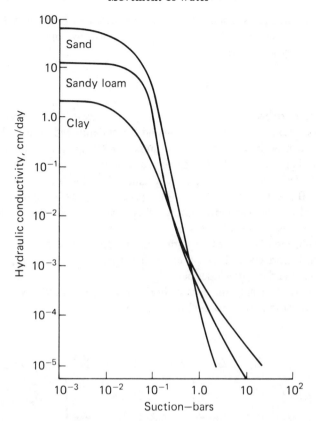

Fig. 2.12. Values of conductivity K for a sand, a sandy loam, and a clay at increasing suctions. Near the vertical axis the soils are water saturated; to the right they become increasingly dry (graph from U.S. Salinity Laboratory)

Flow in Unsaturated Soil

When in Figure 2.11 the inflow of water at the head is stopped, the column drains and out-flow ceases. All large pores are emptied, and in the upper portion smaller pores are likewise drained. Throughout the column water remains in the very small pores and capillaries and is under tension. At the lower end the soil is virtually saturated, the meniscuses facing the atmosphere below having large contact angles. This uneven moisture distribution with stagnant water at the bottom applies for lysimeters and for flower pots as well. It may create reducing conditions.

Water resumes flow when slight suction is imposed at the bottom to overcome the adhesion forces, or when the column is brought in contact with a finer dry soil that sucks it down. For unsaturated flow in horizontal direction it is the gradient in matric potentials that induces water movement. In vertical flow the gravitational potential must be included.

Again Darcy's equation applies but conductivity is strongly modified by water content and tension, as seen by the rapidly declining K values in the right-hand part of

Figure 2.12. At the high suctions of dry soils the curves cross and clays become better conductors than sands. As soils dry out molecular diffusion of moisture in the vapor phase assumes heightened proportions.

Field Conductivities

In pastures of northern Ohio, Taylor et al. (27) encased large, undisturbed tessera blocks of about 1.5 X 1.5 m cross section and 1.6 m depth and measured conductivity K for saturated flow in the presence of grass cover (Fig. 2.13). *Miami silt loam* is moderately well drained, has 27-33% of clay, and bulk densities of 1.5-1.6 in the B and C horizons. *Toledo silty clay* is poorly drained, has 47-59% clay and densities of 1.4-1.5 in the B and C horizons. It has the grayish and mottled appearance of Humic Gley soil or Mollic Haplaquept. During rains, water is ponding and for successful cropping drains must be installed.

In texture-differentiated profiles with a coarse sandy loam A horizon resting on a fine clay loam B horizon, a sharp potential drop spans the contact plane with positive pressures above it and negative ones below.

On a slope, water inside the surface horizon will flow laterally when rainfall rate exceeds the infiltration rate into the lower horizon.

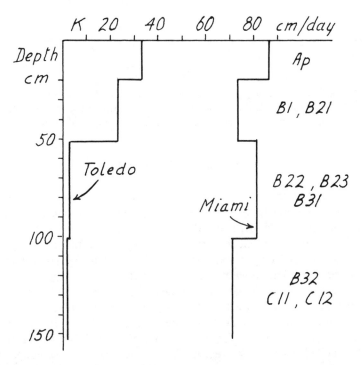

Fig. 2.13. Conductivity values K of horizons in natural soil tessera blocks of Toledo and Miami soils. After Taylor et al. (27)

Temperature-Induced Flow of Water

All preceding formulations postulate constant temperature. In view of the diurnal heat fluctuations plotted in Figure 2.2, complications are to be expected. Jackson (11) observes capillary soil water flowing upward or downward, depending on temperature gradients. While it is current knowledge that rising temperatures augment vapor pressure, conductivity, diffusion, and biological activity and reduce sorption and surface and adhesion tensions, it has been impossible so far to integrate the interactions quantitatively.

I. Flow of Water through the Ecosystem

From soil as source, water moves to roots, to stems, to leaves, and into the atmospheric sink along a path of decreasing potentials. The plant is an efficient transmitter of water, and the removal of brush and trees and the clearcutting of forests multiplies runoff, soil leaching, and stream flow.

Potential and Resistance

In 1928 Gradmann (10) assigned in the soil-plant-atmosphere system (SPA) (Fig. 2.14) zero suction to a soil saturated with water, 50 bars to leaf surface, and 1000 bars to atmospheric air.

Greatest pull and resistance are experienced at the plant-air boundary. Suctions up and down a tree are measurable by subjecting severed twigs to the pressure bomb.

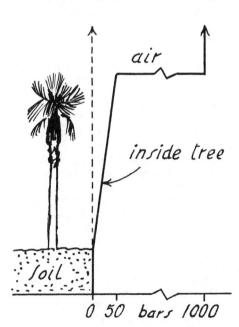

Fig. 2.14. Gradmann's model of flow of water through the soil-plant-atmosphere system along suction gradients

Van den Honert (31) revived the nearly forgotten Gradmann theory and introduced resistances in analogy with Ohm's law of electric current, as $v = (\psi_1 - \psi_2)/R$, where $\psi_1 - \psi_2$ is the potential difference and R the resistance. It is a general flow equation written for steady-state conditions. Discussions and experimentations have been expanded by many workers (12, 13).

Domains of SPA

Pertinent strata of the ecosystem line up as follows:

In atmosphere:	wind turbulence,
	boundary layer
In plant:	stomatal pores in leaves, vascular system of stems, root tissue
In soil:	root zone,
	water table

The horizontal lines mark the sites of important boundary problems and at the bottom a water table allows steady-state conditions of flow which simplifies mathematical treatment.

Environmental Atmosphere. In still air inside a forest dispersal of water molecules proceeds by diffusion, an average molecule traveling 1 cm in a few seconds. In rainy, foggy weather transpiration at an even temperature may be nil.

Above the tree canopy *air turbulence* accelerates the rate of removal, but the millimeter-thin *boundary layer* on the surface of each leaf may become a rate-limiting diffusion step.

Stomatal Pores in Leaves. Botanists single out the microscopic stomata openings on leaves as the main avenues of exit of water vapor and entry of CO_2. The orifices control transpiration by opening and closing their apertures, possibly in response to water deficits of leaf cells. In the stomata, as elsewhere, the conversion of a gram of water to vapor consumes 586 cal of heat and the needed energy is supplied through solar radiation and convection. Transpiration is a cooling process.

Vascular Systems of Stems. Upward water transport in stems and trunks utilizes the xylem vessels, which are cross-walled and have diameters of 20-800 μm (12). Capillary ascent in them could reach 15 m which is too short for supplying tall trees. Water might be propelled upward by root pressure resulting from osmosis, but the speed would never reach the 10-20 m/hr that have been recorded. Kramer (12) prefers the cohesion theory of long standing. It postulates a continuous network of capillary water held together by its own cohesion forces and extending from roots to the transpiring surfaces of the leaves, a sort of giant wick anchored in the soil and "burning" at the top in response to atmospheric suction. Though it has not yet been made clear how water gets to the tree top in the first place, once the vessels are filled the tensile strength transmits the transpirational pull from leaf to root to soil. At a relative air humidity (RH) of 50% the evaporative suction at the leaf surface is 938 atm; inside the stomata, at 97% RH, it is only 42 atm.

Root Tissue and Root-Soil Boundary. The xylem conduits do not reach the root periphery. Unless a metabolic water pump is operative, which is still in doubt, soil water must reach the xylem by osmosis and protoplasmic streaming through the cell interiors or by cohesive pull through ultrafine capillaries or "free spaces" in the cell walls that surround the cytoplasm. Unlike the discontinuous leaf-atmosphere boundary, the root-soil region presents on the submicroscopic scale a continuous network of channels (see Fig. 3.21) in which water flows into and sometimes out of the root in line with the directions of potential gradients, as will now be examined.

J. Soil Water for Plants

Water stress in plants develops when transpiration demands exceed the moisture-supplying power of the soil.

Availability of Soil Water

In filled-up large pores water has nearly zero potential. In spite of high water availability plants may suffer in saturated soils because diminished oxygen supply and high CO_2 pressure interfere with root metabolism and ion uptake. A freshly drained soil having the slight negative potential of 0.1-0.25 bars offers optimum conditions for growth.

 The small soil volume A adjacent to a root (Fig. 2.15) has a high water content (θ) and low suction. Transpirational pull imparts to the root the higher suction ψ_{root} and water flows from soil into root. As the volume A is being depleted the gradient of root-soil potentials declines. Flow may cease altogether unless water in A is replenished from the B area. Temporary wilting may ensue.
 For a simple root model Gardner (9) obtains solutions of the flux equation for soils of various textures. They are in accord with field observations and point to the soil's water-transmitting coefficient K as a determinant of water availability.

As the soil dries out water potential and conductivity assume such low values that the plant is unable to meet its water demands. It wilts permanently, shrivels, and dies, or, in the case of a tree, sheds its leaves. For many domestic plants, even for cacti and young pines, the growth-limiting soil water content θ corresponds to matric potentials of -10 to -20 bars, with a mean of about -15 bars. It is the *permanent wilting percentage* (PWP) proposed by Briggs and Shantz (5) in 1912. This

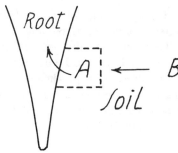

Fig. 2.15. Schematic root with adjacent soil volume A and distant volume B to illustrate water delivery

widely used ecopedological criterion is routinely measured in greenhouse and laboratory. In coarse-textured soils PWP corresponds to less than 5% H_2O by weight and in fine-textured ones to 10-15% or more. McColl (15) observed the high soil suction of 70 bars in a tall eucalyptus forest that showed no visible signs of stress.

Soil Water Storage

Sands have pore space volumes of 30-40% and clay loams and clay range from 40 to 60% and even higher. Soil animals and growing roots continually modify the pore space, but in studies of flow and storage of water, a stable pore geometry is tacitly assumed.

During rains or irrigation the surficial layer of a deep soil has all pores filled with water. When application ceases the large pores keep draining into the dry soil below. In 2-3 days flow comes to a near halt and water hangs in capillaries. The transition from wet to dry soil is visibly sharp. The wet soil is said to be at *field capacity* (FC), measured as percentage water in oven-dry soil. Water is under negative pressure (suction) of approximately one-third bar in medium-textured soils and about one-tenth bar in sands. So important is the concept as a measure of water storage for plants and watershed management that it is being routinely determined on horizon samples brought into the laboratory, e.g., as moisture equivalent (M.E.) by centrifuging, and as one-third atmosphere percentage by the suction plate method.

Field capacity depends foremost on the texture of a soil and the nature of its clay minerals, also on soil structure. On oven-dry basis the span of water contents comprises 4.4% for a gravelly sand, 9.0% for a Bareilly loamy sand, 14.3% for a Yolo sandy loam, 24.4% for a Yolo loam, and 36.2% for a Sacramento clay. Some clays may exceed 50%. As humus builds up in a mineral soil the field capacity is decisively augmented, at the rate of 2.20-2.25% of water for each percent humus for the loams and sandy loams of Figure 2.16 (conventionally, humus = 1.724 · organic carbon).

> For 97 soils Klemmedson (Chapter 12) determined organic carbon (% C), clay, silt, and sand contents and water-holding capacity (M.E.) by centrifuging untreated wet soil at 1000 g. In the collection eight soils had near-identical loam textures, and seven were near-identical sandy loams. The rise in percentage C at constant texture is attributable to increasing mean annual precipitation.

For a given rainfall and in absence of runoff, field capacity controls the depth of water penetration. An old rule of thumb of arid agriculture states that 1 cm of ponded water covering air dry soil will wet a sand to a depth of 6-9 cm or more, a loam to 3-5 cm, and a clay to 2-3 cm. When the soil is initially at permanent wilting percentage the distances are approximately doubled. In many applied soil moisture problems FC is similarly expressed as centimeters of water to a chosen depth of soil; thus, the aforementioned Yolo loam with a bulk density of 1.3 g/cm^3 stores to a soil depth of 100 cm a rainfall of 31.7 cm.

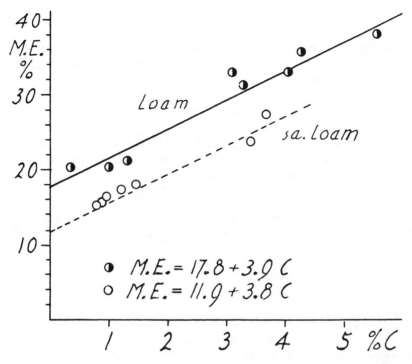

Fig. 2.16. Organic soil carbon (% C) augments field capacity, measured as moisture equivalent (M.E.), illustrated for loams and sandy loams of textural near-identities along a moisture transect (Chapter 13)

Available Water

The moisture difference of field capacity and permanent wilting percentage is the *available water capacity* (AWC), the supply of water to plants in a well-draining soil. In watershed hydrology AWC is the soil's renewable storage volume. For the previously mentioned Yolo loam, PWP is 7.0% and AWC = 24.4% – 7.0% = 17.4% water; hence, a recharge of uniform soil to 1 m depth is accomplished by a rainfall of 22.6 cm and that amount is at the disposal of plants, provided direct evaporation from the soil is banned. Humus augments AWC, according to Rode (21), because it raises FC more than PWP.

The question of how plant growth responds to soil moisture stress between FC and PWP permits a variety of answers. Photosynthesis and respiration appear less sensitive than elongation of organs and production of fresh weights. As flow rates of soil water to the root diminish, the supply of dissolved nutrients to the root likewise is reduced, and nitrate starvation may be the first symptom to accompany plant water stress. The nitrogen-fixing legumes have an advantage.

On gentle clay loam slopes in summer-dry savanna climates the dense mass of fibrous roots of the bunch grasses "goes after the water" and quickly exhausts the A and B horizons. Shrub and tree roots penetrate into deep C horizons and rock crevasses, hence dark-green oaks decorate the rocky knolls and stony ravines that are surrounded by expanses of dry, golden-yellow grasslands.

K. Evapotranspiration and Water Balance

Transpiration returns moisture to the skies, cools the leaves, lifts nutrients, and, combined with evaporation from soils, functions as precipitation in reverse. Evapotranspiration is a salient climatic factor.

Evapotranspiration

Early in this century the team of Briggs and Shantz (5) studied the water requirements of agricultural crops in the Great Plains area by growing plants in well-watered pots. They discovered that dry matter (Y) produced during a season correlates closely with water transpired (Tr) and with water evaporated (E) from a large pan.

In reexamining older and newer studies, Arkley (1) concludes that a given amount of water (Tr) is more growth-effective at high than at low air humidity. Writing Tr/Y instead of his Y/Tr the equation, in milliliters per gram of yield, becomes

$$\frac{Tr}{Y} = k\,(100 - RH)$$

where RH denotes percentage relative atmospheric humidity. In words, transpiration per unit plant weight is proportional to the relative saturation deficit of the air, i.e., proportional to suction at the leaf surface, as would be expected from the cohesional theory of transpiration. For sorghum $k = 5.62$, for corn (*Zea mays*) 6.78, and for oats 12.66.

Viets (32) insists that pots and small field plots experience exaggerated transpiration rates because they are exposed to advective heat carried horizontally by warm winds. Large fields receive their energy of transpiration predominantly in a vertical direction from net radiation, and their transpiration is independent of yield after reasonably complete cover is attained. As it were, the green transpiring layer is simply being elevated as photosynthesis and growth proceed.

At agroclimatic stations, water pans and atmometer cups are placed within grass crop areas or pastures that are extensive enough to minimize border and oasis disturbances caused by advective heat flow. More advantageously, large and deep weighing-lysimeters of several meters diameter sitting on underground delicate balances are planted with grass and set flush with surrounding sod. High soil moisture of low suction is maintained continuously, and the water consumption of the tank ecosystem is termed *"potential evapotranspiration"* (Pet). At semiarid Davis near Sacramento, California, average annual Pet values of about 135 cm per unit area have been registered, whereas the rain gauges fill up to only 42 cm. Compared to grass cover, the rough vegetation surfaces of vineyards and tree crops augment Pet because turbulence of dry air moving over the canopy is enhanced.

Bare soil at field capacity has a high evaporation rate, but as the surface dries, vapor emanation is curtailed to such an extent that during desiccation periods lasting many months the soil does not dry out below 20-30 cm depth. In semiarid regions this conserving of subsoil water at field capacity is the principle of dry-land farming in which land is kept bare for a year or two. Mulches in the form of paper and straw layers or as natural forest floors reduce evaporation from soil.

It has long been a goal to compute evapotranspiration directly from meteorological parameters. In 1948 Penman (17) and Thornthwaite (29) independently published their fruitful procedures. Penman's more theoretical equations utilize vapor pressure, wind, sunshine, and the extraterrestrial radiation and its reflection to calculate the heat available for distilling water, whether from pond, soil, or leaf. These factors are not routinely measured by weather stations. Thornthwaite, like many before him, substitutes air temperatures for the solar energy budget, and his empirical equations, calibrated on United States watersheds, include day-length and latitude but neither wind, sunlight, nor humidity. The variables of his equation are abundantly available and the tedious computations of monthly Pets have been shortened by graphic nomograms (30).

It has since become apparent that Thornthwaite's index tends to overestimate Pet in maritime temperate climates and underestimate it in arid regions. At Sacramento, California, calculated Pet (2) is only 81 cm against the aforementioned 135 cm measured at nearby Davis. Comparisons with Penman's index reveal the following deviations in centimeters: Seattle, Washington, +7.6; Spokane, Washington, ± 0; Sacramento, California, – 17.8; Yuma, Arizona, – 28.0, quoted from Magnuson (14). Perhaps inclusion of the physiologically important relative humidity might bridge the gap but since few stations measure it, the advantage of possessing many data is lost.

Water Balance

The inputs of water, P, into an ecosystem includes all forms of precipitation, fog drip, and irrigation. Outputs consist of actual evapotranspiration, Aet, and through-drainage, Dr. On gentle, vegetated slopes runoff and runin are small and their difference is here assumed to be negligible. The yearly water balance reads

$$P = \text{Aet} + \text{Dr} + \Delta\text{Ms}$$

with ΔMs signifying changes in soil moisture, which can be determined by sampling or in place with a neutron probe. In dry regions Dr and ΔMs are zero for an average year and Aet is simply P. The method is illustrated in Table 2.1 for Davis, California, using a 5- to 7-year period of weighing lysimeter readings for grass.

Row 1:   Long-time mean *monthly temperatures* in $^\circ$C are included for orientation. They are not needed here because at this station Pet is measured, not computed.

Row 2:   Long-time mean *monthly precipitation (P)*.

Row 3:   *Potential evapotranspiration*, Pet, actually measured.

Row 4:   *P minus Pet*. Row 4a has + values with $P >$ Pet. It is surplus water (14.82 cm) available for deep penetration. Arkley (1) designates it as leaching value Li. Row 4b has negative values that indicate water deficits in amounts of 107.73 cm for a 9-month period. To secure maximum crop growth, that much water would have to be supplied by irrigation, or rather more would have to be supplied because of unavoidable losses in watering practice.

Row 5:   *Water storage in soil.* Counting begins here in December, always with the first month of water surplus following the dry period. The monthly surpluses are added until the maximum of 14.82 cm is reached in February. Withdrawal starts in March and all available soil moisture is used up by the end of April.

Table 2.1. Soil and Vegetation Water Balance (Davis, California)

Row	Jan.	Feb.	Mar.	Apr.	May	June	July	Aug.	Sept.	Oct.	Nov.	Dec.	Year
1 Temperature (°C)	8.2	10.7	12.9	15.3	19.0	22.8	24.7	24.0	21.6	17.7	12.6	8.7	16.51
2 Precipitation (P) (cm)	8.05	8.00	6.02	3.10	1.45	0.38	0	0	0.13	2.21	3.58	9.19	42.11
3 Pet (cm)	2.44	5.33	8.59	11.94	15.95	20.02	21.55	18.43	13.75	9.91	4.46	2.65	135.02
4a P-Pet (cm, positive)	5.61	2.67	–	–	–	–	–	–	–	–	–	6.54	14.82
4b P-Pet (cm, negative)	–	–	2.57	8.84	14.50	19.64	21.55	18.43	13.62	7.70	0.88	–	107.73
5 Storage (Ms) (cm)	12.15	14.82	12.25	3.41	0	0	0	0	0	0	0	6.54	–
6 Depth (cm)	54	66										29	–
7 Aet (cm)	2.44	5.33	8.59	11.94	4.86	0.38	0	0	0.13	2.21	3.58	2.65	42.11

Row 6: *Depth of wetting* by surplus water calculated for the aforementioned Yolo loam. If the roots extend to the maximum depth of water penetration (66 cm) then all subsoil moisture is consumed by plants and there is no through-drainage, Dr. Calcareous soils would tend to develop a $CaCO_3$ horizon at this depth. For shallow root systems, through-drainage beyond the root zone must be accounted for separately.

Row 7: *Actual evapotranspiration,* Aet, is a measure of biological activity and to some extent of biomass production. It is calculated by combining P, Pet, and moisture storage. During June, September, October, and November the soil is dry and Pet $> P$. Thence, Aet cannot exceed P. In December, January, and February $P >$ Pet, therefore, Aet = Pet. During March and April again Aet = Pet but is now composed of P and storage withdrawal (2.57 cm in March and 8.84 cm in April). In May, Aet is 4.86 cm, contributed by P (1.45 cm) and final storage depletion (3.41 cm). Annual Aet equals annual P because there is neither runoff nor through-drainage as would take place in wetter and cooler climates.

Arkley and Ulrich's (2) calculation of Aet for nearby Sacramento, using Thornthwaite's method, predicts maximum actual evapotranspiration will occur in May and June instead of March and April. The leaching value Li is less seriously affected by the discrepancy in Pets, being 18.6 cm instead of the 14.8 cm in Table 2.1.

To generalize, actual evapotranspiration (Aet) is limited in arid lands by lack of precipitation and in cold regions by small Pet; in either case, vegetative growth is scant. In the warm, humid tropics Pet and Aet are high and biomass production thrives, provided soil fertility is abundant.

Soil Taxonomy (26) publishes examples of water balances of its five classificatory soil moisture regimes: *aridic* (moisture deficient during most of the year, as normally found in arid climates); *udic* (well supplied with water throughout the year, as is common in humid climates with well-distributed rainfall); *ustic* (intermediate between the two); *xeric* (cool, wet winters alternating with warm, dry summers, typical of Mediterranean climates); and *acquic* (water-saturated by high water tables and capillary fringes and the water depleted of oxygen).

L. Review of Chapter

Soil is the water reservoir for growth of land plants and small fauna. Water is "hanging" in pores and only part of it is available to roots.

Water potentials (ψ) pertain to the free energy of unit mass of water, that is, the capability of doing useful work. At a given temperature, water at any position in the ecosystem flows from higher to lower water potentials.

Gravitational potential refers to energy of vertical position of water. It is highest in tree crowns, lower in A horizons and still lower in B and C horizons. Dry soils sorb or suck in water vapor and liquid water. The energy of unit sorbed water corresponds to the *sorptive potential,* which is the sum of the *osmotic potential,* when salts are present, and the *matric potential,* which deals with water held in capillaries and pores. There it exists under negative pressure or suction, compared to free pool water, which has zero potential. Dry soils have matric potentials below -15 bars and many plants wilt permanently at the corresponding water percentage (PWP). Water will not flow from a wet to a dry site unless the former has a higher potential (a less negative one) than the latter.

For a given gradient of water potentials the velocity of water flow decreases rapidly as the soil dries out.

Potential evapotranspiration occurs at soil moisture contents of near-zero potentials. Computed *actual* evapotranspiration is broadly linked to plant biomass production and relies on soil storage depletion and refill, that is, soil water balances.

References

1. Arkley, R. J. 1963. *Hilgardia* 34: 559-584.
2. Arkley, R. J., and R. Ulrich. 1962. *Hilgardia* 32: 443-462.
3. Babcock, K. L. 1963. *Hilgardia* 34: 417-542.
4. Bodman, G. B., and N. E. Edlefsen. 1934. *Soil Sci.* 38: 425-444.
5. Briggs, L. J., and H. L. Shantz. 1912. *Bot. Gaz.* 53: 229-235.
6. Day, P. R. 1942. *Soil Sci.* 54: 391-400.
7. DeBano, L. F., and J. Letey (Eds.). 1968. Water repellent soils. Proc. Symp. Univ. Calif. Riverside.
8. Delamare Deboutteville, C. 1951. *Microfaune du sol des pays tempérés et tropicaux.* Hermann, Paris.
9. Gardner, W. R. 1960. *Soil Sci.* 89: 63-73.
10. Gradmann, H. 1928. *Jahrb. Wiss. Bot.* 69: 1-100.
11. Jackson, R. D. 1973. In *Field Soil Water Regime,* R. R. Bruce, ed., pp. 37-50. Soil Sci. Soc. Amer., Sp. Publ. 5, Madison, Wisc.
12. Kramer, P. J. 1969. *Plant and Soil Water Relationships.* McGraw-Hill, New York.
13. Lemon, E. R., D. W. Stewart, R. W. Shawcroft, and S. E. Jensen. 1973. In *Field Soil Water Regime,* R. R. Bruce, ed., pp. 57-67. Soil Sci. Soc. Amer. Sp. Publ. 5, Madison, Wisc.
14. Magnuson, M. D. 1968. Evapotranspiration. In *Proc. of West. Reg. Techn. Work-Planning Conf. Coop Soil Survey* (mimegr.). Univ. of Calif., Riverside.
15. McColl, J. G. 1977. *Soil Sci. Soc. Am. J.* 41: 984-988.
16. Nielsen, D. R. (Ed.). 1972. *Soil Water.* Amer. Soc. Agron., Madison, Wisc.
17. Penman, H. L. 1948. *Roy. Soc. London Proc. A* 193: 120-146.
18. Richards, L. A. 1941. *Soil Sci.* 51: 377-386.
19. Richards, L. A. 1949. *Soil Sci.* 68: 95-112.
20. Richards, L. A. (Ed.). 1952. *Saline and Alkali Soils.* U.S.D.A. Agr. Handbook No. 60.
21. Rode, A. A. 1955. *The Moisture Properties of Soils and Underground Strata* (in Russian). Transl. Israeli Progr. Sci. Trans., Jerusalem, 1960.
22. Scholander, P. F., H. T. Hammel, E. D. Bradstreet, and E. A. Hemmingsen. 1965. *Science* 148: 339-346.
23. Schubert, J. 1930. Das Verhalten des Bodens gegen Wärme. In *Handbuch d. Bodenlehre,* E. Blanck, ed., Vol. 6, pp. 342-375. Julius Springer, Berlin.
24. Slatyer, R. O. (Ed.). 1973. *Plant Response to Climatic Factors.* Unesco, Paris.
25. Smith, G. D., F. Newhall, L. H. Robinson, and D. Swanson. 1964. *Soil-Temperature Regimes: Their Characteristics and Predictability.* U.S.D.A. SCS-TP-144.
26. Staff (Soil Survey). 1975. *Soil Taxonomy.* U.S.D.A., S.C.S. Agr. Handb. 436.
27. Taylor, G. S., E. Stibbe, T. J. Thiel, and J. H. Jones. 1970. *Hydraulic Conductivity Profiles of Toledo and Miami Soils as Measured by Field Monoliths.* Ohio Agr. Res. and Dev. Center, Research Bull. 1025.

28. Taylor, S. A., and G. L. Ashcroft. 1972. *Physical Edaphology*. W. H. Freeman, San Francisco.
29. Thornthwaite, C. W. 1948. *Geogr. Rev.* 38: 55-94.
30. United States Dept. Agr., S.C.S. 1960. *Palmer-Havens Diagram for Computing Potential Evapotranspiration by the Thornthwaite Method*. Portland, Oregon.
31. van den Honert, T. H. 1948. *Disc. Faraday Soc.* 3: 146-153.
32. Viets, F. G. 1964. *Plant Food Rev.* 10: No. 2:2-4.

3. Behavior of Ions in Soils and Plant Responses

The chemical elements that function in ecosystems appear as neutral atoms, molecules, and electrically charged ions.

The *vert space* with its plants and animals is dominated by carbon (C) atoms arranged as organic chains, rings, and networks,

$$- \ C \ - \ C \ - \ C \ - \ C \ -$$

with associated oxygen (O), hydrogen (H), and nitrogen (N) atoms, all held together by strong chemical bonds. Total mineral content of vegetation is usually less than 5% of air-dry weight.

The *soil space* is filled with clusters of inorganic ions that have been compared to assortments of tiny billiard balls of diverse sizes and having one or more electric charges. Most sand grains, for instance, are ensembles of large—large as atoms go—negatively charged oxygen ions (O^{2-}) held together by the electrical attraction of small, highly positively charged silicon ions (Si^{4+}). Schematically

$$- \ O^{2-} \ - \ Si^{4+} \ - \ O^{2-} \ - \ Si^{4+} \ - O^{2-} \ - \ Si^{4+} \ - \ O^{2-} \ -$$

Fewer than 1% by weight of the soil ions are mobile. A minute fraction is transported daily to vert space by way of root uptake because green plants cannot create the life-sustaining sugars, starches, and proteins unless critical, soil-derived, mineral elements are present at the biochemical reaction sites. In litter fall and plant remains the vert space ions return to the soil and participate anew in nutrient cycling.

A. Biological Elements

By 1870 the botanist Sachs could announce the "ten essential elements" required for plant growth, namely, C, O, H, N, P, S, K, Ca, Mg, Fe. Except for Fe, they are arranged here as *macroelements* because they are needed by the plant in relatively large amounts (1, 22).

C (carbon), O (oxygen), H (hydrogen). Carbon is the key atom of all forms of life and, along with O and H atoms, constitutes the bulk of organic matter. The three are brought together in photosynthesis.

N (nitrogen). Nitrogen is an integral constituent of all proteins, amino acids, and nucleic acids.

P (phosphorus). Phosphorus is a component of nucleo- and other proteins and of many enzymes; it is involved in phosphorylation and energy transfer.

S (sulfur). Sulfur occurs as sulfur linkage in many proteins and is a component of the vitamin biotin.

K (potassium). Potassium enhances swelling, plays a role in translocations of organic molecules, and operates in the stomatal opening of leaves.

Ca (calcium). Calcium is a stabilizer of macrocmolecular chains and bundles (e.g., pectins).

Mg (magnesium). Magnesium is an integral part of the chlorophyll molecule and cofactor of many enzymes.

Dry plants contain about 1-2% of K and nearly as much Ca, and 0.1-0.3% of P, S, and Mg.

Except for iron, quantities of *micro-* or *trace elements* in plant tissues are so minute that they are counted in parts per million (ppm) or parts per billion (ppb). In water cultures Sachs and later botanists overlooked them because the "macrochemicals" contained them as impurities.

Fe (iron). Iron is a component of respiratory enzymes and of ferredoxin which is active in photosynthesis and nitrogen fixation.

B (boron). Boron is involved in cell division, possibly in sugar translocation, and in building carbohydrate esters and cell wall structures.

Mn (manganese), Zn (zinc), Mo (molybdenum), Cu (copper), Cl (chlorine). Like iron these are involved as cofactors in the enzyme reactions that manipulate the life processes. Manganese and copper serve with oxidation-reduction enzymes, whereas molybdenum is pertinent to nitrate reductase. Zinc, in addition to stimulating auxin formation, stabilizes the association between catalytic and regulatory subunits of enzymes.

Deficiency of the mineral elements is manifested in discoloration of leaves, malformation of fruits, lessened resistance to infections, retarded growth, lack of reproduction, and, finally, death.

Presumably, all higher plants need the 16 elements cited. In addition, blue-green algae and nitrogen-fixing bacteria need *cobalt (Co)*. Diatoms require *Si*, and one of the green algae requires *vanadium (V)*. Livestock must have *selenium (Se)*, which is tied to vitamin E activity, and, together with man, need *iodine (I)* to prevent goiter. *Fluorine (F)* conveys resistance to tooth decay in vertebrates, and *Si* in grasses and cereals conveys resistance to attacks of pathogenic fungi and bacteria. It may become necessary to distinguish the *physiologically essential* elements of plants grown in isolation in greenhouses from *ecologically essential* elements that guarantee survival in a hostile natural environment full of competitors and predators.

At higher concentrations in soil solutions aluminum, manganese, and boron become toxic. Man pollutes ecosystems with *lead (Pb)*, *arsenic (As)*, *mercury (Hg)*, and radioactive ions.

B. Ions in Rock and Soil Crystals and in Solutions

Atoms consist of a nucleus with a positive electric charge surrounded by fast moving electrons (e^-) with a negative charge. The number of electrons equals the charge on the nucleus, and the atom is neutral. When atoms lose one or more electrons they become positively charged cations; when they gain electrons they become negatively charged anions. The two are symbolized by the nutrient ions K^+, NH_4^+, Ca^{2+}, Cl^-, NO_3^-, and SO_4^{2-}. The numbers of + and − signs indicate the net charges per ion, or its valence. Roughly, ions may be considered fairly rigid spheres, with like charges repelling and unlike charges attracting one another. Sizes of ions in Angstrom units (1 Å = 10^{-8} cm) are plotted in Figure 3.1. They are the Goldschmidt radii for soil minerals.

Crystal Lattice Energy

Ions interact according to Coulomb's law, which says that in air the *force (F)* of repulsion or attraction between two monovalent ions A^+, A^+ or B^-, B^- or A^+, B^- varies as the product of their charges (ϵ) divided by the square of the distance (d) of separation, or $F = \pm \epsilon \cdot \epsilon / d^2$. For ions touching each other, d is the sum of their radii.

The *work (W)* done in pairing A^+ and B^- is $-\epsilon_A \cdot \epsilon_B / (r_A + r_B)$, the negative sign implying that the two ions give up free energy in coming together. For sodium (Na^+) and chloride (Cl^-) the 118,000 cal thus generated for 1 mole (6.06 × 10^{23}) of pairs is viewed as the energy needed to break the ionic bonds. To arrange these ion pairs into the crystal lattice structure of table salt (NaCl), shown in Figure 3.6, involves further attractions as well as localized repulsions, which bring the calculated *lattice energy* to 184,000 cal. The experimental value is 183,000 cal. Most crystals are more complicated than NaCl but the calculations invoke Coulomb's law.

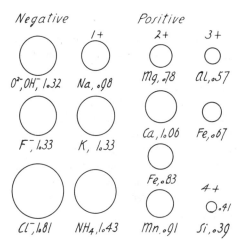

Fig. 3.1. Goldschmidt radii of ions in silicate crystals, in Angstrom-units

The greater the lattice energy, the more stable, harder, and chemically resistant is a crystal. Stability is favored by small ions with high charges, such as silicon (Si^{4+}), aluminum (Al^{3+}), and iron (Fe^{3+}). Combined with the large oxygen (O^{2-}) and hydroxyl (OH^-) anions the three cations constitute the aluminosilicate networks of rocks and soils.

Aluminosilicate Crystals

The basic building units are the four-sided *tetrahedron* with small Si^{4+} in the central cavity of $4O^{2-}$ and the eight-sided *octahedron* with Al^{3+} or Mg^{2+} in the interstice of six touching O^{2-} and OH^-, as displayed in Figure 3.2 as clusters of spheres and as their

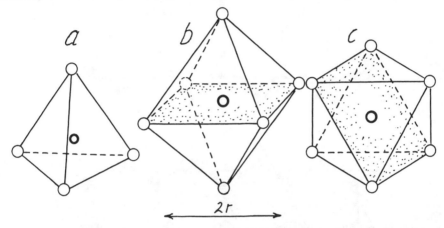

Fig. 3.2. Building units of aluminiosilicates: the tetrahedron and the octahedron, displayed as spheres of O^{2-} ions and hidden Si^{4+} and Al^{3+} ions. In their geometrical presentation the spheres are not drawn to size. In b the octahedron is "standing up," in c it is "resting"

corresponding geometrical images. (The reader is encouraged to construct these models with rubber balls, marbles, and cardboard.)

Tetrahedra are aligned as isolated units, as chains, ribbons, and sheets (Fig. 3.3) and as three-dimensional crystals (see Fig. 4.1). A sheet of OH^- octahedra with Mg^{2+} in each cavity, or Al^{3+} in two out of three, is drawn in Figure 3.4.

The combination structure in Figure 3.5 is a layered aluminosilicate crystal (pyrophyllite). A sheet of O^{2-} and OH^- octahedra with Al^{3+} in two-thirds of the interstices is sandwiched between two sheets of Si^{4+}-O^{2-} tetrahedra. The interior of the platy

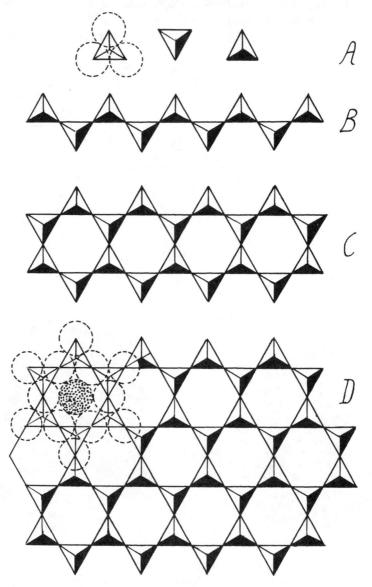

Fig. 3.3. Line up of SiO_4 tetrahedra in a plane. (A) Individuals; (B) a chain; (C) a band or ribbon; (D) a net

crystal and the exposed planar surfaces are electrically balanced, hence neutral. Electric stray forces reside at exposed, unsatisfied ions of corners, edges, and vertical faces. They attract other ions, water molecules, and organic substances.

A Note on Chemical Bonds

Coulomb's law applied to the NaCl crystal typifies the strong *ionic bond,* illustrated in Figure 3.6. Neutral atoms and molecules may attract each other weakly if they are dipoles having unsymmetrical internal distributions of protons and electrons. The resulting attractive forces contribute to the *van der Waals* bonds of cohesion, important in orienting and holding together inert organic carbon chains as in paraffin and humus. Ions attract dipoles too, as when Na^+ surrounds itself with a shell of neutral water molecules. The small H^+ ion, a mere proton, is especially effective and when placed between two oxygen ions or between oxygen and nitrogen holds them together as *hydrogen bond.* One of the strongest bonds is the *covalent bond* or electron-pair bond in which two atoms share two electrons. It is the foremost force among the carbon atoms in organic molecules. Calculation of its molecule energy utilizes the methods of quantum mechanics. In many compounds and with such nutrient ions as Cu^{2+} and Zn^{2+} several of the bonds, sketched in Figure 3.6, operate simultaneously (52).

Ions in Solution

When a crystal dissolves, the released ions gain freedom of action and become chemically highly reactive. Some of the polyvalent ions may reassociate to neutral ion pairs.

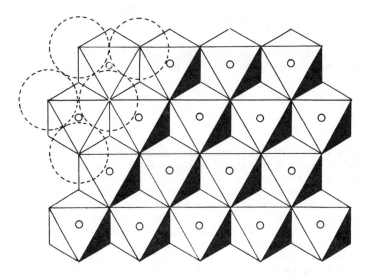

Fig. 3.4. A layer of octahedra. In hydroxide crystals the corners are occupied by OH^- ions. When the center of each octahedron contains a Mg^{2+} ion, the crystal is brucite, $Mg(OH)_2$; when two out of three—the third remaining empty—carry Al^{3+} ions, the crystal in gibbsite ($Al(OH)_3$). The size of the interstice sphere is drawn in smaller proportion

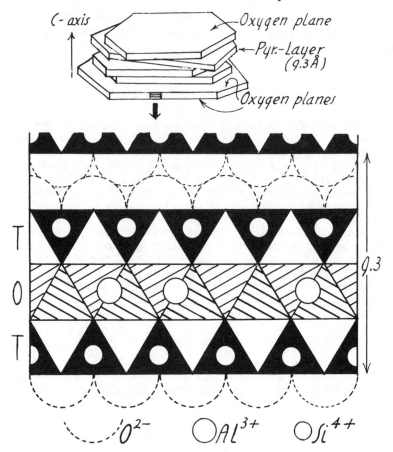

Fig. 3.5. Above: five layers of the mineral pyrophyllite ($Al_2 Si_4 O_{10} (OH)_2$) stacked irregularly to form a platy colloidal clay particle. Below: arrangement of sheets of (black) tetrahedra (T) and (striped) octahedra (O) in a pyrophyllite layer (Pyr), 9.3 Å thick. The neutral oxygen planes correspond to Fig. 3.3D, viewed from below

Hydration of Ions. In nature the water molecule (H_2O) is the most prominent dipole and its lineup with a cation is shown in Figure 3.6, as ion-dipole bond. An ion's electric field strength, which attracts the water molecules, is inversely proportional to r^2; hence, it is the small ions with high charges that are most strongly hydrated. The first and rather rigid hydration shell often contains 6 H_2O. Farther away from the ion surface the water molecules are held less strongly. Wandering ions in solution carry their water envelopes with them. At exposed crystal surfaces ions may be but partially hydrated.

The tiny hydrogen ion H^+ is a cardinal constituent of acids and attaches itself to a water molecule forming the monovalent hydronium cation $(H_3O)^+$. It has an approximate effective radius of about 1.4 Å. For typographical brevity the symbol H^+ is retained in ensuing chemical equations and the hydration shells of other ions are omitted.

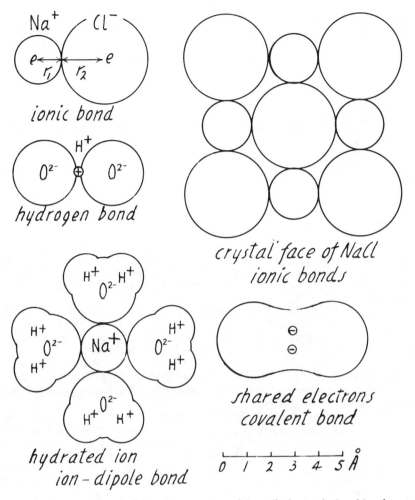

Fig. 3.6. Prominent chemical bonds encountered in soil-plant relationships (see text)

Chemical Potential. Work is required to remove a molecule or a pair of ions from a solution. The change in free energy, when 1 mole of substance i, such as sugar or Na^+Cl^-, is removed or added to a large volume of solution, is its chemical potential μ_i written

$$\mu_i = \mu_i^\circ + RT \ln \gamma_i m_i$$

where μ_i° is the chemical potential in a standard reference state, m_i is the concentration of i as moles per kilogram of water, and γ_i is the activity coefficient. The product $\gamma_i m_i$ is the activity or active mass a_i. At high dilutions ions are far apart, do not interfere with one another, and $\gamma_i = 1$. Very little is known about γ_i in living tissues or in soil pores of high water stress, but inferences are drawn from ion concentrations in extracts of plants and soils.

Law of Mass Action. The multitude of ionic and organic reactions in soils and organisms is encompassed by the versatile law of mass action, written for a simple case as

$$A + B \quad \rightleftharpoons \quad C + D$$
$$\text{reactants} \quad\quad\quad \text{products}$$

One mole of substance A reacts with one mole of B to be converted to 1 mole each of C and D. In many soil and organism processes conversion is not complete and A, B, C, and D coexist. Equilibrium may then be approached from either side indicated by the two opposing arrows. The tendency for a reaction to occur is greatest when it is farthest from equilibrium.

Applying the chemical potential μ leads to the temperature-dependent equilibrium constant K. In its simplest form it is

$$K = \frac{[C] \cdot [D]}{[A] \cdot [B]}$$

the brackets denoting activities. The constant lends itself to calculating the important free energy change ΔG and, therewith, the maximum work that could be obtained from the reaction under ideal conditions. For 1 mole of each of the substances, $\Delta G = -RT \ln K$.

An important corollary is the "solubility product" of a *saturated* solution. When crystal $A^+ B^-$ dissolves into A^+ and B^- ions, $A^+ B^- \rightleftharpoons A^+ + B^-$, the product $[A^+] \cdot [B^-]$ is a constant at a given temperature. Adding to the solution either A^+ or B^- ions from a second salt (e.g., $A^+ C^-$) reduces the solubility of the first.

Note. Analytically, most elements are determined as concentrations; when used in the mass action equation they require conversion to activities by multiplication with γ, the activity coefficient.

Chemical symbols vary in different areas of research. Phosphorus appears as P in leaf testing literature, as pentoxide (P_2O_5) in rock analyses, and as PO_4-P, viz. P present as PO_4, in soil solution work. Potassium in any form is K; when stressed as ion it is K^+; the oxide is K_2O. N-NH_4 indicates nitrogen as an ammonium ion, N-NO_3 as a nitrate ion.

C. Exchangeable Ions and Ion Exchange

In aquatic ecosystems the ions are dissolved and obey the laws of diffusion and dilute solutions. In soils most ions have limited mobility because they are intimately allied with solid particles, including clay and humus. A whole new set of rules governs their conduct.

Soil, an Ion Exchanger

In Way's experiment of 1850 a dilute solution of potassium chloride (KCl) is percolated through a column of soil, as shown in Figure 3.7. In the effluent the anion Cl^- comes through, whereas the cation K^+ stays behind. In its place the element Ca^{2+} accompanies Cl^-. Today we write in law-of-mass-action style:

Fig. 3.7. Way's experiment (1850) of adding dilute KCl
solution to soil. K is retained by the soil particles and in
its place Ca appears in the filtrate; Cl is not affected

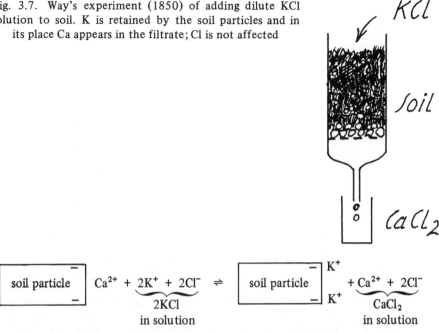

Calcium attached to or adsorbed on soil particles is replaced by two potassium ions. The reaction is termed base exchange, an older term, cation exchange, or simply ion exchange (29). It is prominent on the extensive surfaces that characterize the colloidal realm.

Most soil chemists write the above equation:

$$Ca^{2+} - \text{particle} + 2K^+ \ \rightleftharpoons \ K_2^+ - \text{particle} + Ca^{2+}$$

The Cl^- is left out because it does not participate in the reaction. Either formulation will be used in ensuing sections depending on what features are stressed.

Ion Exchange with Clay Particles

Clays are assemblages of tetrahedra and octahedra, arranged either at random, as in amorphous clays, or as crystallized layer lattices. Many are conceptually related to pyrophyllite.

Charge-Dependent Exchange Capacity. During synthesis of a clay platelet (Fig. 3.5) a trivalent Al ion instead of a tetravalent Si may be caught in one of the tetrahedral interstices. The otherwise neutral oxygen plane acquires a surplus negative charge in direct proportion to the number of Al^{3+} substituting for Si^{4+}. Likewise, charge imbalance may have its source in the interior of the octahedral layer where small Mg^{2+} may substitute for Al^{3+}. The pyrophyllite clay then becomes a smectite type, beidellite in the first instance, montmorillonite in the second. Nature rectifies the excess negative charge by placing extra cations—K^+, Na^+, or Ca^{2+}—on the oxygen plane (Fig. 3.8). These balancing cations are held weakly and in water they try to diffuse away from the plane

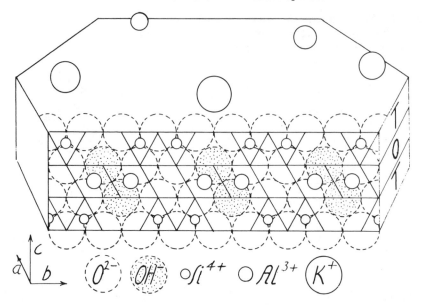

Fig. 3.8. Pyrophyllite layer modified by Al^{3+} substituting for Si^{4+} in tetrahedra or Mg^{2+} for Al^{3+} in octahedra (neither shown). The resulting excess negative charge on the oxygen plane is balanced by exchangeable K^+ (solid circles)

but are pulled back again electrostatically, continually oscillating to and fro, in a sort of thermal motion. They are the exchangeable cations. Their number characterizes the cation, charge-dependent, *exchange capacity* (CEC). It may exceed 100 me/100 g of clay. It is largely independent of pH, as seen in Figure 3.9 for Yolo clay loam soil.

Electric Double Layers. An exposed plane of negative O^{2-} ions and its balancing swarm of exchangeable cations constitute an electric double layer, schematized in Figure 3.10, left, for Na-clay. Beyond the outer cation layer no electric field exists, and, to an observer, a bacterium for instance, the face appears neutral. Regardless of whether the particle settles out or rises, the ion cloud is dragged along.

Addition of the salt NaCl to a Na-clay suspension allows individual Na^+ to join the ion swarm, whereas Cl^- is repelled by the negative inner wall. The interaction forces salt to concentrate in the bulk of the solution. This is negative adsorption or anion exclusion.

Based on the mathematical double layer theories of Helmholtz, Gouy, and later workers (25) the concentration profiles of ions near the surface are of the kind shown in Figure 3.10, right. The thickness of the double layer (< 100 Å) corresponds to the distance from the wall to the electrical center of gravity of the exchangeable ions, and it varies with their nature.

Ion Exchange Model. When KCl is dissolved in a suspension of Na-clay, a wandering K^+ may perchance slip between the wall and an oscillating Na^+, take its position, and repel it into the solution phase where it associates with the lone Cl^-. The process is *cation exchange*, viewed as a statistical phenomenon (33) but written in mass action form as

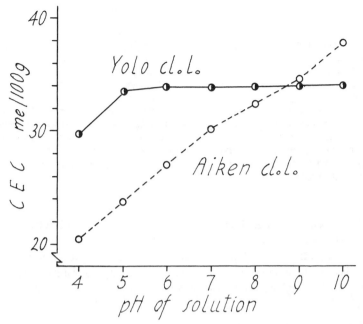

Fig. 3.9. Cation exchange capacity (CEC) related to pH of replacing solution (NH₄ -
acetate). Yolo clay loam soil has montmorillonitic minerals; Aiken clay loam is rich
in kaolinitic and oxide clays that render CEC pH-dependent

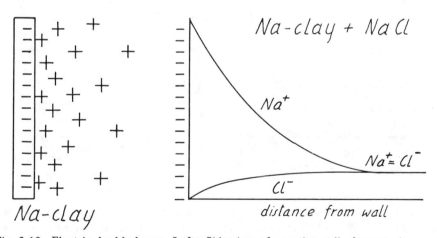

Fig. 3.10. Electric double layers. Left: Side-view of negative wall of oxygen ions and
positive swarm of exchangeable Na⁺ ions. Right: Concentration profiles of Na⁺ and
Cl⁻ in relation to distance from wall

$$Na^+\text{-clay} + K^+Cl^- \rightleftharpoons K^+\text{-clay} + Na^+Cl^-$$

The reaction may be initiated from the left or from the right, and at equilibrium the ion swarm of a clay particle embraces both K^+ and Na^+ ions.

As long as the exchange sites on the oxygen plane are not too far apart the oscillating cations may displace one another sideways, inducing *surface migration*. It provides a jumping mechanism of ion diffusion along clay plates. A further variant is *contact exchange* between two planar surfaces forced together face to face so that their double layers intermingle. Cations exchange without an electrolyte solution necessarily being present. An extension is *exchange diffusion* of cations inside a gel in charged pores and channels without solution anion partners (36). These mechanisms are shown in Figure 3.11. They are operative in clay soils, humus bodies, and organic membranes (59).

The pH-Dependent Exchange Sites. Soils in low latitudes abound in clay lattices that are high in Al or Fe ions surrounded by sheets of hydroxyl (OH⁻) anions (illustrated in Fig. 3.12 for a platelet of kaolinite). Although it lacks substitutions in tetrahedra and octahedra layers, the mineral possesses exchange capacity, explained here as follows: on the vertical faces of the crystal the broken bonds of exposed Si and Al ions attract water molecules that dissociate and cover the surface with OH ions. For an Al site:

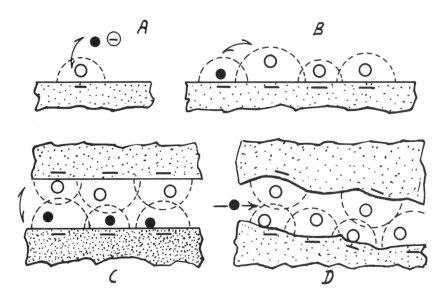

Fig. 3.11. Behavior of exchangeable cations, the dashed semicircles suggesting the domains of oscillation. (A) Cation exchange with a salt solution composed of ● cation and ⊖ anion. (B) Surface migration on a clay surface. (C) Contact exchange of cations between two negative surfaces. (D) Exchange diffusion in a pore of a gel. No soluble anions need be present in B, C, or D

$$\begin{array}{ccccccc}
\text{H} & & \text{H} & & & \text{H} & & \text{H}\\
\text{---} & \text{O} & \text{---} & \text{O} & & \text{---} & \text{O} & \text{---} & \text{O}\\
& & & | & & & & & |\\
& & & \text{Al} + \text{H}_2\text{O} & \rightleftharpoons & & & & \text{Al} - \text{OH}\\
& & & | & & & & & |\\
& & & \text{O} & & & & & \text{OH}
\end{array}$$

At high pH some of the H are exchangeable according to the reaction $-\text{Al}-\text{OH} + \text{NaOH} \rightleftharpoons -\text{Al}-\text{O}\cdot\text{Na} + \text{H}_2\text{O}$. The particle is negatively charged and carries exchangeable Na ions. The displacement intensifies with rising pH and the extent achieved is the *pH-dependent exchange capacity* (Fig. 3.9).

Anion Exchange. Many soils and seepage waters contain particles of Al and Fe oxides and hydroxides composed of Fe^{3+}, Al^{3+}, O^{2-}, and OH^- ions. Their surfaces may be covered with OH ions as in the above case of kaolinite. In presence of salt (e.g., NaCl) at low pH the entire OH^- anion may be replaced by Cl^-: $=\text{Fe}-\text{OH} + \text{NaCl} \rightleftharpoons =\text{Fe}\cdot\text{Cl} + \text{NaOH}$. The particle is now positive and possesses anion exchange capacity, Cl^- being replaceable by nitrates (NO_3^-), bicarbonates (HCO_3^-), and sulfates (SO_4^{2-}). More sophisticated models of synthesis of + charge, also involving H_2O and OH, have been discussed by various authors (15). As part of a double layer the exchangeable anions in soils possess surface mobility, yet are protected from being leached by rains (64) unless first exchanged into the percolating soil solution.

High valency anions of the type PO_4^{3-} (phosphate) effectively replace surficial lattice OH^-, and they are held with great tenacity on the positive sites of exposed Fe^{3+} and Al^{3+}. Unsatisfied negative valences of sorbed PO_4 may raise CEC. Strong phos-

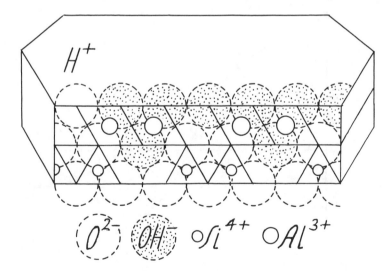

Fig. 3.12. Front view of a kaolinite-type of clay platelet showing an exposed horizontal sheet of hydroxyl (OH^-) anions. The OH^- ion at the upper left corner is dissociated into an oxygen ion (O^{2-}) and an exchangeable hydrogen ion (H^+)

phoric acid (H_3PO_4) actually decomposes kaolinite and hydrous oxides and is precipitated as sparingly soluble strengite ($FePO_4$) and variscite ($AlPO_4$) minerals.

In the iron-rich Oxisols and lateritic soils phosphate retention and concomitant unavailability exerts tremendous pressure on natural selection of plant species (4), and it poses wide-spread adversity in tropical and subtropical agriculture. In rice paddies prolonged flooding initiates reducing conditions, and P of strengite becomes available (51).

Amphoteric Exchangers. When two platinum wires connected to the poles of a battery are inserted in a suspension, the particles migrate visibly to either the negative or the positive wire depending on whether exchangeable cations or anions dominate. The observed "electrophoretic" velocity of particles is often recalculated as an electric potential difference, the zeta-potential (ζ). In a Belgian Alfisol colloidal particles of A horizons move faster toward the + pole than particles of B horizons, suggesting greater transportability of the former (14).

Many particles are two-sided or *amphoteric*. At low pH they are positive and migrate to the − pole; at high pH they are negative and move to the + pole. Montmorillonite clay partially coated with $Fe(OH)_3$ performs in this manner; so do the proteins. The reaction (pH) or the ion activity of the liquid at which no movement is observed is the *isoelectric point* of the colloid, illustrated in Figure 3.13. The particle may then have no charge at all and no exchange properties, or, it may carry an equal number of + and − charge sites with combinations of cation and anion exchange. Negative smectite clay particles turn positive when thorium (Th^{4+}) is added as $ThCl_4$, seen in Figure 3.13, and the ion becomes practically nonexchangeable. The "zero-point charge" of clays has been keyed to degree of horizon development of Spodosols (30).

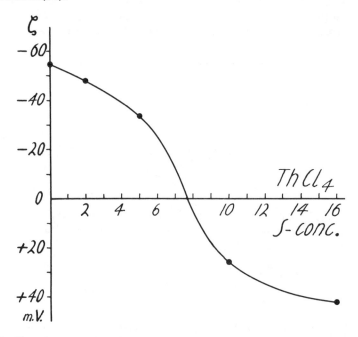

Fig. 3.13. Negative smectite clay turns positive when thorium salt; $ThCl_4$, is added. S, symmetry concentration; mV, millivolts. Zeta (ζ)-potential is proportional to electrophoretic velocity

A relevant historical vignette is Aristotle's claim that sea water is made potable by passing it through soil. Removal of both Na^+ and Cl^- is required, hence the soil must be amphoteric. Some of the Mediterranean red earths may satisfy the criterion.

Organic Exchangers

Organic acids (Fig. 3.14) carry carboxl —COOH groups that are weakly dissociated $(-COO^-H^+)$ and are neutralized by a base (Fig. 3.13).

$$R - COOH + NaOH \rightleftharpoons R - COO^-Na^+ + HOH$$
$$\text{dissociated} \quad H_2O$$

R denotes the remainder of the molecule. When carboxyl groups are incorporated in colloidal particles, as in soil organic matter or in micelles of plant cell walls, they impart negative charges with balancing cation swarms and function as seats of cation exchange. The capacity (CEC) is of the pH-dependent type. With bacteria, fungi, yeasts, algae, lichen, and mosses, Knight et al. (42) record a parallelism of CEC with uronic acid content. In roots of angiosperms carboxyl groups of galacturonic acid are identified as exchange sites (40).

The long list compiled by Drake et al. (19) cites *root capacities* of 100 me/100 g for delphinium, 10-20 me/100 g for grasses, and double that or more for legumes. The amphoteric proteins in roots and soil fauna tissues are cation and anion exchangers that accommodate up to 100 me/100 g of either type of ion, depending on solution pH. The anion sites reside at $-NH_2$ groups (Fig. 3.14), the reaction $-NH_2 + HCl \rightleftharpoons -NH_3^+ \, Cl^-$ furnishing exchangeable Cl^- ions.

In *soil humus* cation exchange sites are contributed by carboxyls, hydroxyls (—COH) of alcohols and phenols, and by other groups (Chapter 5).

> For irrigated Ramona sandy loam having a range of organic carbon contents from about 0.4 to 1.3%, Pratt (57) obtains for 14 manured plots the regression CEC = 2.4 + 4.9%C and a correlation coefficient of 0.98. The contribution of C to capacity (CEC) at pH 7 is contained in the slope of the curve, which is 4.9. It corresponds to 490 me/100 g of C, or 284 me/100 g of humus, using a carbon to humus conversion factor of 1.724.
>
> In Klemmedson's (41) set of 52 California soils derived from granitic rocks are 13 loams that have near-identical quantities of sand, silt, and clay. Correlation analysis between CEC of these soils and their percentages of carbon, 0.4-5.5%C, produces a CEC at pH 7 of 199 me/100 g for humus *in situ*.
>
> Klemmedson's samples are from humid coniferous forests. Those of Pratt are from cultivated, semiarid lands.

Ion Exchange Equations

For equilibrium studies clay particles or organic exchangers are processed to have uniform ion swarms containing one species only, say NH_4^+. To aliquots of a NH_4^+-clay suspension are added various amounts of a salt, often in S-units, which are fractions or

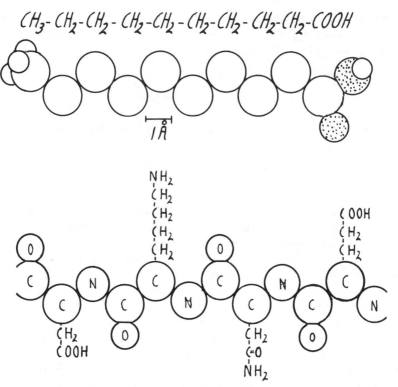

Fig. 3.14. Components of organic ion exchangers. Above: A fatty acid chain having a COOH group at one end. In the scale model the H_2 are not drawn. Below: Short segment of a protein molecule with side groups for cation ($-COOH$) and anion ($-NH_2$) exchange

multiples of equivalents of the exchangeable NH_4^+ originally present. After prolonged shaking at constant temperature the clays are sedimented and the clear supernatant solutions are analyzed for cations and anions.

In Figure 3.15 the percentages of the exchangeable NH_4^+ replaced from the particles by ingoing Na^+, K^+, and Ca^{2+} are plotted as four exchange isotherms (33). The NH_4^+-K^+ exchange is written

$$NH_4^+\text{-clay} \quad + \quad K^+ Cl^- \quad \rightleftharpoons \quad K^+\text{-clay} \quad + \quad NH_4^+ Cl^-$$

| solid | in | solid | in |
| phase | solution | phase | solution |

Experimentally, the reaction runs from left to right. Had initially K^+-clay been prepared and subsequently $NH_4^+ Cl^-$ added, the event would follow the right to left arrow. In either case the same equilibrium situation is achieved.

Because the chloride anion Cl^- does not participate in the reaction, the equilibrium constant K of the law of mass action is simplified to

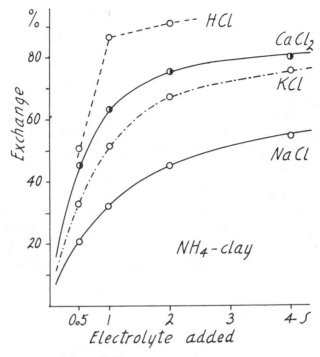

Fig. 3.15. Ion exchange curves for NH_4-clay suspensions (1.5%). Salts and HCL added in proportion to exchangeable NH_4 in 500 ml suspension. (The HCl curve is confounded with Al^{3+} dissolved by the acid)

$$K = \frac{(K^+\text{-clay}) \cdot [NH_4^+]}{(NH_4^+\text{-clay}) \cdot [K^+]}, \text{ which is also } \frac{(K/NH_4) \text{ in exchanger}}{[K/NH_4] \text{ in solution}}$$

known as selectivity coefficient. The solution ions are written as activities (brackets), the exchangeable ions as concentrations (parentheses) because their activities are not readily obtained. For the potassium curve in Figure 3.15 the coefficient is 1.10, for the sodium curve it is 4.56; therefore, the swarm of Na^+ is more diffuse than that of K^+ or NH_4^+, meaning that Na^+ is held loosely by the negative surface, compared to K^+ and NH_4^+. It is in accord with the idea that Na^+ on exposed clay surfaces is more hydrated than K^+.

The preference of negative surfaces for K^+ over Na^+ is widespread among clays, soil, and peats and has vast geochemical implications because the selectivity tends to leave Na^+ in the soil solution phase to be carried by percolating waters into creeks, rivers, and oceans where it accumulates. This trend is reinforced by vegetation that consumes more K than Na. Whereas in igneous rocks the ratio Na/K averages 1.68/1, it widens to 4.63/1 in rivers and lakes and to 47.18/1 in the sea. In confirmation, the sedimentary rocks, which may be ancient soil materials, have the narrow ratio of 0.31/1.

When divalent cations, as $CaCl_2$ or $MgCl_2$, replace NH_4 the reaction is written

$$2\,NH_4^+\text{-clay} + Ca^{2+} + 2\,Cl^- \rightleftharpoons Ca^{2+}\text{-clay} + 2\,NH_4^+ + 2\,Cl^-$$

The formal mass action law no longer yields a constant and often the left-hand and right-hand approaches give discordant results. The discrepancy is aggravated by trivalent and higher valent cations.

A simple criterion for *comparisons of ion behavior* is the percentage of exchange achieved when the number of equivalents of added salt matches the NH_4^+ equivalents in the exchanger, that is $S = 1$. In Figure 3.15 the percentages of exchange are 32 for Na^+, 51 for K^+, and 63 for Ca^{2+}. Such values are graphed in Figure 3.16 against the crystal lattice radii of the ingoing cations. The higher the valency of a cation the more readily it displaces NH_4^+ and the more tightly it is held on the surface. At the same valency, the more hydrated cations (Li^+, Na^+, Mg^{2+}) are the less effective replacers of NH_4^+ and are loosely held. The cations cesium (Cs^+) and strontium (Sr^{2+}) are rare in ecosystems but are assuming prominence as radioactive polluters. Figure 3.16, and what has been said about it, pertains to the beidellite type of clay. In other exchangers the ion sequences or "lyotropic series" are somewhat different. Further, the nature of the anion, whether Cl^-, NO_3^-, SO_4^{2-}, or $H_2PO_4^-$, also guides the extent of cation exchange (69). Thermodynamic equations are discussed by Babcock (2) and Sposito (60).

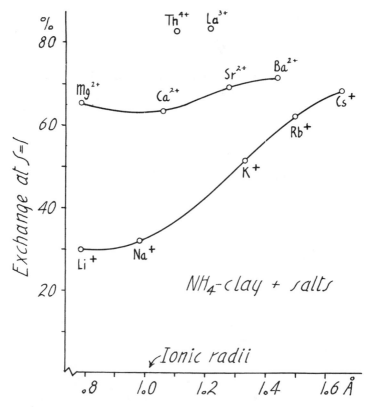

Fig. 3.16. Percentage of NH_4 ions released from NH_4-clay by salts (chlorides) of cations of the first two periods of the periodic table (rare gas type ions). Lyotropic series. (Courtesy The Williams and Wilkins Co., Baltimore, MD)

Concerns of waste disposal direct attention to exchange reactions of ions of heavy metals: Zinc (Zn^{2+}), copper (Cu^{2+}), nickel (Ni^{2+}), cadmium (Cd^{2+}), etc. In NaCl solutions of saline soils the metal cations attract the Cl^- anions and unite to complexes having reduced charges. Thus, the ion pair ($Cd^{2+}Cl^-$)$^+$ behaves as a monovalent cation that is less strongly held on clay particles than the normal, divalent Cd^{2+}. Lessened adsorbability implies heightened mobility in the soil (16). The anions and molecules that are coordinated with a cation are named ligands.

D. Ion Currents by Diffusion

Water flowing up a tree or down a soil column along gradients of water potentials sweeps solution ions with it, which is mass flow of ions. In water resting in pores, dissolved salts move by diffusion, which is random walking of ions driven by gradients of concentrations (39).

Diffusion Visualized

The small segment of a long cylinder in Figure 3.17 is filled with watery agar jelly, an organic network with large pores in which ions diffuse as readily as in pure water. The left-half compartment contains dissolved KCl of uniform concentration C^0, the right-half water only. The net movement of salt is to the right. Water proceeds to the left to fill the spaces vacated by the ions. When equilibrium is reached KCl is evenly distributed over both compartments. At any time during the process the agar plug may be pushed out of the tube and sliced into thin discs that are analyzed for KCl.

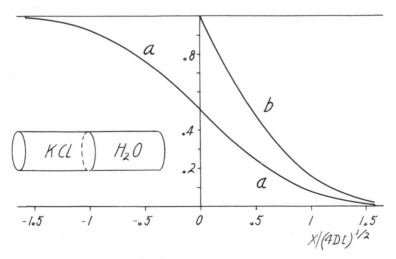

Fig. 3.17. Sketch of diffusion cell (left) and curves a and b for transitory diffusion described in text

Transitory Diffusion

Prior to equilibrium the sigmoidal curve a in Figure 3.17 traces the KCl concentrations C after a chosen diffusion time t. At the boundary "O" the relative concentration is always 0.5. In the left portion of curve a the quasi-triangular area *above* the sigmoid branch is the quantity of KCl that moved to the right side. It is shown again in the right compartment as the same area *under* the sigmoid branch.

In the equation of the curve,

$$C = \frac{C^0}{2} \left(1 - \text{erf}\ \frac{x}{2\sqrt{Dt}}\right)$$

where erf is the normal probability or error function, x is the distance in centimeters from the central boundary plane of $x = 0$, t is the time in seconds, and D is the *diffusion coefficient*. The quantity of ions that passes in 1 sec through 1 cm^2 of boundary area is the flow, current, or flux, usually written as $J = -D(\partial C/\partial x)$. D is the flux per gradient in square centimeters per second.

For KCl in agar, D of Cl$^-$ is 1.6; in a paste of red Aiken soil Moreno (47) finds a value of 1.0; and in dead roots Charley (9) observes a value of 0.10; all three values are to be multiplied by 10^{-5} cm^2/sec. For phosphates (K$_2$HPO$_4$), Moreno obtains values of 0.65 in agar and 0.062 in a silty clay loam at field capacity; at 6 bar tension Olsen et al. (49) find the value to be 0.004; all three D's are to be multiplied by 10^{-5} cm^2/sec.

The diffusion coefficient D measures the mobility of a substance in a given medium. When a formula by Einstein is applied to the above values, Cl$^-$ advances 1.7 cm/day in agar and 0.4 cm/day in the roots. The travel distance of the phosphate of Olsen et al. is less than 1 mm/day.

Curve b in Figure 3.17 portrays diffusion from a constant source at the central boundary plane, as when a large, sparingly soluble crystal placed there slowly dissolves, maintaining a constant ion concentration at its surface while ions are diffusing away from it. The current of ions entering the right compartment is exactly double that of curve a.

Exchangeable cations migrate along surfaces and porewalls, as previously explained. Diffusion coefficients of monovalent cations in moist clay slabs (43) average 0.0075 × 10^{-5} cm^2/sec, corresponding to a migration velocity of about 1 mm/day. Movement is much slower for divalent and trivalent ions. Since the number of exchangeable ions in clay pastes is large, transport is substantial and can be greater than in an equal volume of equilibrium solution.

Diffusion is fast over short distances and can become spectacular over long periods of time. In exchanger gels it is strongly temperature-dependent (44).

Periodic Bands and Rings and Soil Profile Implications

When diffusing substances interact and precipitate, bands and rings delineate the path of migration. To follow Popp's procedure (56), pour fine sand into a test tube, fill the pores with saturated magnesium chloride (MgCl$_2$) solution, and top the mixture with a layer of ammonia (NH_4OH). Within days, white bands of solid magnesium hydroxide ($Mg(OH)_2$, brucite) appear in zebra-like pattern. On the other hand, when the sand pores are filled with manganese sulfate, black discs of dioxide (MnO_2) precipitate.

When the diffusion matrix is placed as a shallow layer in a dish and a drop of the diffusing reagent is put at the center, a family of concentric rings emerges, the Liesegang rings (61). Micromorphologists are well aware of diffusion phenomena in soil pores, and in millennia of soil genesis even slow migration rates bring forth macroscopic prominence. Periodic or rhythmic layering and structuring is treasured as aesthetic designs in agate and opal stone. It commonly appears as weathering rings and shells in rocks and as clay and iron bands in soil profiles. Photographs of large ellipses in a sand deposit at the U.S. Army Base in Giessen (Germany) are shown in Figure 3.18.

The forest floor performs as a perpetual source of ions and humus molecules, and their downward diffusion profiles imitate the b curve in Figure 3.17 turned $90°$ to the right. Both leaching and diffusion proceed in the same direction. The thin layers of Figure 3.18 might be interpreted as rhythmic bandings of Fe, Mn-humus.

E. Oxidation-Reduction Potentials

Elements differ in their combining powers, denoted *valence* or oxidation number. An atom of hydrogen (H) will combine with one—but only one—atom of another element. It is univalent. In the water molecule (H_2O) oxygen associates with 2H and is divalent. Nitrogen in ammonia (NH_3) is trivalent.

Iron, sulfur, nitrogen, and many micronutrients can shift their valency from lower to higher, and vice versa; the shifting processes are known as oxidation and reduction, respectively. In soils the conversions are triggered by drying and wetting, and the pedogenic consequences are far-reaching.

Redox System Defined

In moist air metallic iron combines with oxygen and rusts, undergoing *oxidation*. Prominent iron oxides in soils are the greenish-white FeO of divalent (Fe^{2+}) and the "rouge" colored powder Fe_2O_3 of trivalent iron (Fe^{3+}), oxygen in either case being divalent (O^{2-}). In a blast furnace Fe_2O_3 is *reduced* to FeO and then to metallic iron.

The oxides dissolve in acids, e.g., HCl, producing $FeCl_2$ (ferrous chloride) and $FeCl_3$ (ferric chloride). In an experiment (Fig. 3.19) conducted by Stieglitz (62) concentrated solutions are put into two beakers or half-cells and are connected by a salt bridge containing KCl in agar jelly. A platinum (Pt) foil-electrode is inserted into the solution in each beaker. When the wires are joined an electric current passes spontaneously through the metal, the electrons (e^-) flowing from left to right. In the left beaker or half-cell Fe^{2+} donates electrons to electrode and wire and is oxidized to brownish Fe^{3+}. In the right half-cell Fe^{3+} accepts e^- from the electrode and is reduced to Fe^{2+}. No oxygen participates.

Fig. 3.18. Rhymthic banding of iron and manganese in North European sands. (A) Ellipses (entire shovel is 68 cm long). (B) Bands under forest (the vertical arrow is 10 cm
long) (Photo A by author; photo B by R. Bach)

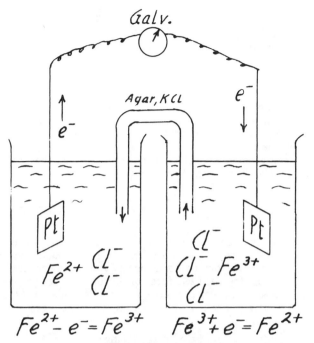

Fig. 3.19. Double cell with iron chlorides portraying electron (e^-) flow in wires and Cl^- migration in agar-KCl bridge (62). Galv., Galvanometer; Pt, platinum electrodes

Oxidation and reduction are electron transfers, symbolized for iron as

$$Fe^{2+} \xrightleftharpoons[\text{reduction}]{\text{oxidation}} Fe^{3+} + e^-$$

ferrous	ferric	to e^- acceptor or
iron	iron	from e^- donor

In soils oxygen, nitrates, sulfates, and many organic acids are good e^- acceptors and are, thereby, reduced. Schematically, a neutral oxygen atom, ½ O_2, admitting 2 e^-, becomes the oxygen ion O^{2-}. Sugars, alcohols, and many humus substances are e^- donors and are oxidized during the transaction.

Redox Potentials

When the two wires in Figure 3.19 are connected to the poles of a storage battery that sends a current opposite to the instantaneous flow and exactly balances it, the voltage (*v*) read off is the *redox potential*. It expresses the tendency of oxidation and reduction to occur (53). It is divided among the two half-cells as single electrode potentials, E, specifically, as reduction (+) and oxidation potential (−). Since neither can be measured separately, the left-hand half-cell is replaced by a hydrogen (H_2) electrode that is taken as a reference electrode and is assigned a voltage of zero. Hence

$$\begin{array}{ccc} Eh & = & E & - & E \\ \text{of total cell} & & \text{right-hand} & & H_2 - \text{cell} \\ \text{(measured)} & & \text{cell} & & V = 0 \end{array}$$

The right-hand cell contains a "couple," a mixture of the oxidized and reduced forms, and the voltage read is assigned to it as reduction potential, e.g., $Fe^{3+} + e^- = Fe^{2+}$. Various couples, e.g., Fe^{3+} and Fe^{2+}; Mn^{3+} and Mn^{2+}; S and H_2S, produce characteristic voltages. Under conditions of near-molal quantities (unit activities) and a pH of zero, the *standard potentials* E^0 of couples obtain and a few are plotted on the vertical axis in Figure 3.20.

Single electrode potentials (Eh) depend on E^0 and on the activity ratio of the two ions involved as given by the Nernst equation. For iron it is

$$Eh = E^0 - \frac{RT}{nF} \ln \frac{Fe^{2+}}{Fe^{3+}} = E^0 - \frac{0.059}{n} \log \frac{Fe^{2+}}{Fe^{3+}}$$

where R is the gas constant, T the absolute temperature, nF the quantity of electricity involved, and E^0 the standard potential. For some couples the pH has to be included. High Eh means strong oxidizing power, an urge to attract e^-. The higher Eh is, the more readily the oxidized form of an ion oxidizes the reduced form of another ion located at a lower peg. Low and negative values of Eh indicate high e^--escaping tendencies, a power to reduce the ions of a higher peg.

Free Energy Change

Alignment of two half-cells creates a spontaneous electron flow, as in Figure 3.19 or in a commercial battery. The current may be harnessed to drive a motor in the cell's environment, hence *an oxidation-reduction reaction can do work.* Under ideal, reversible conditions the maximum useful work equals the voltage difference ΔEh times the quantity of electrons (nF) transferred, or $\Delta G = -nF\Delta Eh$. In organisms the free energy changes of redox reactions are tied to growth processes.

Manganese and Iron

In comparison to the ferric ion Fe^{3+} the manganic ion (Mn^{3+}) more readily accepts e^- from outside sources, i.e., Fe^{2+} is oxidized in preference to Mn^{2+}. Further, $Mn^{3+} + Fe^{2+} \rightleftharpoons Mn^{2+} + Fe^{3+}$, the Fe^{2+} donating e^- directly to Mn^{3+}. Hydrogen sulfide (H_2S) reduces trivalent Fe and Mn. The reduced forms are much more soluble and mobile than the oxidized states, and they participate in ion exchange.

Nonmetals

The *oxidation* of odorous hydrogen sulfide (H_2S) to yellow sulfur (S) and to sulfuric acid (H_2SO_4) initiates the genesis of acid-sulfate soils (Chapter 11), and that of ammonia (NH_3) to nitrate (NO_3^-) is vital in plant nutrition. *Reductions* of nonmetals essential to ecosystem maintenance comprise the fixation of atmospheric nitrogen (N_2) to ammonia and the conversion of nitrate anions to gaseous N_2, to be discussed in later chapters.

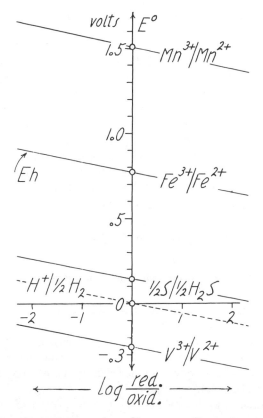

Fig. 3.20. Selected redox couples (e.g., $Fe^{3+} + e^- = Fe^{2+}$) showing standard reduction potentials (E^0) on the vertical axis and the variation of Eh with the logarithms of the concentration ratios of reduced and oxidized forms. The couple $H^+/\frac{1}{2}H_2$ pertains to the hydrogen reference electrode (at pH zero, $E^0 = 0$)

Organic Systems

The burning of coal, $C + O_2 \rightarrow CO_2$, is the classic organic oxidation reaction. Chemists also consider the removal of H atoms from *organic molecules* (dehydrogenation) an oxidation even though no oxygen participates. Addition of H atoms to organic compounds is reduction. Since the H atom consists of the proton H^+ and the electron e^-, dehydrogenation may be viewed as a removal of electrons. Conceptually, it unifies organic and inorganic oxidations.

To illustrate, the ring molecule quinone is reduced to hydroquinone as follows:

The couple has a redox potential of 0.70 V, and the H^+ ions make the Eh voltage pH-dependent.

Incorporation of redox couples into surfaces of clay and humus particles creates *electron exchangers* (29) in analogy with ion exchangers. The couples oxidize or reduce electrolytes in circulating solutions, but they themselves remain attached to the solids. For humus the formulation is Humus + H^+ + e^- $\rightarrow$ Humus (H). Its E^0 is unknown. In fact, most organic couples are electrode inactive and E^0 is calculated from the equilibrium constant K, because $\Delta E^0 = (0.059/n) \log K$.

In the photosynthetic process of leaves and green cells, light rays decompose neutral water at 0.81 V to oxygen and hydrogen (H^+ + e^-) and drive electrons down the scale to -0.40 V where CO_2 is reduced to carbohydrate and energy is stored (58). Respiration or breathing of soil flora and fauna is oxidation of organic molecules by means of dehydrogenation. Free energy becomes available and CO_2 and H_2O are liberated. These master reactions are examined in Chapter 5.

Hydromorphic Soil Conditions

Reductions in wet soil pores are initiated by aerobic microbes that consume all oxygen. The succeeding anaerobic bacteria that can function without free oxygen dehydrogenate soil organic matter, donating e^- to Mn^{3+}, Fe^{3+}, and SO_4^{2-}, which become reduced and Eh is lowered. In the absence of either microbes or organic matter the extent of reduction is negligible (7). In water-saturated soils low Eh persists because diffusion of O_2 to the depleted sites is extremely slow.

When a suspension of kaolinite clays with natural coatings of reddish Fe_2O_3 is infused (8) with soil bacteria and sugar, Eh is slowly lowered by fermentation to 0.25 and 0.15 V, and the color bleaches from bright red to gray and white. As much as 20-60% of the total iron is removed from the clay and its CEC increases in proportion. In immersed paddy (rice) soils having Eh 0.5 V and containing near-insoluble iron and manganese oxides, the Mn compound is reduced first followed by Fe at Eh of 0.15 to -0.10 V and both are solubilized (63). Soils inundated by sea water (7) foster the appearance of Mn^{2+}, Fe^{2+}, and S^{2-} (as sulfide), in that order, and subsequently the soil blackens from precipitated iron sulfide (FeS), as detailed in the chapter on topography (Chapter 11).

F. Aspects of Plant Nutrition in Soils

Roots are surrounded by a diversified ionic environment from which they draw nutrients. In conventional plant nutrition studies, roots are bathed in stirred, ion-rich media (e.g., Hoagland solution), and the rate of uptake is governed by what the inside tissues can transport and metabolize (22). Most soil solutions have notoriously low concentrations of crucial ions, and nutrient delivery from soil particle to root becomes a growth-limiting factor.

Nature of Soil Solution

Some workers identify the soil solution with the liquid phase of the soil. It includes the exchangeable ions. Isolating the liquid phase requires suction or pressure that leaves the solid phase, its sorbed water, and the ion swarms behind. Operationally, the soil solution is set at par with a soil extract or filtrate.

Solubilities. Chlorides (e.g., NaCl, $MgCl_2$), nitrates (e.g., $(Ca(NO_3)_2)$), and magnesium sulfate ($MgSO_4$) dissolve readily and bring forth high ion concentrations. As many as 3000 me of Cl have been detected in a liter of extract of saline soil (17). While desert salt bush (*Atriplex* sp.) may endure in its leaves 40% of salt by weight, domestic plants suffer fatal plasmolysis.

Calcium sulfate as gypsum ($CaSO_4 \cdot 2H_2O$) is less soluble (28.0 me Ca/liter or 561 ppm) and calcium carbonate ($CaCO_3$) still less, even though its solubility is raised by the high CO_2 pressures of soil air (see Table 7.9). Phosphorus in soil solutions rarely exceeds 1 ppm, and soils that are P-deficient for field crops have solution concentrations of less than 0.10 ppm P. Notoriously inert are rock phosphates ($Ca_3(PO_4)_2$) and allied apatites with solubilities as low as 0.005 ppm P. Aluminum and iron phosphates likewise are sparingly soluble.

Under evapotranspiration, sparingly soluble salts maintain their low concentrations because excesses are relieved by precipitation in line with the solubility products of the law of mass action. When soils of arid regions dry out, the escaping CO_2 changes soil reaction from slightly acid to mildly alkaline and $CaCO_3$ precipitates (see Table 7.9). The much more soluble $MgCO_3$ reacts with silica to become insoluble magnesium silicate (20). In turn, the proportion of Na^+ to ($Ca^{2+} + Mg^{2+}$) remaining in the soil solution rises and may harm roots and soil structure.

Exchange Contributions to the Soil Solutions. Ion swarms service the soil solution by way of "hydrolysis" with water ($H_2O \rightleftharpoons H^+ + OH^-$):

$$Na, Ca\text{-clay} + 3HOH \rightleftharpoons H_3\text{-clay} + \begin{cases} NaOH \\ Ca(OH)_2 \end{cases}$$

$$\text{water} \qquad\qquad\qquad \text{hydroxides} \\ \text{(pH 7-11)}$$

More common is exchange with biologically produced carbonic acid because CO_2 gas dissolved in water furnishes H^+ ions in the sequences $CO_2 + H_2O \rightarrow H_2CO_3 \rightarrow H^+ + HCO_3^-$, recorded in Table 7.9. For a clay particle with a coat of Na^+ and Ca^{2+} the exchange is written schematically, with ion symbols omitted, as

$$\boxed{\text{clay}}\begin{array}{l} Na \\ \\ Ca \end{array} + 3\,H_2CO_3 \rightleftharpoons \boxed{\text{clay}}\begin{array}{l} H \\ H \\ H \end{array} + \begin{array}{l} NaHCO_3 \\ \\ Ca(HCO_3)_2 \end{array}$$

nearly	carbonic	acid	bicarbonates
neutral	acid from	clay	in soil solution
clay	respiration		(pH 6-9)

Hydrolysis is less effective than CO_2 exchange and yields a solution of high pH (8-11). Nitric acid (HNO_3) is the most efficient reagent.

Displacement of an ion from a mixed swarm is facilitated by its relative abundance, viz. a high degree of saturation DS, based on ion equivalents. In the above examples, DS is 33% for Na and 67% for Ca. When DS of sodium exceeds 15%, as observed in alkali soils, clay structures and root tissues begin to collapse because of Na-induced swelling of clay and cell wall constituents. Malnutrition ensues.

Nutrient Solution Profiles. The instruments assembled in the inset of Figure 3.21 measure precipitation (rain gauge), soil water potentials (tensiometers), soil moisture contents (neutron probe), and soil solutions, the latter by extraction with tubes of 3.5 cm diameter and fitted with ceramic cups.

In a clay loam soil (Argixeroll) with a B2t horizon the moisture devices record less soil water and higher water potentials in a 60-year-old California eucalyptus forest (*E. globulus* (Labill.)) than in an adjacent, recent clearcut area covered with grasses in the spring (46).

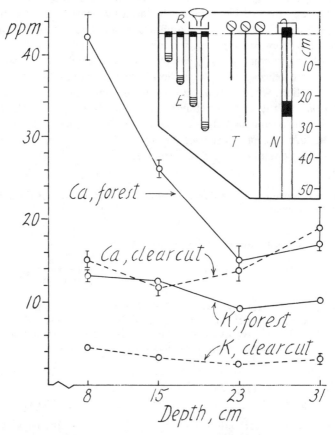

Fig. 3.21. Calcium (Ca) and potassium (K) concentrations of soil solutions from eucalyptus forest and grassy clearcut area. Vertical line segments through circles denote standard errors of means, which are very small for K. Inset: R, rain gauge; E, extraction tubes; T, tensiometers; N, neutron probe (46)

The buried extraction tubes are slightly evacuated and kept at a tension of 0.5 bar throughout the rainy season. The soil solutions sucked from the larger pores are siphoned off 9-10 times during January to May, yielding each time 200-330 ml per tube. From the dozens of ions analyzed, the means of Ca and K concentrations (ppm) during the season are plotted as depth functions in Figure 3.21. The anions are predominantly HCO_3 of metabolic origin, followed by SO_4 and Cl furnished by rainfall and leaf wash.

In the surface soil ($<$ 23 cm depth) the solution concentrations of Ca and K and of most of the other macronutrients are much lower in the clearcut area than in the forest soil where leaf wash and litter decay enlarge the influx of elements. Ammonium (NH_4) is totally absent; its transform, the nitrate (NO_3) ranges from 3-8 ppm in the clearcut area to 5-16 ppm in the forest. For comparison of sites McColl (46) computes the products of concentrations times the solution volumes extracted.

Root-Soil Boundary Zone

Figure 3.22 is a highly magnified segment of an epidermis cell at the root surface. At the far left the cytoplasm of the cell interior shelters organelles such as Golgi bodies, mitochondria, and endoplasmic reticulum. The wiggly double line on its right, the plasma membrane (Pl), is believed to be the site where enzymes engage the nutrient ions for intracellular transport. The broad region labeled "cell wall" is permeated by strands of cellulose fibrils (m) that, according to the works of Frey-Wyssling and others, provide the framework of the wall. The empty spaces between them are "free space" (f)

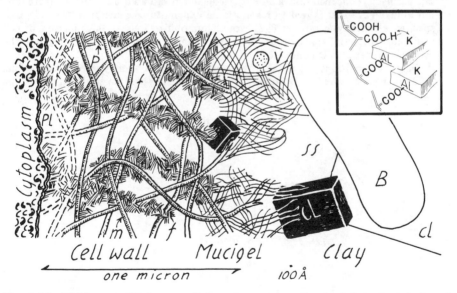

Fig. 3.22. Highly magnified root-soil boundary region. On the left is the interior of an epidermal cell with cytoplasm and plasma membrane (Pl). The cell wall region depicts cellulose microfibrills (m), pectic (p) gel zones, and freespace (f), with its channel widths exaggerated. The cell wall is fringed by the mucigel that is contacting clay (cl) particles, a bacterium (B), a virus (v) and is permeated by soil solution (ss). The 100-Å dot attests to the relative smallness of enzyme molecules and nutrient ions

serving as avenues of ion diffusion and flow of transpiration water. Portions (p) of the interfibrillar space are filled with hemicelluloses and pectic substances rich in —COOH groups, also with protein molecules, all contributing to the ion exchange capacity of the root. The space occupied by their electric double layers is designated Donnan free space.

On the right of the cell wall region is the mucigel (27, 28, 35), which is an open mesh of solidified exudate possessing cation exchange properties. Its irregular shape mirrors the region between cell wall proper and adjacent soil grains. The mucigel is permeated by soil solution (ss), and bacteria (B) and clay platelets (cl) are seen enmeshed in it, the latter possibly forming bonds between —COO radicals and Al and Fe ions as suggested in Figure 3.22 (inset). Root solution, soil solution, organic and inorganic solids, and microbes belong to a plant-soil continuum.

Passive and Active Ion Uptake by Roots

Ions flowing to the root passively with the transpirational *mass flow* may be partly rejected at the plasma membrane barrier and pile up outside the root forming tube-like incrustations of calcium carbonate or sulfate. At the low transpirational rates during foggy weather and at night, entry of nutrients is supplemented by *diffusion* of ions into the root. A zone of ion depletion surrounds the root that has been "photographed" as autoradiographs (37).

To explain how ions are able to pass from a dilute soil solution into the salty cell sap against a concentration gradient, Hoagland postulated an *active* or metabolic ion uptake (22) that is tied to respirational expenditure of energy in the root. Ion-specific, chelate-type *carriers* (32) would immobilize ions at the junction of cell wall and cytoplasm, transport them across it, and release them irreversibly into the vacuole, all steps being controlled metabolically.

In stirred solutions ion uptake by excised roots follows a hyperbolic curve (Fig. 3.23) that is described by Epstein (22) as the Michaelis-Menten equation of enzyme

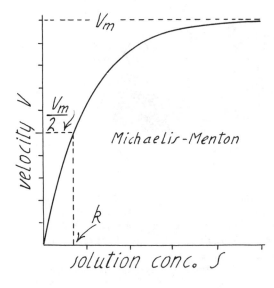

Fig. 3.23. Michaelis-Menton curve of velocity v of ion uptake related to ion concentrations in the nutrient solution

kinetics, $V = V_m \cdot S/(k + S)$, in which V is the velocity of uptake, S is the external ion concentration, V_m is the maximum velocity, and k is the ion concentration at which V is half of V_m.

Role of Root Exchange Capacity (CEC)

Ions taken up by plants are held in tissues as soluble, exchangeable and non-exchangeable ions, the latter being closely tied to metabolic processes. All three types perform essential functions. A linkage or pathway between exchangeable and non-exchangeable ions has long been implied (34) and is invoked by Volz and Jacobson (66) to account for the superior metabolic uptake of potassium by vetch roots compared to barley roots. The former have higher root CEC than the latter.

Positive correlations between CEC of roots and total cations (Ca, Mg, K, Na) in the above-ground parts have been observed for many crops. Biomass of sugar cane and leek varieties rises with CEC of their roots (12).

Double layer theory predicts for high charge density of a surface layer a greater replaceability of monovalent than divalent cations because the latter are held more tightly as the negative attraction spots are set closer together (Fig. 3.24). From a Na, Ca-clay *barley* roots of low CEC remove 2.71 times more Na^+ than Ca^{2+}, whereas high CEC pea roots acquire only one-third as much (21). The various root surfaces that confront clay colloids may have a bearing on the compatability of species.

Modes of Transport of Ions from Soil Domains to Roots

In addition to mass flow with the transpiration stream, ions approach the root surface by diffusion.

Boundary Film. A solution streaming past a root creates a hydrodynamic boundary layer around it, conceptualized by Nernst (29) as a stagnant film 10-100 μm thick. When the root depletes it of ions resupply must diffuse through the Nernst film, which gets thinner the more vigorously the liquid flows. Hence, roots accumulate ions faster from a moving or stirred solution than from one at rest (55), and at high dilution plants may prosper in the one but not in the other.

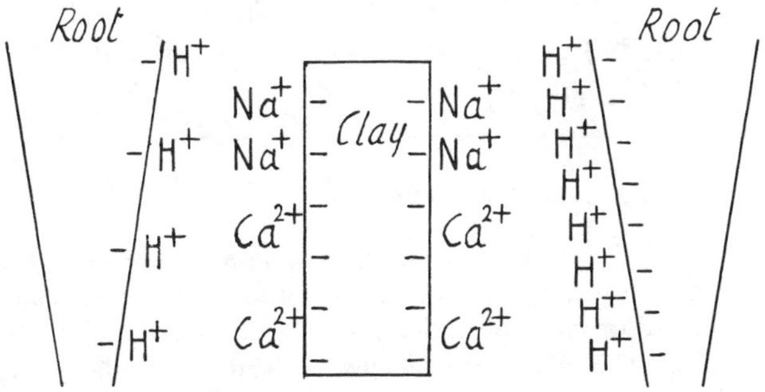

Fig. 3.24. Clay particle confronted by roots of low (left) and high (right) cation exchange capacities. Preferentially, Na^+ will appear in the left root, Ca in the right one

Stationary or Steady-State Diffusion. When a large crystal of sparsely soluble aluminum phosphate dissolves at its surface, and when a nearby root segment grabs every phosphate anion that reaches it, the first acts as a source, the second as a sink of diffusing ions. In the simple case sketched in Figure 3.25 the concentration profile is linear and ion flux $J = (C^0 - C)/d$. The shorter the distance d between source and sink, the steeper is the gradient and the larger the ion flux to the root surface.

Nonstationary Diffusion. A root cell wall of thickness τ is portrayed in Figure 3.25 by carboxyl groups. It faces a soil solution of thickness d containing Na^+, H^+, HCO_3^-, and small amounts of NO_3^-. The bicarbonate salt is in equilibrium with respiratory CO_2 and with an extensive gel of Na-clay platelets. The clay is the source of the Na^+ current and the plasmalemma between cell wall and cytoplasm is the sink. H^+ ions emanating from the root replace Na^+ on the clay. No steady state is achieved because Na^+ concentration in the clay phase and in the adjacent soil solution at $X=0$ are continually diminishing. Figure 3.25 gives the Na^+ concentration profile (34) at arbitrary time t.

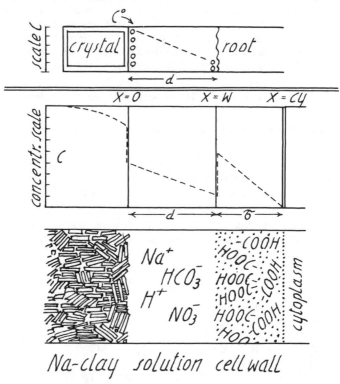

Fig. 3.25. Ion currents in soil-root microregions. Top (above double lines): steady-state diffusion of ions from a crystal surface having constant ion concentration, C^0, to sink at root surface. Bottom: Nonstationary diffusion flow of Na^+ from clay phase through soil solution and cell wall H-exchanger to sink in cytoplasm. Dashed lines are ion concentration curves; Na^+ moves from left to right, H^+ from right to left. Electric double layers at $X=0$ and $X=W$ are not shown

At $X=0$, Na^+ is released from the clay phase and enters the dilute soil solution, at $X=W$ it is taken up by the root exchanger, and at $X=CY$ it is irreversibly captured by the cytoplasm.

As the root edges closer to the clay the distance d shortens. The Na^+ current accelerates and maximizes when the double layers of root and clay touch each other. A logical counterpoint to this contact uptake is contact depletion of a root brought in juxtaposition with acid clay particles. Roots lose copious quantities of ions, as elaborated in the section on soil acidity.

In soils the tortuosity of paths, the varied configurations of roots, and the changes in moisture tension impose complicated situations (48).

In pots of soil holding exchangeable Ca^{2+} and Sr^{2+}, Barber's team (e.g., 3) cultivates tomato and soybean plants that do not discriminate between the two cations. During 3 weeks the transpirational mass flow from *muck soil* supplies all of the Ca and Sr found in the plants and their Ca/Sr ratios are those of the soil solution. On *Sidell soil* mass-flow furnishes less than 20% of ion uptake. The remainder comes from diffusion, probably exchange diffusion, because the plants' Ca/Sr ratios are akin to those of the exchangeable ions. (The ions in the equilibrium solution have wider ratios.) Corn plants in K-Rb-soils also consume the cations in proportion to their ratios on clay, not in solution.

In these experiments the amount of water that has transpired multiplied by the nutrient concentration of the soil solution measures the mass flow of ions. Its difference from the total uptake is attributed to ion diffusion, either in solution or along clay surfaces. Soil ions move to the root surface and roots advance to the exchange sites by elongation and root hair proliferation.

Root Interaction with Inert Solids

Roots may overcome deficient solutions by locally tapping practically insoluble compounds.

Iron Oxide. In a calcareous sand medium with continually circulating, aerated, *iron-deficient* nutrient solution of pH 8, alfalfa plants and citrus shoots are stunted and yellow-leafed, that is, chlorotic. Malnutrition is overcome by amending the bed of quartz sand with grains of practically insoluble iron oxide (Fe_2O_3) or iron hydroxide ($Fe(OH)_3$).

The procurement of Fe by alfalfa and the extent of disappearance of chlorosis in the double column experiment of Figure 3.26 is directly proportional to the number of grains of iron oxide that touch 1 cm^2 of root surface (26). Fe uptake appears localized at the junction of root and oxide, and the idea of an overall soil solution of iron ions or chelates is negated. Bacteria of the mucigel are not involved because iron gets into the plant under sterile growth conditions. The lower plants in Figure 3.26 fail to register any benefits of iron-solubilizing root exudates that might have been washed downstairs.

The localized interaction is explained either as Fe exchange with root $-COOH$ groups, or as root surface chelation, or as root-induced reduction of Fe^{3+} to the more soluble Fe^{2+}. "Contact reduction" has been cited (50) as a modus of availability of insoluble MnO_2 to plants.

Phosphates. Scarcely soluble rock phosphate $(Ca_3(PO_4)_2)$ obeys the solubility product $(Ca^{2+})^3 \cdot (PO_4^{3-})^2 = k$. Roots trade $2H^+$ for each Ca^{2+}, and, to keep k invarient, more PO_4^{3-} enters the solution and encourages plant growth (18). Beyond solution supply the rye grass (*Lolium italicum*) of Blanchet et al. (5) utilizes phosphorus tenaciously adsorbed on kaolinite. Matar's (45) barley plants accumulate 150 times more P from a slurry of red Aiken soil than from its equilibrium solution. The roots acquire a coat of red clay particles that resists being washed off, as if it were held in chemical bondage. The interaction releases sorbed PO_4^{3+} to the soil solution.

From 3 to 90% of total soil phosphorus is located in insoluble organic fractions, and up to half of it exists as the ring-structured inositol hexaphosphate (13). Mineralization of organic soil P to the available inorganic forms is accomplished by extracellular, dephosphorylizing enzymes of microbes, and P liberated in laboratory incubation tests is judged as being significant for plant growth. The humus-rich soils of East Africa have 86% of total P in organic form and fertilizer response correlates highest with total organic P (24). In concert with others, Friend and Birch (24) are inclined to attribute P-mobilization to enzymes in root gels that operate at the interface of root

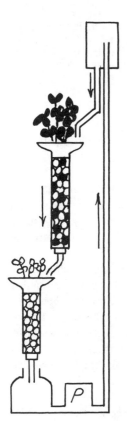

Fig. 3.26. Circulating iron-free nutrient solution at pH 8 induces chlorosis of alfalfa in the lower column but not in the upper one, which contains insoluble iron oxide grains (black dots) that touch the roots. P, pump

and humus. Estermann and McLaren (23) demonstrate enzyme phosphatase activity of root caps and epidermis mucigels under sterile conditions. Electron micrographs by Herrera et al. (31) show mycorrhiza hyphae contacting particles of leaf litter.

Fixed Potassium. Ramona soil (Haploxeralf) originals from granitic alluvium and contains 78 ppm of NH_4-exchangeable K and 20,000 ppm of non-exchangeable K locked up inside lattices of mica and related clay minerals. Lattice confinement is not absolute; month-long leaching with water saturated with CO_2 having pH 4.0 removes 15-27 ppm of non-exchangeable K in addition to the 78 ppm of exchangeable K (70).

A dense mat of rye seedlings growing on the same soil by "Neubauer's technique" acquires in 18 days all of the exchangeable K plus 188 ppm of non-exchangeable K, a feat that is approached only by long-time soil extraction with strong acids ($<$ pH 2) or shaking the soil with commercial ion-exchanger resins or with K-capturing tetraphenyl-boron reagent.

In micaceous minerals K is lodged between charged sheets of oxygen ions (Fig. 3.27), and K at the edges is displaced by external H^+ that advance into the interior of the sheets by slow exchange diffusion. The H-exchange sites of the root that touch a biotite mica crystal are effective in donating H^+ and in capturing K^+ as fast as it surfaces. Clay exchangers perform similarly. In securing K, the roots may transform the biotite lattice to clay minerals (Chapter 4).

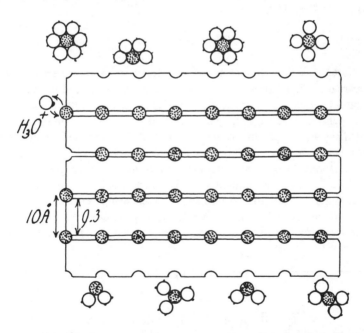

Fig. 3.27. A micaceous clay particle composed of a stack of modified pyrophyllite layers (see Fig. 3.8). The K^+ (dotted circles) are imprisoned in the interlayer space. At the edge a hydronium ion replaces a K^+ and will exchange-diffuse into the interior. Surface Ks are partially hydrated

G. Problems of Soil Acidity

The ecosystem casts soil reaction to acidic, neutral, or basic. In turn, changes in reaction evoke biological responses. Some pedologists single out reaction as the preeminent soil property.

Acidity and Alkalinity

The intensity of sour or soapy taste is characterized by pH which is the negative logarithm of the concentration or, precisely, activity of hydrogen ions (H^+). Neutral reaction has pH 7, acid reaction has a lower pH, and alkaline reaction has a higher pH (Fig. 3.28). Most natural soils are confined to pH 3-10.

> The H^+ activity is measured by inserting into a solution a H^+-sensitive glass electrode tied to a calomel reference electrode, the latter contacting the liquid through a KCl junction (Fig. 3.29). On commercial pH meters the voltage arising in the circuit is calibrated directly as pH. Tests with color indicators furnish approximations.
>
> A strong base (NaOH) neutralizes strong acids such as hydrochloric (HCl) and sulfuric (H_2SO_4) to a salt. The *titration curve* (Fig. 3.30) is steep and the endpoint is reached at pH 7. Weak acids such as vinegar (acetic acid, CH_3COOH) produce "flattened" curves that are "buffered." Their endpoint of titration is at pH above 7 because hydrolysis of the salt, CH_3COONa, furnishes alkalinity: CH_3COO $Na + HOH \rightleftharpoons CH_3COOH + NaOH$.

Suspension Effect

In fine-textured soils the pH of a slurry or paste is often lower, more acid, than the pH of the supernatant liquid (Fig. 3.30), or water extract, or the soil solution, indicating that soil particles seemingly carry acidity not shared with the water phase.

A 1% suspension of acid clay may have a pH of 3.6 and the equilibrium supernatant liquid may have a pH 5.2. This "suspension effect" is interpreted either as a response of the glass electrode to a swarm of exchangeable H^+ ions on clay and humus particles (68), or as a Donnan membrane potential (25), or as a junction potential at the KCl junction (37). The three incompatible explanations inject ambiguity into the fundamental meaning of ion activities in soils and other exchanger systems.

Practice circumvents the conflict by relating pragmatically the pH meter recording of a soil slurry or paste with whatever soil and plant properties are being investigated.

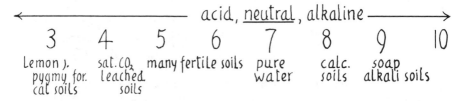

Fig. 3.28. Chart of soil reactions expressed as pH. j., juice; for., forest; sat., saturated; cal., calcareous

Fig. 3.29. Suspension effect. Glass electrode (G) and calomel electrode (C), with KCl junction as black strip, dipping in supernatant liquid (I) record higher pH than in sediment (II). Connecting the two Gs gives a reading of zero mV, connecting the two Cs gives 120 mV, which corresponds to Δ pH=2, that is, pH 6-pH 4 (25,37)

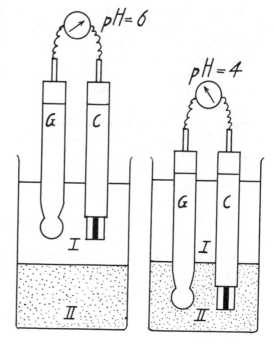

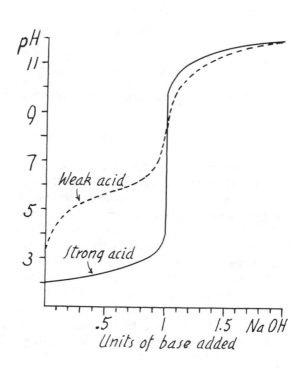

Fig. 3.30. Examples of titration curves (acid + base) of a strong and a weak acid

Such empirical correlations are reproducible and possess utility. Thus, for micronutrients the availability of Mo rises with increasing soil pH, whereas that of B, Cu, Zn, Fe, and Mn falls. However, the stability diagrams of clays in Chapter 4 demand pH of soil solutions, not of suspensions.

The suspension effect is eliminated by shaking soil with strong, neutral salt solution (KCl, $CaCl_2$), which releases the exchangeable acidity into the liquid phase. The pH of slurry then matches the pH of supernatant liquid. For soils of the temperate region in which silicate clays dominate, the pH of a soil slurry in water is lowered by KCl addition, e.g., 5.0 in KCl versus 6.2 in H_2O. In tropical soils that have a preponderance of iron oxide particles the reverse is found, e.g., 6.5 in KCl versus 5.5 in H_2O. In the one case K^+ exchanges acidity, in the other Cl^- is believed to replace OH^- of amphoteric clays.

Titratable Soil Acidity

Either suspensions of soils or their salt extracts may be titrated with a base. Titratable acidity in a KCl extract is assigned to sites of permanent-charge CEC. In the slightly alkaline extraction procedure of Mehlich, soil is leached with a $CaCl_2$ triethanolamine buffer of pH 8.1. This extract gives much higher titration values than KCl because the reagent also reacts with loci of pH-dependent CEC such as aluminum compounds and weak organic acids.

For the 72 coarse-textured soil samples of the Mt. Shasta mud flow area of Figure 9.3 the overall regression furnishes

$$aci = 0.42 + 3.333\% \, C \; (r = 0.991)$$

in which aci is the Mehlich exchange acidity in me/100 g of soil and C the organic carbon content. Over 98% of the aci-variation is accounted for by the spread in carbon from 0.05 to 9.7%. Each gram of humus ($1.724 \cdot \%C$) brings to the soil 1.93 me of titratable acidity. KCl exchange acidity of these samples is very low.

Role of Aluminum Ions

Natural smectite clay is transformed to pure H-clay by passing a suspension through a column of commercial H-resin exchanger. Titration of this "clay acid" shows it to be *strong*, that is, highly dissociated, and leaching it with KCl releases 90 me/100 g of exchangeable H^+ ions according to the equation: H-clay + KCl → K-clay + HCl.

In a suspension of H-clay at rest for months, its exchangeable H^+ ions, slowly diffuse into the interior of the clay crystals and displace in the tetrahedral and octahedral sheets aluminum and magnesium ions that move outward and occupy the vacated H^+-sites as exchangeable cations. The pH rises. Resulting Al-clay titrates as a *weak acid*. Salts react with Al-clay as: Al-clay + 3 KCl ⇌ K_3-clay + $AlCl_3$.

As seen, a salt (KCl) extract of Al-clay contains aluminum chloride ($AlCl_3$), which is acid. Its titration with NaOH furnishes insoluble $Al(OH)_3$ and titratable acidity (HCl) that is about equal to the KCl-exchangeable acidity of the original H-clay.

On clay surfaces a variety of Al-hydroxy complexes or polymers may reside adjacent to H^+ and Al^{3+} ions depending on pH:

$$[Al(OH)_3]^0, [Al(OH)_2]^+, [Al(OH)]^{2+}, [Al \overset{OH}{\underset{OH}{\diamond}} Al]^{4+}, \text{etc.}$$

For brevity, Al^{3+} only is used in equations. Acid humus may support Al and Fe ions attached to carboxyl groups. In nature, acid soils low in humus and high in clay content are rich in exchangeable aluminum ions rather than H^+ ions (11).

Physiological Action

In a leaching regime at high rainfall the proportions of exchangeable Al^{3+} and H^+ on clay and humus increase and those of Mg^{2+}, Ca^{2+}, Na^+, and K^+ decrease, which brings on nutritional deprivation.

The question of toxicity of H^+ and Al^{3+} on exchange sites is ascertained (54) by immersing roots in suspensions of pure H^+-clay (pH 3.6) or Al^{3+}-clay (pH 5.1) free of salts. The roots lose copious amounts of K, Ca, and Mg in the former and small ones in the latter. In the filtrates, which have pH 5.2 and 5.6, respectively, the roots remain intact. Evidently, H^+ ions in exchange position are much more debilitating than adsorbed Al ions.

Soil water contains dissolved acids, bases, and salts, and their concentrations may escalate by fertilizer practice and influx of acid rain of industrial origin. The cations displace exchangeable Al ions into the soil solution in which they stunt roots at concentrations of very few ppm. Dissolved Al, Fe, and Mn ions are thought to be the main cause of biologically harmful soil acidity.

In agriculture acid soils are reclaimed by neutralization with powdered limestone:

$$\boxed{\begin{array}{c}\text{clay} \\ \text{humus}\end{array}}\begin{array}{l}H^+ \\ Al^{3+}\end{array} + 2\,CaCO_3 + H_2O \rightleftharpoons \boxed{\begin{array}{c}\text{clay} \\ \text{humus}\end{array}}\begin{array}{l}Ca^{2+} \\ Ca^{2+}\end{array} + Al(OH)_3 + 2CO_2$$

| acid particle | calcium carbonate | neutral particle | insoluble aluminum hydroxide | carbon dioxide gas |

Decomposing leaf litter high in Ca, Mg, K, and Na likewise neutralizes soil acidity, but in the long run the biological cycling of bases merely delays the Al escalation of climatic leaching.

H. Mineral Cycling

Quantitative studies of nutrient flow through soil and vegetation are of recent origin (38). The object of Gessel's work group in Seattle, Washington, is a 36-year-old, second-growth Douglas fir forest (*Pseudotsuga menziesii*) growing in Everett soil (Inceptisol) on a loamy sand terrace (12,000 years B.P.). The subordinate vegetation includes salal (*Gaultheria shalon*), Oregon grape (*Berberis nervosa*), huckleberry (*Vaccinium parvifolium*), and bracken fern (*Pteridium aquilinum*). Mean annual precipitation is 136 cm, nearly all of it as rain.

Procedurally, an area of 40 m² occupied by 10 trees is monitored during 1 year for influxes, litterfall, crown drip, stem flow, and effluxes. Soil percolates are collected continually by placing large porcelain tension plates directly beneath the forest floor and at 90 cm soil depth. At the end of the year the entire area is sacrificed, all vegetation harvested, roots dug up quantitatively, and samples of soil horizons secured. The flux values from soil to tree are based on analysis of current year foliage, branches, and wood increment, the latter estimated from widths of rings. The combined quantity of organic accruement is 10,000 kg/ha, or 1 kg/m², during the year of observation.

The condensed summary of the ecosystem inventory in Table 3.1 is composed of standing vegetation and soil reserves and assigns 51% of the system's organic matter to vert space and 49% to soil space, the latter including roots and forest floor. The bulk of the elements N, P, K, and Ca congregates in the soil space even though K and Ca are merely the exchangeable fraction of the mineral soil.

In Table 3.2 the influxes and effluxes are slight compared to the signal flows within the ecosystem. The collection of the individual transfer rates invites drawing diverse combinations of balances and flows, the one for Ca being shown in Figure 3.31. For the ecosystem as a whole the outflux to nonsoil slightly exceeds the influx from rainfall, leading eventually to depletion after weathering has taken its course. The rate of loss is accelerated by logging and shipping the lumber from the site to cities.

I. Limiting Factor or Performance Equations

Any variable that affects growth is a growth factor. Nutrients are growth factors, as are water and light. To avoid circular reasoning, such variables are called growth factors only if they can be manipulated from the environment, particularly by man. Countless workers have attempted to formulate a general law of the action of growth factors on biomass or yield of crops. High on the list are the names of Liebig and Mitscherlich.

Table 3.1. Inventory of Organic Matter (OM)[a] and Nutrient Elements of Douglas Fir Ecosystem (kg/ha)[b]

Components	OM × 10⁻³	N	P	K	Ca
Trees					
Above ground	171.5	288	60	196	296
Roots	33.0	32	6	24	37
Subordinate vegetation	1.0	6	1	7	9
Forest floor	22.8	175	26	32	137
Mineral soil (0-60 cm)	111.6	2809	3878	234[c]	741[c]
Total ecosystem	339.9	3310	3971	493	1220

[a] Oven-dried at 70°C.
[b] After Cole et al. (10).
[c] Exchangeable ions only.

Table 3.2. Annual Fluxes of Elements of Douglas Fir Ecosystem (kg/ha)[a,b]

Number	Pathways	N	P	K	Ca
1	Soil to trees	38.8	7.2	29.4	24.4
2	Soil to nonsoil below	0.6	0.02	1.0	4.5
	To forest floor:				
3	From litter fall	13.6	0.2	2.7	11.1
4	From crown drip and stem flow	1.7	0.4	12.3	4.6
5	From rainfall	1.1	trace	0.8	2.8
6	To soil from forest floor	4.8	0.95	10.5	17.4
7	Net loss of soil (1+2−6)	34.6	6.27	19.9	11.5

[a]Subordinate vegetation excluded.
[b]From Cole et al. (10).

Liebig's Law of the Minimum (1855)

Yield, said Liebig, is controlled by the factor that is present in minimal quantity or intensity, and growth Y is directly proportional to the input of this limiting factor X, as graphed in Figure 3.32. At some level of supply the factor ceases to be limiting and the rising line turns sharply to a horizontal direction. A new factor then assumes the limiting role and the curve may start climbing again.

Law of Diminishing Returns

More common is the gradually bending curve of Figure 3.32. The first unit of input, say 100 kg/ha of nitrogen fertilizer to a N-free soil, produces the largest return or increment ΔY in yield, the second unit a somewhat lesser one, and each succeeding unit of

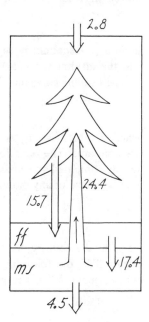

Fig. 3.31. Annual fluxes of calcium (kg/ha) into and out of a conifer ecosystem, and cycling within it (10). (1 kg/ha = 0.1 g/m²). ff, Forest floor; ms, mineral soil

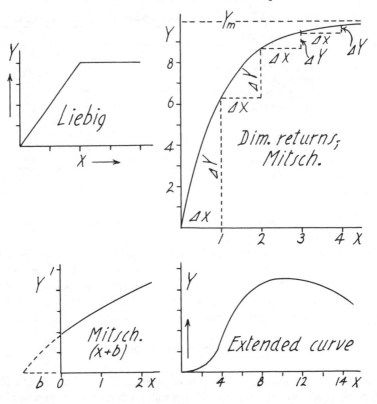

Fig. 3.32. Limiting factor curves: Liebig's law of minimum, diminishing returns, and Mitscherlich curves; extended sigmoid curves with stimulation at low and toxicity at high inputs

input a correspondingly smaller return ΔY until finally the inputs become ineffective and the maximum yield Y_m for a given constellation of growth factors is reached.
  The bending curve may be described and fitted to observations with the polynomial

$$Y = a + bX + cX^2 + \ldots$$

where Y is biomass or yield (used here synonymously), X the growth factor, and a, b, c are constants. The equation appeals to statistically oriented researchers because it is readily computed and amenable to significance tests. The equation is strictly empirical, that is, devoid of any biological content or argument.

Mitscherlich Equation

Early in the century, Mitscherlich proposed for any point Y on the growth curve of Figure 3.32 the following action model: the yield increase ΔY brought about by a unit of input ΔX is the greater the farther away the yield Y is from the maximum yield Y_m.

Written as differential and integrated equations:

$$\frac{dY}{dX} = k\,(Y_m - Y), \quad \text{and} \quad Y = Y_m\,(1 - e^{-kX})$$

where k is the efficiency coefficient of the growth factor and e the base of logarithms. All other controllable growth factors are supposed to remain constant.

The equation satisfies the biological certainty that $Y = 0$ for $X = 0$, or "no nutrient, no yield." In the usual situation the soil contains some of the nutrient X, say, the quantity b, prior to fertilization and the exponent is then written $-k(X + b)$. In Figure 3.32, b is appraised by graphic extrapolation.

A large number of field and greenhouse experiments, but by no means all, fit the equation within the experimental errors that are substantial (10% and more). No physiological reason is given why a plant or, more precisely, an ecosystem should choose to behave in this simple mathematical fashion, but inasmuch as metabolic ion uptake and photosynthetic response to CO_2 pressure follow the very same curve, the equation has achieved a measure of credibility. However, the claim of universality of the efficiency coefficient k is invalid. Actually it can be proven that the Mitscherlich equation, the well-founded sorption isotherm of Langmuir, the Michalis-Menten equation of enzyme kinetics and ion uptake, and the law of mass action for a monomolecular reaction are mathematically one and the same.

The Extended Yield Curve

In field and pot experiments the yield curve may have a sigmoidal initial branch caused by soil chemical or microbial nutrient fixation, and a final declining branch caused by toxicity of high salt concentrations (Fig. 3.32). Mitscherlich had extended his equation to include the declining branch, and a more recent version has been published by von Boguslawski and Schneider (6). The polynomial equation also describes the extended curve.

Functions of Leaf Analysis

An "internal" curve correlates crop yield with the chemical composition of an organ of the crop such as K in petioles. The approach offers promise in assessing the nutritional status of plants in natural ecosystems (65).

K. Review of Chapter

Soils are immense stockpiles of ions, randomized in amorphous gels, highly ordered in crystals, and wandering in solutions. Ions behave individualistically, being controlled by their size and charge, but obeying the chemical law of mass action.

The bulk of mobile ions resides on clay and humus particles as exchangeable ions that diffuse along solid surfaces and enter the soil solution by ion exchange.

Iron and manganese are multivalent and participate in oxidations and reductions. The reduced forms (Fe^{2+}, Mn^{2+}) are much more mobile than the oxidized partners (Fe^{3+}, Mn^{3+}).

Roots take up nutrient ions passively with the transpiration stream and by diffusion and actively by metabolic processes. Roots touching sparingly soluble minerals may acquire ions by various interaction mechanisms. The nutrients are transported to leaves and return to the soil by way of litterfall, which is nutrient cycling. High soil acidity is harmful because of toxicity of dissolved Al^{3+} ions.

In fertilizer experiments, yield augmentations tend to line up along a performance curve in which the first dosage produces the highest increment and subsequent dosages induce lowered responses.

References

1. Allaway, W. H. 1975. *The Effects of Soils and Fertilizers on Human and Animal Nutrition.* U.S.D.A. Agr. Inf. Bull. No. 378, Washington, D.C.
2. Babcock, K. L. 1963. *Hilgardia* 34: 417-542.
3. Barber, S. A., S. M. Elgawhary, and G. L. Malzer. 1972. In: *Characterization of nutrient supply mechanisms to plant roots using double labelling and the ratio of Ca:Sr absorbed.* Int. Atom. Energy Ag. Proc. Isotopes and radiation in soil-plant relationships including forestry. pp. 11-18. Vienna.
4. Beadle, N. C. W. 1966. *Ecology* 47: 992-1007.
5. Blanchet, R., C. Chaumont, and R. Studer. 1962. *Ann. Agron.* 13: 21-29.
6. Boguslawski, E. von. 1972. Wachstum und Ertragsgesetze. In *Pflanzenernährung,* H. Linser, ed., Vol. 1:2, pp. 739-788. Springer, New York.
7. Brümmer, G. 1974. *Geoderma* 12:207-222.
8. Carroll, D. 1958. *Geochim. Cosmochim. Acta* 14: 1-28.
9. Charley, J. L., and H. Jenny. 1961. *Agrochimica* 5: 99-107.
10. Cole, D. W., S. P. Gessel, and S. F. Dice. 1967. In *Symposium on Primary Productivity and Mineral Cycling in Natural Ecosystems,* H. E. Young, ed., pp. 198-232. Univ. Maine Press, Orono.
11. Coleman, N. T., and G. W. Thomas. 1967. In *Soil Acidity and Liming,* R. W. Pearson and F. Adams, eds., pp. 1-41. Am. Soc. Agron., No. 12, Madison,Wisc.
12. Crooke, W. M., and A. H. Knight. 1971. *J. Sci. Food. Agric.* 22: 389-392.
13. Dalal, R. L. 1977. Soil organic phosphorus. In *Advances in Agronomy,* N. C. Brady, ed., Vol. 29, pp. 83-117. Academic Press, New York.
14. Dixit, S. P. 1978. *Agrochimica* 22: 25-31.
15. Dixon, J. B., and S. B. Weed (eds.). 1977. *Minerals in Soil Environments.* Soil Sci. Soc. Am., Madison, Wisc.
16. Doner, H. E. 1978. *Soil Sci. Soc. Am. J.* 42: 882-885.
17. Doner, H. E., and W. C. Lynn. 1977. Carbonate, halide, sulfate, and sulfide minerals. In *Minerals in Soil Environments,* J. B. Dixon and S. B. Weed, eds., pp. 75-98. Soil Sci. Soc. Am., Madison, Wisc.
18. Drake, M., and J. E. Steckel. 1955. *Soil Sci. Soc. Am. Proc.* 19: 449-450.
19. Drake, M., J. Vengris, and W. G. Colby. 1951. *Soil Sci.* 72: 139-147.
20. Eaton, F. M., G. W. McLean, G. S. Bredall, and H. E. Doner. 1968. *Soil Sci.* 105: 260-280.
21. Elgabaly, M. M., and L. Wiklander. 1949. *Soil Sci.* 67: 419-424.
22. Epstein, E. 1972. *Mineral Nutrition of Plants: Principles and Perspectives.* Wiley, New York.

23. Estermann, E. F., and A. D. McLaren. 1961. *Plant Soil* 15: 243-260.
24. Friend, M. T., and H. F. Birch. 1960. *J. Agr. Sci.* 54: 341-347.
25. Gast, R. G. 1977. Surface and colloid chemistry. In *Minerals in Soil Environments*, J. B. Dixon and S. B. Weed, eds., pp. 27-73. Soil Sci. Soc. Am., Madison, Wisc.
26. Glauser, R., and H. Jenny. 1960. *Agrochimica* 4: 263-278.
27. Greaves, M. P., and J. F. Darbyshire. 1972. *Soil Biol. Biochem.* 4: 443-449.
28. Guckert, A., H. Breisch, and O. Reisinger. 1975. *Soil Biol. Biochem.* 7: 241-250.
29. Helfferich, F. 1962. *Ion Exchange*. McGraw-Hill, New York.
30. Hendershot, W. H., and L. M. Lavkulich. 1978. *Soil Sci. Soc. Am. J.* 42: 468-472.
31. Herrera, R., T. Merida, N. Stark, and C. F. Jordan. 1978. *Naturwissenschaften* 65: 208-209.
32. Jacobson, L., and R. Overstreet. 1947. *Am. J. Bot.* 34: 415-420.
33. Jenny, H. 1936. *J. Phys. Chem.* 40: 501-517.
34. Jenny, H. 1966. *Plant Soil* 25: 265-289.
35. Jenny, H., and K. Grossenbacher. 1963. *Soil Sci. Soc. Am. Proc.* 27: 273-277.
36. Jenny, H., and R. Overstreet. 1939. *J. Phys. Chem.* 43: 1185-1196.
37. Jenny, H., T. R. Nielsen, N. T. Coleman, and D. E. Williams. 1950. *Science* 112: 164-167.
38. Jordan, C. F., and J. R. Kline. 1972. *Ann. Rev. Ecol. Syst.* 3: 33-50.
39. Jost, W. 1952. *Diffusion in Solids, Liquids, Gases*. Academic Press, New York.
40. Keller, P., and H. Deuel. 1957. *Z. Pfl. Düng. Bod.* 79: 119-131.
41. Klemmedson, J. O. 1959. Influence of pedogenic factors on availability of nitrogen, sulfur, and phosphorus in forest and grassland soils of California. Ph.D. Thesis, Univ. of Calif., Berkeley.
42. Knight, A. H., W. M. Crooke, and R. H. E. Inkson. 1961. *Nature (London)* 192: 142-143.
43. Lai, T. M., and M. M. Mortland. 1961. *Soil Sci. Soc. Am. Proc.* 25: 353-357.
44. Lopez-Gonzales, J. de, and H. Jenny. 1959. *J. Colloid Sci.* 14: 533-542.
45. Matar, A. E. 1966. Film replenishment and contact interaction in ion transfer with soil phosphates. Ph.D. Thesis, Univ. of Calif., Berkeley.
46. McColl, J. G. 1978. *Soil Sci. Soc. Am. J.* 42: 358-363.
47. Moreno, E. 1957. Diffusion rates of soil phosphorus from solid to liquid phase. Ph.D. Thesis, Univ. of Calif., Berkeley.
48. Nye, P. H., and P. B. Tinker. 1977. *Solute Movement in the Soil-Root System*. Univ. Calif. Press, Berkeley.
49. Olsen, S. R., W. D. Kemper, and J. C. Van Schaik. 1965. *Soil Sci. Soc. Am. Proc.* 29: 154-158.
50. Passioura, J. B., and G. W. Leeper. 1963. *Nature (London)* 200: 29-30.
51. Patrick, W. H., Jr., S. Gotoh, and B. G. Williams. 1973. *Science* 179: 564-565.
52. Pauling, L. C. 1967. *The Chemical Bond*. Cornell Univ. Press, Ithaca, New York.
53. Pimentel, G. C., and R. D. Spratley. 1971. *Understanding Chemistry*. Holden-Day, San Francisco.
54. Polle, E. O., and H. Jenny. 1968. *Soil Sci. Soc. Am. Proc.* 32: 528-530.
55. Polle, E. O., and H. Jenny. 1971. *Physiol. Plant.* 25: 219-224.
56. Popp, K. 1925. *Kolloid Z.* 36: 208-215.
57. Pratt, P. F. 1957. *Soil Sci.* 83: 85-89.
58. Rabinowitch, E., and Govindjee. 1969. *Photosynthesis*. Wiley, New York.
59. Sollner, K. 1969. *J. Macromol. Sci.* A3, 1-86.
60. Sposito, G. 1980. Cation exchange in soils. In *Chemistry in the Soil Environment*, D. E. Baker, ed. (In Press). Soil Sci. Soc. Am., Madison, Wisc.
61. Stern, K. H. 1954. *Chem. Rev.* 54: 79-99.
62. Stieglitz, J. O. 1917. *The Elements of Qualitative Chemical Analysis*. Century, New York.
63. Takai, Y., and T. Kamura. 1966. *Folia Microbiol.* 11: 304-313.

64. Thomas, G. W. 1970. In *Nutrient Mobility in Soils: Accumulation and Losses,* D. P. Englestad, ed., pp. 1-20. Soil Sci. Soc. Am. Spec. Publ. No. 4.
65. Ulrich, A., and P. L. Gersper. 1978. In *Vegetation and Production Ecology of an Alaskan Arctic Tundra,* L. L. Tieszen, ed., pp. 457-480. Springer-Verlag, New York.
66. Volz, M. G., and L. Jacobson. 1977. *Plant Soil* 46: 79-91.
67. Walker, J. M., and S. A. Barber. 1961. *Science* 133: 881-882.
68. Wiegner, G. 1931. *J. Soc. Chem. Ind.* 50: 103T-112T.
69. Wiklander, L. 1975/76. *Grundförbättring* 27: 125-135.
70. Williams, D. E., and H. Jenny. 1952. *Soil Sci. Soc. Am. Proc.* 16: 216-221.

4. Origin, Transformation, and Stability of Clay Particles

Hard rocks weather and turn into substrates for higher plants. The mineral assemblies of rocks are fragmentized to gravels, sands, and silts and transform themselves chemically to clay particles, molecules, and ions. Clay formation is a cardinal pedogenic process.

A. Six Common Primary Silicate Minerals

As outlined in Chapter 3 rock minerals have a skeleton of O^{2-} and OH^- anions arranged as tetrahedra and octahedra with small, multivalent cations in their interstices.

Olivine (olive green) is composed of isolated $(SiO_4)^{4-}$ tetrahedra bonded mainly to Mg^{2+} and Fe^{2+} and formulated chemically either as $Mg_2 SiO_4$ or $Fe_2 SiO_4$ or mixed $(Mg,Fe)_2 SiO_4$.

Augite (black, prismatic) contains individual chains of linked Si-O tetrahedra having the units $(SiO_3)^{2-}$. The strands are held together by Ca^{2+}, Mg^{2+}, and Fe^{2+} (formula: $Ca(Mg,Fe)Si_2 O_6$).

Hornblende (black, prismatic-fibrous) is made up of double chains of Si-O tetrahedra with units of $(Si_4 O_{11})^{6-}$, joined to Ca^{2+}, Mg^{2+}, Fe^{2+}, OH^-, with the formula $Ca_2 (Mg,Fe)_5 (OH)_2 Si_8 O_{22}$.

These three "ferromagnesian" minerals serve as hosts for minor elements, and they are often associated with the mineral apatite which contains phosphorus.

Quartz (colorless, with greasy luster) represents a three-dimensional framework of SiO_4 tetrahedra, all corners being shared (Fig. 4.1) except, of course, those protruding at the outer surfaces. The formula is SiO_2 (silicon dioxide) and the mineral is highly resistant to weathering. Randomized SiO_2 frameworks are attributes of the noncrystalline opals, agates, flints, cherts, and quartzites.

Feldspars, whitish, are the most common of all rock minerals. Like quartz they have a framework of linked tetrahedra, but in one-fourth of them Al^{3+} substitutes for Si^{4+}. To overcome the resulting deficiency of positive charges additional cations, mostly K^+, Na^+, and Ca^{2+}, are incorporated in the open spaces. Common feldspars are *orthoclase* ($K Al Si_3 O_8$), *albite* ($Na Al Si_3 O_8$), and *anorthite* ($Ca Al_2 Si_2 O_8$). The last two may appear as mixed crystals called *plagioclase*. The formulas are idealized because the crystals are never pure. Orthoclase is the most resistant to weathering.

Micas are flaky minerals with layer lattices that are akin to those of *pyrophyllite,* $Si_4 O_{10} Al_2 (OH)_2$, the symbol in italics pertaining to cations in the octa-

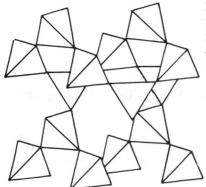

Fig. 4.1. Three-dimensional arrangement of SiO_4 tetrahedra in quartz. Note large cavities. After Berry and Mason, *Mineralogy: Concepts, Descriptions, Determinations*. W. H. Freeman and Company. Copyright © 1959

hedral sheet (Fig. 3.5). The many types of mica found in nature result from partial substitutions of ions of similar size and dissimilar charge, Si^{4+} and Al^{3+} in tetrahedra and Mg^{2+}, Fe^{2+}, Fe^{3+}, and Al^{3+} in octahedra.

When all tetrahedra are filled with Si^{4+} and all octahedra with Mg^{2+}, the mineral is *talc*, with the formula $Si_4O_{10}Mg_3(OH)_2$. In the common micas one-fourth of tetrahedral Si^{4+} is replaced by the slightly larger Al^{3+} which distorts somewhat the ideal hexagonal pattern and leaves a + charge deficiency as in some of the clays. The electrical imbalance is alleviated by placing between the layers K^+ ions, one for each Al^{3+}, such that they are partly sunk in the hexagonal cavities of the tetrahedral sheets, which are slightly forced apart. The K^+s hold adjacent layers tightly together, as shown in Figure 3.25. Along a stack of layers (*C*-axis) the period of structure repetition of C-spacing is 10 Å. For *muscovite* mica the resulting formula is $K(AlSi_3O_{10})Al_2(OH)_2$. In *biotite* mica the octahedral layer corresponds to that of talc with Fe^{2+} substituting partially for Mg^{2+}, producing the formula $K(AlSi_3O_{10})$ $(Mg,Fe)_3(OH)_2$.

The six groups of minerals described are primary silicates, the feldspars and micas also being primary aluminosilicates. The idealized formulas given omit the biologically important micronutrient accessories Zn, Mn, S, Co, F, and others.

B. Weathering and Neogenesis of Clays

Grinding quartz and feldspar crystals to fine powders does not produce a clay that swells and shrinks, is plastic and sticky, and specifically reactive. Physical break-up must be accompanied by chemical change.

Initial Hydrolysis Step

Long ago, O. Tamm in Sweden drilled a cylindrical cavity into a large orthoclase crystal and filled it with water and small chips of the feldspar. Long-time shaking abraded the fragments without contamination by foreign surfaces. He found, as did many others, the solution to be alkaline from a surface reaction with water:

$$\langle Si, Al, O \rangle K^+ \quad + \quad H^+ OH^- \quad \rightarrow \quad \langle Si, Al, O \rangle H^+ \quad + \quad K^+ OH^-$$

fresh feldspar	water	hydrolyzed	potassium
		feldspar	hydroxide in
			solution, pH 8-11

Not depicted are the H_2O molecules sorbed by the exposed crystal lattice ions.

This cation exchange reaction is "mineral hydrolysis" and it applies to all species mentioned, except quartz, which has no bases. The trend is accelerated by increasing the H^+ concentration in water by microbially produced carbonic, sulfuric, phosphoric, and nitric acids, acid clays (23), organic acids, and H-humus.

The K concentration in solution is critical because small additions of K, as in litterfall, drive the reaction to the left and retard weathering. With mica, 2 ppm of K in solution prolong the creation of a weathering rim of 1 mm width from 33 to 10,000 years (29). K is *removed* in outward diffusion, in leaching by rainfall, and in uptake by organisms.

Dissolution

The H^+ ions acquired by the feldspar diffuse into the crystal, convert O^{2-} to OH^-, and impart local instability. Si and Al tetrahedra peel off. Silicon in solution remains tetrahedral as $Si(OH)_4$, which is silicic acid, H_4SiO_4. Aluminum in water tends to surround itself octahedrally with six ligands that are either H_2O molecules, or OH^- and O^{2-} ions, or combinations thereof. The element appears as $Al(OH)_3$ molecules and as $Al(OH)_4^-$, $Al(OH)_2^+$, $Al(OH)^{2+}$, and Al^{3+} ions, depending on pH (Fig. 4.8). The solution is dilute and equations of solubility products may be applied.

Ferromagnesian minerals add ionized forms of $Ca(OH)_2$, $Mg(OH)_2$, and ferrous $Fe(OH)_2$. The oxidation of Fe^{2+} to Fe^{3+} creates $Fe(OH)_3$ and the ferric ions $Fe(OH)_4^-$, $Fe(OH)_2^+$, $Fe(OH)^{2+}$, and Fe^{3+} in relation to pH (Fig. 4.9).

Particles of Small Dimensions

As the feldspar crystal continues weathering new solid substances appear, but seldom in macroscopic size. In soils, diameters of micrometers and nanometers, the colloid range, are the preferred ones.

Colloidal Silica. Several $Si(OH)_4$ molecules may agglomerate (polymerize) to colloidal "silica," the word denoting any form of Si sharing O and OH. The polymer is written in Wiegner's (38) rectangular, *colloid notation* as $\boxed{Si(OH)_4}$. Its slight negative charge contributes to pH-dependent CEC. Within the particles the tetrahedra are either ordered as crystalline quartz varieties (40) or randomized as watery flocs. In the latter case drying leads to amorphous opal or slightly organized chalcedony.

Colloidal Aluminum Oxide Group. The colloidal particles of the various hydroxides $(Al(OH)_3)$, oxyhydroxides $(Al(OH))$, and oxides (Al_2O_3) listed in Table 4.1, are symbolized as Al, O, $\overset{+-}{OH}$. In neutral soils all varieties are but sparingly soluble. The

Table 4.1. Sesquioxide Clays[a]

Aluminum compounds (colorless, white)	Iron compounds (red, yellowish and brown)
Amorphous hydroxides $Al(OH)_3$ or $\frac{1}{2}(Al_2O_3 \cdot 3H_2O)$	Amorphous hydroxides $Fe(OH)_3$ or $\frac{1}{2}(Fe_2O_3 \cdot 3H_2O)$
Crystalline hydroxides Gibbsite and bayerite $Al(OH)_3$ or $\frac{1}{2}(Al_2O_3 \cdot 3H_2O)$	Paracrystalline hydroxide Ferrihydrite
Crystalline oxyhydroxides Boehmite and diaspore $AlOOH$ or $\frac{1}{2}(Al_2O_3 \cdot 1H_2O)$	Crystalline oxyhydroxides Goethite and lepidocrocite $FeOOH$ or $\frac{1}{2}(Fe_2O_3 \cdot 1H_2O)$
Crystalline oxide Corundum Al_2O_3 —	Crystalline oxide Hematite and maghemite Fe_2O_3 Magnetite $FeO \cdot Fe_2O_3$ or Fe_3O_4

[a] From various sources (e.g., 10).

+ – sign on the rectangle marks the amphoteric behavior including anion and cation exchange. The particles are either amorphous or crystalline, gibbsite (Fig. 3.4) being the most common mineral. Corundum is not found in soils. Bauxite deposits abound in $Al(OH)_3$ and are the source materials for the aluminum industry.

Colloidal Iron Oxide Group. The octahedral clusters of Fe^{3+} with OH^- and O^{2-} ions are generalized as $\boxed{Fe^{3+}, O, \overset{+-}{OH}}$ and are described in Table 4.1. At pH 5-8 all phases are practically insoluble. Crystalline forms assume the shapes of spheres, needles, and platelets. Ferrihydrate (32) is a forerunner to the ubiquitous goethite and to lepidocrocite of wet places. Both minerals are oxyhydroxides. Magnetite in soils is extractable with a magnet. Limonite and bog iron are poorly defined mixtures. Goethite—as well as gibbsite and boehmite—were identified by Hénin and Pedro (17) in a brief, 2-year-long weathering experiment of basalt rock.

Heating expels OH ions from the lattice as H_2O, leaving the oxide behind. Two $Al(OH)_3$ units are then written $Al_2O_3 \cdot 3H_2O$. The combination of Al^{3+} or Fe^{3+} with O^{2-} is in the proportion of 2:3 or 1:1.5 (1½ = sesqui), which explains the common name *sesquioxides.*

Colloidal Secondary Aluminosilicates. The dissolved Si and Al tetrahedra and octahedra may recombine to new types of aluminosilicates or silicate clays. Wiegner (38) symbolized the colloidal particles as $\boxed{nH_2O \cdot SiO_2 \mid Al(OH)_3}$; now, the simpler version $\boxed{Si, Al, O, \overline{OH}}$ seems more appropriate. Whether crystalline or not, the particles are most often negatively charged, act as huge anions, and possess CEC.

Weathering products of volcanic ash abound in noncrystalline and paracrystalline *allophanes* having SiO_2/Al_2O_3 ratios of 1.3-2.0, isoelectric points below pH 7, and large surface areas of 1000 m^2/g. A better crystallized variant is threadlike *imogolite* (SiO$_2$ · Al$_2$O$_3$ · 2.5H$_2$O) that may convert to halloysite and kaolinite (37). With dilute solutions of silica and Al-salts and with ingenious mixing techniques French and Belgian soil mineralogists (7) succeeded in synthesizing at room temperature the silicate clays kaolinite (Fig. 3.9) and montmorillonite (Fig. 3.8). These and illite, vermiculite, and chlorite will be described in the next section.

When aluminum hydroxide sol is poured over sodium silicate solution in a test tube, vigorous shaking produces a stiff gel. The white coprecipitate is an aluminosilicate. Wiegner (38), Gedroiz (15), Mattson (24), and many others believed that these amorphous flocs would provide a rational explanation of world-wide clay formation, a view that fell into disrepute with the discovery of clay crystallinity by X-ray (19) in the early 1930s.

Summary of Weathering and Neogenesis of Clays

A rock containing a mixture of the 6 primary minerals gives rise to the following major products:

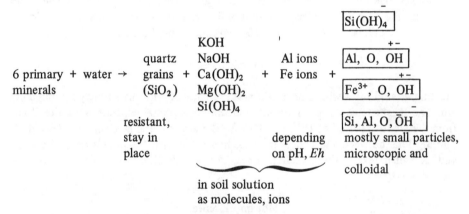

The ubiquitous carbonic acid (CO$_2$) of air converts the hydroxides of K, Na, Ca, and Mg to carbonates and bicarbonates, those of Ca and Mg being rather insoluble.

The multitude of soils on the surface of the Earth is keyed to the fate of these weathering products, whether they stay in place or move on, are stable or disintegrate, or transform to new syntheses in a variety of microenvironments.

Traditionally, American pedologists call all four groups "clays" as long as the diameters are less than 2 μm. In this book the terms "oxide clays" and "silicate clays" are also employed. Mineralogists confine the word clay to the aluminosilicate group and speak of "clay minerals."

For the transformation of orthoclase to kaolinite clay the time-honored overall equation is written:

$$2K\,Al\,Si_3O_8 + 11H_2O \rightleftharpoons 2KOH + 4Si(OH)_4 + Si_2O_5\,Al_2(OH)_4$$

orthoclase   kaolinite

The equilibrium constant turns out to be $[Si(OH)_4]^2 \cdot [K^+]/[H^+]$. The three components in solution control the reaction, but nothing is divulged about mechanisms, intermediate fabrics, colloidality, or speed of weathering.

C. The Structural Kinship of Micas and Clays

Clay formation by neogenesis postulates that all polyhedra of primary minerals, especially of olivine, augite, hornblende, and feldspars are liberated by hydrolysis and then reassembled to clays either inside or adjacent to the parent crystal or after movement to a different horizon site.

In the case of mica flakes the transformation is facilitated because reagent (mica) and reaction product (clay) have similar layer lattices.

Illites

Leaching mica with dilute acids, including CO_2, slowly replaces some of the interlayer K^+ by exchange diffusion with H^+ (or H_3O^+), as visualized in Figure 3.25. The flakes curl up at the edges and undergo diminution in size (10). The altered mica is *hydromica* or *illite*. It retains the characteristic 10-Å spacing along the C-axis.

Vermiculites

Treating ground mica with a magnesium salt replaces K with the strongly hydrated Mg ion. The layers separate and hydrate, and the mica crystal expands along the *C*-axis from the identity period of 10 Å to 14-15 Å. The new mineral is *vermiculite* (Fig. 4.2). The Mg ions between the layers are reexchangeable with Na and Ca and, of course, with K which reverts the crystal to mica. Soil vermiculites have a CEC of 100-200 me/100 g (10).

Montmorillonites

Further transformation of mica reduces CEC to 80-100 me/100 g. The forces holding the layers together are now so weakened that the crystal strongly hydrates and expands to 18-40 Å; in fact, with exchangeable Na and in agitated suspension complete

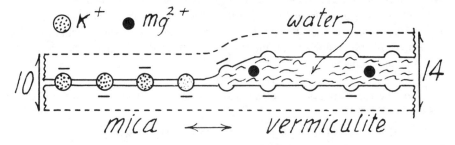

Fig. 4.2. Transition from mica to vermiculite. Numbers denote angstrom units (Å)

separation of individual plates is observed. The strongly swelling minerals constitute the *smectite group* of which *montmorillonite* is the classic member. Its swelling and exchange behavior is attributed to Mg replacing Al in the octahedral sheet. Associates are beidellite, with partial Al substitution in the tetrahedral sheet, iron-rich nontronite, and saponite and hectorite (10).

Interstratification

Chlorites are mica-related minerals in which K of interlayer space is replaced by a positively charged sheet of Mg, Al octahedra (Fig. 4.3) which reduces CEC severely. Interstratification with $Al(OH)_3$ is common in vermiculites and montmorillonites. Either Al is incorporated directly from the soil solution or exchangeable H^+ between layers switches positions with octahedral Al. In older soils complicated, regular, and irregular stacking sequences may develop that render clays rather stable to solution attack.

How these various clays may arise in a soil profile is documented by Wilson (41) in Scotland for an uncultivated brown forest soil on biotite-hornblende rock weathered to over 1 m depth. In the C horizon (pH 5.8) biotite becomes hydromica with loss of K, and all iron is oxidized. In the B horizon (pH 6) interstratified vermiculite-chlorite appears, having a Mg octahedra sheet between the mica layers, the Mg possibly coming from nearby weathering hornblende crystals. In the A horizon (pH 5.2) the proportion of vermiculite to chlorite is augmented and in this process of "vermiculitization" the $Mg(OH)_2$ sheet is removed, probably by the high soil acidity.

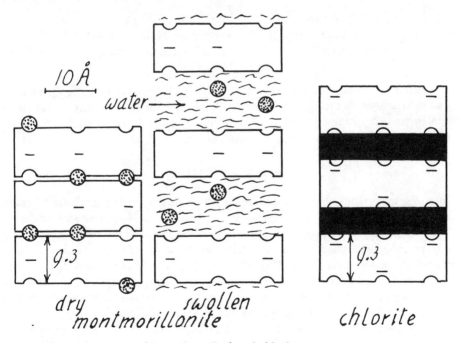

Fig. 4.3. Montmorillonite (dry and swollen) and chlorite

Kaolinites

In a drastic alteration of pyrophyllite one of the two sheets of Si tetrahedra is removed and the lost O^{2-}s of the octahedral plane are replaced by OH^- of water.

The truncated remainder is *kaolinite* (Fig. 3.9). It has a C-spacing of 7.1 Å and the formula is $Si_2 O_5 Al_2 (OH)_4$. Heating pyrophyllite or kaolinite to high temperatures expels the OH ions as water, and their formulas are often written $4 SiO_2 \cdot Al_2 O_3 \cdot H_2 O$ and $2 SiO_2 \cdot Al_2 O_3 \cdot 2H_2 O$, respectively, the $SiO_2/Al_2 O_3$ ratios being 4 and 2.

The interstices of the hydroxyl sheet are occupied by Al, those of the tetrahedra sheet by Si. There are few substitutions and the layers are neutral. Weak hydrogen bonds between O and OH planes hold the layers together, yet the mineral does not swell because it lacks hydrating interlayer exchangeable cations. At pH 7 the CEC is low, 3-15 me/100 g, depending on particle size.

In the closely related *serpentine* minerals Mg^{2+} and some Fe^{2+} rather than Al^{3+} reside within the octahedral cavities. The formula is $Si_2 O_5 Mg_3 (OH)_4$. *Halloysites* have a plane of water molecules interposed between two kaolinite layers. The sheets tend to roll up into submicroscopic tubes which are readily identified on electron micrographs.

D. Clay Sequences

The principal clays that might come forth in weathering of mica are graphically summed up as:

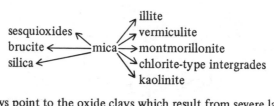

The left arrows point to the oxide clays which result from severe lattice disruptions. The right arrows lead to the silicate clays. Opinions differ on the dominance of either neogenesis or transform processes. If orthoclase is put into the center of the above scheme all clay minerals are products of neogenesis.

Mineral Environments

In a rock mineral the clays may synthesize within its microscopic cracks and fissures and eventually convert the entire crystal, yet preserve its shape (pseudomorphism). The specific clay to be generated is governed by the milieu of the weathering primary mineral, whether flushed by acid or alkaline solutions. Broadly, clay genesis is a function of the soil-forming factors (25). Tardy et al. (35) find biotite converted to montmorillonite in arid climates, to vermiculite in temperate climates, and to kaolinite in the humid tropics. Plagioclase feldspar, on the other hand, progresses directly to kaolinite in all climates, perhaps because it lacks a source of Mg.

In "biologically induced weathering," the biotite lattice is altered by wheat plants to vermiculite (26) and by white cedar, hemlock, and other tree seedlings directly to kaolinite (34).

Transformation of Clays

Omitted in the above sketch are the many arrows that connect the clay species among themselves. Much publicized is the forward and backward conversion: mica ⇌ illite ⇌ vermiculite ⇌ montmorillonite ⇌ kaolinite. Pedogenetically, the sequence is multifactorial. It may not materialize unless the state factors are modified, such as a change in climate to raise pH or to initiate a leaching regime, or a shift to a new topographic position to supply bases or encourage erosion. Thus, the three conversions: gibbsite → kaolinite, kaolinite → montmorillonite, and gibbsite → montmorillonite are *resilications* and require a localized influx of silica.

Quantitative depth functions of clay mineralogy in soils have been made available by Loughnan (22). Schellmann's (31) soil, formed on serpentine rock in a dense tropical forest in Borneo, is copied in Figure 4.4. In the horizons at 7 m depth montmorillonite and chlorite stand out, whereas the surface layer is mostly hydrous oxide clay. Should the soil surface slowly erode and montmorillonite approach the surface the mineral is expected to convert to the kaolinite-gibbsite suite. Seen in this light, the existing depth sequence reflects a steady-state clay profile, the more so if the rate of rock weathering at the base should match the rate of erosion at the surface.

Indices of Clay Suites

Every soil and every horizon carries a commixture of clay types. For a population of soils the *integration of clay species* into a singular, numerical index has been attempted by computation of principal components (PC) which are statistical blends. For 97 soil

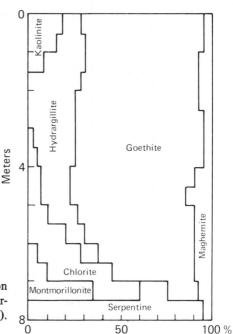

Fig. 4.4. Schellmann's (3) depth function of clay types in a red soil derived from serpentinite in Borneo (courtesy Geol. Jahrb.). Hydrargillite is the same as gibbsite

sites in a climatic transect the first PC, the most definitive amalgam of species, comprises the four clay minerals arranged in Table 10.5 in order of statistical correlation significance. The role of parent rock on preponderance of clay types is striking even though the soils are very old.

Years ago, workers (2) discovered many striking correlations between the ratio $SiO_2/(Al_2O_3 + Fe_2O_3)$ of natural clays and properties like CEC, sorption of NH_3 gas and dyes, and heat of wetting, the latter shown in Figure 4.5. Rather than proving an amorphous clay continuum, as once believed, each point around the line must be interpreted as the mean value of wetting properties of the suite of clays in a single soil sample.

E. Solubilities and Stability Fields of Clays

Keenly felt is the need for a fundamental approach to clay stability based on thermodynamic criteria, as advocated by Garrels and his school (14). When the "free energies of formation," ΔG_f^0, of clays in a standard state are known, transformation and coexistence of clays are predictable. The approach relies on solubilities and calorimetric measurements of mineral reactions but is handicapped by analytical demands and by the uncertainty of equilibrium attainment.

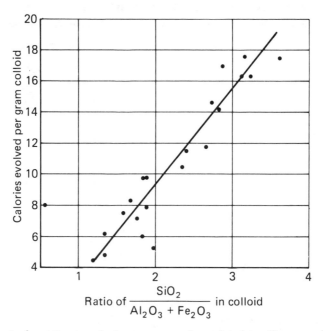

Fig. 4.5. Heat of wetting in calories per gram clay related to silica-sesquioxide ratios of natural soil clays (2)

Solubility of Pure Silica

Crystals of quartz $(SiO)_2$ dissolve in water slowly in amounts ranging from 5-20 ppm of SiO_2. Hydration, as $SiO_2 + 2H_2O$, produces $Si(OH)_4$ molecules. Amorphous silica precipitated in the laboratory has a porous gel structure and delivers to water 110-150 ppm of SiO_2 at pH 7. The high solubility marks it as a thermodynamically unstable intermediate because the gel is expected to turn into quartz. The conversion rate is too slow to have much pedological significance. Because $Si(OH)_4$ is adsorbed on sesquioxide clays many soils have less than 10 ppm of soluble silica (40). Only 0.67 ppm SiO_2 is detected in a water extract (12) of a ferruginous soil in Hawaii. Since acids dissolve sesquioxides, soil solution content of silica may rise sharply (3) as pH is lowered from 7 to 2. Metabolic ATP readily dissolves quartz (11).

Correns's (9) extractions of silica gel with dilute HCl and NaOH are compared in Figure 4.6 with the solubility-product calculations of Alexander et al. (1). In confirmation of the trend a natural alkali soil of pH 10.4 supports 375 ppm SiO_2 in a water extract (18). Temperature augments solubility linearly (28), the ppm SiO_2 being 100 at $0°C$, 180 at $22°C$, 260 at $50°C$, 330 at $58°C$, and 380 at $73°C$.

The essentially neutral $Si(OH)_4$ molecule readily enters plant roots passively in the transpiration stream and accumulates in stems and leaves as amorphous plant opal. In bamboo species it reaches sizes of 3-4 cm (tabasheer). The silica richness of tropical plants has been attributed to high weathering rates and high silica solubility. Annual SiO_2 additions to soil in leaffall are estimated in Oregon (27) as 9 g/m^2 for fir forest and 11 g/m^2 for a natural grassland. Albic A2 horizons have been found enriched in "phytoliths" and Riquier (30) describes a fake tropical podsol whose A2 is white plant opal.

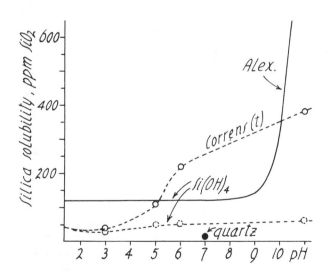

Fig. 4.6. Solubilities of quartz and silica gels as a function of pH. After Correns (9), Alexander et al. (1), and other sources

Brucite Stability

Brucite, $Mg(OH)_2$, is composed of a stack of layers of Mg octahedra (Fig. 3.4). The mineral is slightly soluble in water, about 6 ppm, and imparts a reaction of about pH 10. Higher alkalinity decreases the solubility and adding acid augments it.

For the dissociation reaction $Mg(OH)_2 = Mg^{2+} + 2\ OH^-$ the solubility product is

$$k_{sp} = \frac{Mg^{2+} \cdot (OH^-)^2}{Mg(OH)_2}\ , \text{ or for activities, } k = [Mg^{2+}] \cdot [OH^-]^2$$

In logarithms, $\log k = \log [Mg^{2+}] + 2 \log [OH^-]$, and Garrels' thermodynamic data (14) provide a value of -11.06. Replacement of OH^- by H^+ of dissociation of water, which is $H^+ \cdot OH^- = 10^{-14}$, transforms $\log k$ to 16.9. Therefore, $\log [Mg^{2+}] = 16.9 - 2\ pH$.

To interpret Figure 4.7 select a pH of interest, say 8.5, and draw a vertical line (dashed). Along it the activities of dissolved Mg^{2+} ions are read off. Below the slanted solid line the solutions are unsaturated; on the line Mg saturation prevails and solution and solid coexist, like water and ice at 0°C. Above the line unstable supersaturation precipitates solid $Mg(OH)_2$, that is, above the solid line brucite is stable, below it it dissolves, the more so the lower pH. The instability extends, qualitatively at least, to Mg octahedra sheets in interstratified minerals such as chlorite. Looking at the situation differently, the analysis of a soil solution for Mg and pH tells whether or not the brucite mineral is likely to be present in the solid phase.

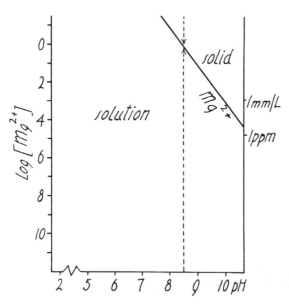

Fig. 4.7. Stability field of brucite, $Mg(OH)_2$ as a function of pH. Logarithm of Mg^{2+} activities (left), and concentrations of Mg in millimoles/liter and ppm

Gibbsite Stability

In pure water a trace of $Al(OH)_3$ dissolves and dissociates into Al^{3+} and $3OH^-$ ions. Acidity dissolves gibbsite, as indicated by the Al^{3+} equilibrium line in Figure 4.8.

The product of the ions is $k_1 = [Al^{3+}] \cdot [OH^-]^3$, and Frink and Peech (13) obtained the small value of log $k_1 = -33.5$. The Al^{3+} ions attract OH^- ions of water and become complex ions of the types $(Al\,OH)^{2+}$ and $(Al(OH)_2)^+$. Sillén and Martell (33) provide numerous "hydrolysis constants," and choosing for $(Al\,OH)^{2+}$ Frink and Peech's (13) value of log $k_2 = -5.0$, and for $(Al(OH)_2)^+$ van Schuylenborgh's (36) choice of log $k_3 = -4.7$, Figure 4.8 is constructed. It relates the active mass of the various aluminum ions to the pH of the soil solution. Owing to the amphoteric nature of $Al(OH)_3$ the aluminate anion $(Al(OH)_4)^-$ is detectable at high pH. Its logarithmic hydrolysis constant k_4 is -12.7 (36).

In acid soil solutions of pH less than 4, Al^{3+} and $(Al\,OH)^{2+}$ ions are prominent and may become toxic to roots. With increasing pH aluminum precipitates, but dissolves again in alkaline soils because of $Al(OH)_4^-$. Gibbsite is least soluble, ormost stable, at pH 5-6 and a kind of "stability field" is fenced in by the solid equilibrium lines in Figure 4.8. Crystallization lowers the solubility.

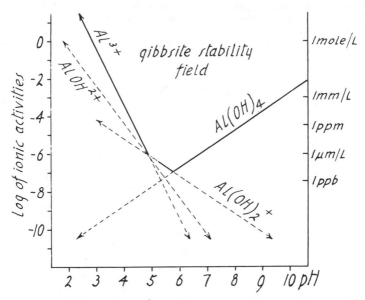

Fig. 4.8. Stability field of gibbsite, $\overset{\cdot}{A}l(OH)_3$. Various Al ions are plotted in relation to pH. Left: logarithms of activities; right: positions of absolute concentrations in millimoles/liter and micromoles/liter

Iron Oxide Stability

The iron oxide group of Fe^{3+} is still more insoluble than the aluminum oxide group, as seen by comparing Figures 4.8 and 4.9.

For fresh $Fe(OH)_3$ in aerated soil suspensions Bohn (6) obtains the solubility product $\log k_1 = -38.8$, representing $[Fe^{3+}] \cdot [OH^-]^3$. The curves for his set of hydrolysis constants are plotted in Figure 4.9, solid lines, but only the dominant iron species at various pH values are graphed.

The curve labeled G in the lower left part of Figure 4.9 is based on the Garrels-Christ equations for aged, crystalline hematite (6) employing $\log k_1 = -42.72$. Except for pH < 2, the Fe concentrations are vanishingly small. Hematite is considered thermodynamically the most stable of the oxides (4) and thus the surviving one, but Schwertmann and Taylor (32) point to soil disequilibria that allow coexistence with other oxides during pedologic time periods.

Reduced Iron

In striking contrast to trivalent iron is the greenish-white hydroxide of divalent iron, $Fe(OH)_2$, of reducing conditions. It furnishes relatively large quantities of Fe^{2+} and $Fe(OH)^+$ in neutral and acid soils (Fig. 4.9, dashed lines).

In the absence of oxygen the solubility product of $Fe(OH)_2$ is $\log k_1 = 14.6$ (36) and the pertinent equations include the redox potential Eh. For hematite at $25°C$ ($k_1 = -42.72$), dissolved Fe^{2+} follows the equation:

$$\log [Fe^{2+}] = -16.95\,Eh - 3\,pH + 12.34$$

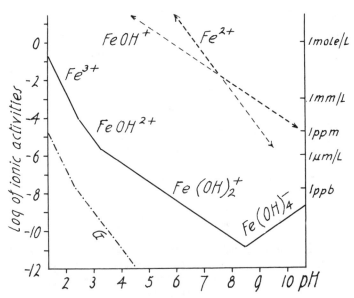

Fig. 4.9. Stability fields of iron oxides as ferric (Fe^{3+}) and ferrous (Fe^{2+}) hydroxides

At pH 5, for instance, under the oxygen-saturated, strongly oxidizing regimes of *Eh* of 0.6 and 0.4 V the Fe^{2+} concentrations are $10^{-12.8}$ and $10^{-9.4}$ moles/liter, both likely too low to have pedological or ecological significance. For the strongly reducing categories of *Eh* of zero and -0.2 V, the Fe^{2+} concentrations of $10^{-2.6}$ and $10^{0.73}$ moles/liter are likely to be pertinent.

Manganese

Manganese is capable of assuming many more oxidation states (valencies) than iron, and it displays more complex reactions (21). Its prominent precipitates in soils are $Mn(OH)_2$, $MnOOH$, MnO_2, Mn_2O_3, and Mn_3O_4, all stable at neutral to alkaline reactions. These compounds impart to soil aggregates a blackish color, foremost along cracks that admit air. Manganese vigorously oxidizes hydrogen peroxide (H_2O_2) with a "fizz," a reaction that is utilized for qualitative Mn testing in the field.

Kaolinite Stability

Stripping off in water the silica layer of kaolinite (Fig. 3.9) leaves as remainder an Al-OH-sheet, which is gibbsite. Quantitatively,

$$Si_2O_5Al_2(OH)_4 \quad + \quad 5H_2O \quad = \quad 2\,Al(OH)_3 \quad + \quad 2\,Si(OH)_4$$

$$\text{kaolinite} \qquad\qquad\qquad\qquad \text{gibbsite} \qquad\quad \text{silica}$$

The solubility product of the dissolved species reduces to $k = [Si(OH)_4]^2$, and with the aid of Garrels and Christ's (14) collection of ΔG_f^0 values log k is calculated as -9.311. It is equal to a $Si(OH)_4$ solubility of $10^{-4.66}$ moles/liter or 2.1 ppm, which is 1.3 ppm SiO_2.

Importantly, in this equilibrium solution kaolinite and gibbsite coexist. If less $Si(OH)_4$ is present kaolinite dissolves to gibbsite, if more, gibbsite and $Si(OH)_4$ transform to kaolinite. The reaction pertains to pH 5-6 at which gibbsite solubility is minimal.

At higher $Si(OH)_4$ concentrations kaolinite resilicates to montmorillonite and the two coexist in a solution of 96 ppm SiO_2, according to Kittrick's (20) years-long equilibrations. Below that threshold montmorillonite converts to kaolinite. No predictions on mechanisms or rates are promised by the thermodynamic approach.

The central position of kaolinite on a silicate axis at neutral reaction and with adequate Mg and Al concentrations appears as follows:

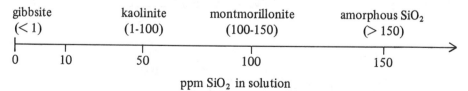

The ppm quantities are taken from Kittrick's work (20) but should not be considered final because other authors get somewhat different answers, depending on pH, bases, and clay subspecies involved. Coexistence of more than two of the four minerals is forbidden, but exceptions in field soils are noted and, unfortunately, can always be "explained" as disequilibria or the minerals not sharing an identical solution.

The minerals serpentine and brucite are structurally related to kaolinite and gibbsite:

$$Si_2O_5Mg_3(OH)_4 \ + \ 5H_2O \ = \ 3\,Mg(OH)_2 \ + \ 2\,Si(OH)_4$$

$$\text{serpentine} \qquad\qquad\quad \text{brucite} \qquad \text{silica}$$

They coexist likewise, as seen in the three-dimensional stability fields calculated by Wildman et al. (39) and depicted in Figure 4.10.

The solution of coexistence is characterized by $\log\,[Si(OH)_4] = -8.6$ which is 0.15 ppb SiO_2, a very low value. Below it only brucite is stable, above it only serpentine. The left-front face is akin to Figure 4.7 and the back face is put at the saturation solubility of silica, taken as $10^{-2.6}$ moles/liter or 151 ppm SiO_2.

It is immediately apparent that both serpentine and brucite are unstable over most of the pH range of soils (<7) unless very high Mg salinity prevails.

F. Clays as Initiators of Organic Evolution

During early abiotic Earth times electrical discharges generated amino acids, as has been proven in laboratory experiments, but their dilution in the sea must have been so high that the probability of collisions leading to organized macromolecules was practically nil. Goldschmidt and others propose that the organic molecules were adsorbed on clay

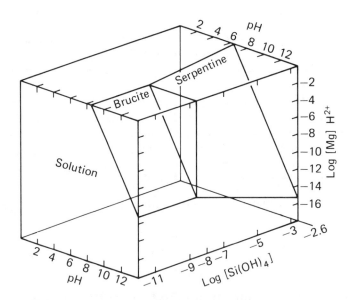

Fig. 4.10. Stability fields of brucite and serpentine; after Wildman et al. (39). (Courtesy The American Mineralogist, Washington, D.C.)

surfaces, came spacially together, and polymerized. Cairns-Smith (8) further argues that adsorbed organic monomers line up along crystal edges and become thread molecules, as seems to have been demonstrated (San Francisco Chronicle, Oct. 15, 1977), and assemble between planes to two-dimensional networks that are precursors to membranes. In Cairns-Smith's picture, primitive life materials and evolution were initiated in soils, not in the sea.

Hartman (16) conjectures an autotrophic origin of metabolism from a simple environment and again invokes clays to serve as templets of orientation and reproduction.

G. Review of Chapter

Soil clays ($<$ 2 μm) isolated by mechanical analysis are inherited from parent materials and originate in horizons from weathering of feldspars and ferromagnesian minerals, and from transformations of mica. Prominent among the *oxide clays* are the oxides, hydroxides, and oxyhydroxides of silicon, aluminum (e.g., gibbsite), and iron (e.g., goethite). Neogenesis and alteration of mica generate *silicate clays* ("clay minerals") such as illite, vermiculite, smectites, chlorites, and kaolinites. Depending on their immediate environments the clays may convert from one to the other. Stability of clays is related to their solubility products in soil solutions.

Chemists speculate that clays played a vital role in the origin of life.

References

1. Alexander, G. B., W. M. Heston, and R. K. Iler. 1954. *J. Phys. Chem.* 58: 453-455.
2. Anderson, M. S., and S. Mattson. 1926. *Properties of the Colloidal Soil Material.* U.S.D.A. Bull. 1452. Washington, D.C.
3. Beckwith, R. S., and R. Reeve. 1963. *Aust. J. Soil Res.* 1: 157-168.
4. Berner, R. A. 1971. *Principles of Chemical Sedimentology.* McGraw-Hill, New York.
5. Berry, L. G., and B. Mason. 1959. *Mineralogy Concepts, Descriptions, Determinations.* Freeman, San Francisco.
6. Bohn, H. L. 1967. *Soil Sci. Soc. Am. Proc.* 31: 641-644.
7. Caillère, S., and S. Hénin. 1965. In *Experimental Pedology,* E. G. Hallsworth and D. V. Crawford, eds., pp. 99-112. Butterworths, London.
8. Cairns-Smith, A. G. 1971. *The Life Puzzle.* Toronto Univ. Press, Toronto.
9. Correns, C. W. 1940. *Chem. Erde* 13: 92-96.
10. Dixon, J. B., and S. B. Weed (eds.). 1977. *Minerals in Soil Environments.* Soil Sci. Soc. Am., Madison, Wisc.
11. Evans, W. D. 1965. In *Experimental Pedology,* E. G. Hallsworth and D. V. Crawford, eds., pp. 14-27. Butterworths, London.
12. Fox, R. L., J. A. Silva, D. R. Younge, D. L. Pluckett, and G. D. Sherman. 1967. *Soil Sci. Soc. Am. Proc.* 31: 775-779.
13. Frink, C. R., and M. Peech. 1963. *Inorg. Chem.* 2: 473-478.
14. Garrels, R. M., and C. L. Christ. 1965. *Solutions, Minerals, and Equilibria.* Harper and Row, New York.
15. Gedroiz, K. K. 1931. *Kolloidchem. Beih.* 317-448.
16. Hartman, H. 1975. *J. Mol. Evol.* 4: 359-370.
17. Hénin, S., and G. Pedro. 1965. In *Experimental Pedology,* E. G. Hallsworth and D. V. Crawford, eds., pp. 29-39. Butterworths, London.

18. Kelley, W. P., and S. M. Brown. 1939. *J. Am. Soc. Agron.* 31: 41-43.
19. Kelley, W. P., W. H. Dorc, and S. M. Brown. 1931. *Soil Sci.* 31: 25-55.
20. Kittrick, J. A. 1977. In *Minerals in Soil Environments*, J. B. Dixon and S. B. Weed, eds., pp. 1-25. Soil Sci. Soc. Am., Madison, Wisc.
21. Krauskopf, K. B. 1967. *Introduction to Geochemistry.* McGraw-Hill, New York.
22. Loughhann, F. C. 1969. *Chemical Weathering of the Silicate Minerals.* Elsevier, Amsterdam.
23. Marshall, C. E. 1977. *The Physical Chemistry and Mineralogy of Soils.* Wiley, New York.
24. Mattson, S. 1930. *Soil Sci.* 30: 459-495.
25. Mitchell, B. D. 1962. In *Genèse et synthèse des argiles*, pp. 139-147. Centre Nat. Res. Sci., Paris.
26. Mortland, M. M., K. Lawton, and G. Uehara. 1956. *Soil Sci.* 82: 477-481.
27. Norgren, J. A. 1973. Opal phytolyths as indicators of soil age and vegetative history. Ph.D. thesis, Oregon State Univ., Corvallis.
28. Okamoto, G., T. Okura, and K. Goto. 1957. *Geochim Cosmochim.* Acta 12: 123-132.
29. Rausell-Colom, J. A., et al. 1965. In *Experimental Pedology*, E. G. Hallsworth and D. V. Crawford, eds., pp. 40-72, 92-96. Butterworths, London.
30. Riquier, J. 1960. *Int. Congr. Soil Sci. 7th* 4: 425-431.
31. Schellmann, W. 1964. *Geol. Jahrb.* 81: 645-678.
32. Schwertmann, U., and R. M. Taylor. 1977. In *Minerals in Soil Environments*, J. B. Dixon and S. B. Weed, eds., pp. 145-180. Soil Sci. Soc. Am., Madison, Wisc.
33. Sillén, L. G., and H. E. Martwell. 1964. *Chem. Soc.*, London, Spec. Publ. 17, 754 pp.
34. Spyridakis, D. E., G. Chesters, and S. A. Wilde. 1967. *Soil Sci. Soc. Am. Proc.* 31: 203-210.
35. Tardy, Y., G. Bocquier, H. Paquet, and G. Millot. 1973. *Geoderma* 10: 271-284.
36. Van Schuylenborgh, J. 1965. In *Experimental Pedology*, E. G. Hallsworth and D. V. Crawford, eds., pp. 113-125. Butterworths, London.
37. Wada, K. 1977. In *Minerals in Soil Environments*, J. B. Dixon and S. B. Weed, eds., pp. 603-638. Soil Sci. Soc. Am., Madison, Wisc.
38. Wiegner, G. 1921. *Boden und Bodenbildung.* T. Steinkopff, Dresden.
39. Wildman, W. E., L. D. Wittig, and M. L. Jackson. 1971. *Am. Min.* 56: 587-602.
40. Wilding, L. P., N. E. Smeck, and L. R. Drees. 1977. In *Minerals in Soil Environments*, J. B. Dixon and S. B. Weed, eds., pp. 471-552. Soil Sci. Soc. Am., Madison, Wisc.
41. Wilson, M. J. 1970. *Clay Minerals* 8: 291-303.

5. Biomass and Humus

Starting with rock, water, and air, only the blue-green algae and the purple bacteria are able to put together a viable combination of C, H, O, and N atoms in the form of biomass. The higher plants assimilate carbon dioxide of the air but depend on microbes to supply available nitrogen. All soil organisms respire and consume plant materials, directly or indirectly. During metabolic conversion dark humus substances appear and accumulate in the soil. The path from greenery to blackness is reconnoitered in ensuing sections.

A. Carbon Fixation

By 1850 learned men had accepted the then revolutionary idea that in sunlight green plants transform the invisible gas CO_2 to edible sugars and starches. This type of carbon (C) assimilation is photosynthesis and is confined in higher plants to green, chlorophyll-containing chloroplast bodies (13).

Participants

For creating grape sugar (glucose) the overall process is written

$$6 CO_2 \; + \; 6H_2O \; \xrightarrow[\text{chloroplast}]{\text{light } (hv)} \; C_6H_{12}O_6 \; + \; 6 O_2 \qquad (5.1)$$

| carbon dioxide gas | water | | glucose (high in chemical energy) | oxygen gas (from water) |

The quantities are moles which are one for glucose (180 g) and six each of CO_2 (44 g), H_2O (18 g), and O_2 (32 g). Light energy is expressed as hv, the product of Planck's universal constant h and the frequency v of light vibrations.

Of the sunlight that impinges on a single leaf 5-15% is used in photosynthesis. The remainder is partitioned among reflection (ca. 20%), heat of transpiration (ca. 60%), and heat radiation (ca. 30%). In an alfalfa field the gross photosynthetic utilization of incident light is about 1% or less.

Besides the chlorophyll molecules with their magnesium cores the photosynthetic apparatus of land plants harbors a battery of enzyme catalysts with inorganic cofactors such as Fe, Mn, and possibly Cl. A further participant is phosphorus (P) present in the organic ester adenosine (A) as ATP^{4-} and ADP^{3-}:

$$
\begin{array}{ccc}
\overset{\displaystyle O}{\overset{\displaystyle \|}{}} \ \ \overset{\displaystyle O}{\overset{\displaystyle \|}{}} \ \ \overset{\displaystyle O}{\overset{\displaystyle \|}{}} & & \overset{\displaystyle O}{\overset{\displaystyle \|}{}} \ \ \overset{\displaystyle O}{\overset{\displaystyle \|}{}}
\end{array}
$$

$$
\text{A–O–P–O–P–O–P–OH} + \text{HOH} \ \rightleftharpoons \ \text{A–O–P–O–P–OH} + H_3PO_4 \qquad (5.2)
$$

$$
\underset{OH\ \ \ OH\ \ \ OH}{} \qquad\qquad \underset{OH\ \ \ OH}{}
$$

| adenosine triphosphate | water | adenosine diphos- | phosphoric |
| (ATP) | | phate (ADP) | acid (Pi) |

Adenosine triphosphate (ATP) is "energy-rich," a kind of fuel, because it releases by hydrolysis 7-10,000 cal/mole of free energy, $-\Delta G$, that may be harnessed for driving biochemical reactions. Sensitivity of C-fixation to phosphorus status of sugar beet leaves is revealed by the curve in Figure 5.1.

Processes

When a tiny bundle of light, a photon or quantum, strikes an atom it may eject an electron (e^-) and impart to it high velocity, as Einstein reasoned in 1905. Applying the effect to a chlorophyll molecule, Arnon (30) sees the expelled electron being captured by the enzyme ferrodoxin, an Fe-S-protein, and then hopping back to its source along a redox chain of cytochrome enzymes. On the way it transduces part of its energy to the creation of one (30) or two ATP molecules by way of reversing hydrolysis [Eq. (5.2)].

In complementary, more involved pathways water molecules in chloroplast are split by light to give off electrons and oxygen ($H_2O = 2\ e^- + 2H^+ + \frac{1}{2} O_2$) and the two e^-s are donated along with the $2H^+$ to the complex molecule NADP which reduces CO_2 in the dark by way of hydrogenation. During e^- flow additional ATPs appear.

In 20-30 discrete reactions with as many intermediary compounds CO_2 is converted to glucose sugar and other carbohydrates, carboxyl groups, fatty acids, and amino acids (Calvin cycle). When the first detectable molecule is a 3-carbon acid the plants are referred to as C_3-plants; when a 4-carbon acid they are C_4-plants. The latter are frequent among species of tropical origin.

Among the lower green plants, algae also contain diverse pigments and complex membranes that enable them to photosynthesize. Algae may become integral parts of the surface soil, developing green and brown felts, matts, scums, and crusts. In deserts they may grow under and adjacent to translucent stones (e.g., quartzites) because of light and slightly improved moisture conditions (10).

To render photosynthesis ecologically and pedologically meaningful it must be related to a season or a year and to unit area of land, which assigns to soil a key role as reservoir of water and ions, as outlined for the leaf in Figure 5.2. Biomass production is often proportional to transpiration (Chapter 2) and the ascending sap carries phosphates, sulfates, nitrates, essential cations and, possibly, high bicarbonate anions acquired from the soil atmosphere which is richer in CO_2 (3000 ppm) compared to ordinary air (330 ppm).

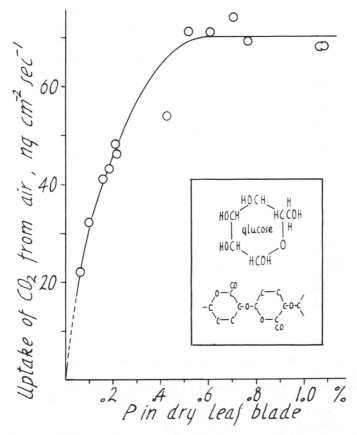

Fig. 5.1. Uptake of CO_2 by intact sugar beet leaves of variable total phosphorus (P) content, courtesy N. Terry (71). Inset: structure formula of glucose sugar, and segment of the carbon skeleton of a long anhydro-glucose chain in the cellulose framework of a cell wall

B. Respiration, the Energy Supplier of Soil Organisms

When cigarette ashes are sprinkled on a sugar cube and a lighted match is put to it the cube burns with a hot, visible flame. Sugar is being oxidized, the ash acting as a catalyst. Two centuries ago, Lavoisier conceptualized life processes (metabolism) as oxidations and he detected that breathing animals consume O_2 and expel CO_2.

Energy Aspects

For an entire mole of glucose sugar the equation of burning in a calorimeter reads:

$$C_6H_{12}O_6 + 6\,O_2 \rightarrow 6\,CO_2 + 6H_2O, \quad \Delta H = -673{,}000 \text{ cal} \quad (5.3)$$

glucose sugar	oxygen gas	carbon dioxide gas	liquid water	heat evolved

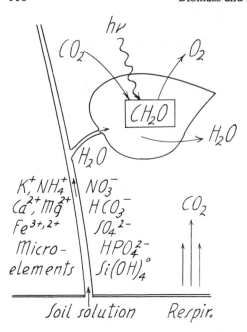

Fig. 5.2. Scheme of transport of soil components to sites of photosynthesis and metabolism. CH_2O, Carbohydrate (1/6 glucose); Respir., soil respiration

This oxidation equation is the exact reverse of photosynthesis [Eq. (5.1)], which is a reduction process, but the pathways are not the same.

In a reversible glucose oxidation 686,000 cal of free energy $(-\Delta G)$ are offered to organisms to carry out respiration, ion accumulation, locomotion, and the making and breaking of covalent bonds among the "big four" C, O, H, N. The amount includes 13,000 cal $(T\Delta S)$ of energy or 43.6 cal/degree of entropy ΔS taken in from the environment that is also available for work.

When glucose, or any other organic compound, oxidizes irreversibly without doing any useful work, as in the sugar burn described above or in a forest fire, the entire free energy is converted to entropy yielding, in the words of Prigogine (e.g., 56), the internal entropy production $\Delta H/T$ of 2258 cal/degree.

Respiratory Electron Flow and Biosynthesis

In organic oxidation (Chapter 3) the act of detaching hydrogen atoms ($H=H^+ + e^-$) from carbon atoms is viewed as a removal of electrons by oxygen atoms which are thereby reduced to anions (O^{2-}). They attract the trailing H^+ ions and form water $2e^- + \frac{1}{2} O_2 + 2H^+ = H_2O$. The electron flow generates 38 ATP molecules that participate in over 70 linked enzyme-catalyzed molecular steps (Krebs cycle), releasing the oxidation energy in sequential small packets. Only a fraction of the free energy change $(-\Delta G)$ is irretrievably lost as wasted, frictional heat; the bulk is immediately coupled to vital processes.

All sorts of carbohydrates, fatty acids, and amino acids have their covalent C–C bonds broken. The intermediate molecules (metabolites) are linked to multitudes of individualized proteins and other macromolecules. The speed with which synthesis takes place is enormous. Reportedly, a single cell of the bacterium *Escherichia coli* will create 12,500 lipid molecules and 1400 highly coded protein particles in 1 sec.

In order to function properly most enzymes require suitable, local ionic environments, and the cations K^+, Rb^+, Mg^{2+}, Ca^{2+}, Mn^{2+}, Cu^{2+}, Zu^{2+}, Ni^{2+}, Co^{2+}, Cr^{3+}, Al^{3+} have been identified as enzyme activators. In soils of low mineral variety, crucial metabolic pathways in microbes and roots may be blocked for lack of proper ions, leading to impoverished microflora and to deficiency symptoms in plants, manifest as curled, discolored, and dying leaves.

C. Vegetation Productivities

Since plants breathe, inhaling O_2 and exhaling CO_2 and oxidizing some of the photosynthates they manufacture, it is the *net* photosynthesis that provides the plant biomass available to other organisms, including man. C_4-plants are more conservative respirers than C_3-plants and they are often higher yielding.

Net Productivity

Net primary productivity (NPP)—to distinguish it from animal productivity, which is secondary—is the harvested yield or biomass of *annual* field crops, provided dead leaves and live and dead roots are weighed too, and insect grazing is accounted for. Dry weights of potato crops may reach 2200 g/m² and those of maize (grain, stalks, leaves) 3200 g/m² (13). Growth and decay of root hairs and root exudates escape detection.

Sampling difficulties increase with perennial plants because growth increments must be separated from older standing crop or stock. In trees the differentials are tallied from enlargements of diameters and annual rings. Root mass in forests is measured in randomly distributed cores and pits collected throughout the season. In the 50-year-old deciduous tulip poplar (*Liriodendron*) forest at Oak Ridge National Laboratory (ORNL) in Tennessee (Fig. 5.3) seasonal maxima and minima of root mass (< 0.5 cm diam.) reveal a production of 450 g C/m^2/year, and a similar quantity of root dieback, as reported by Edwards and Harris (22). For the same forest the ORNL workers (31) compute for vert space an annual net productivity (NPP) of 318 g C/m^2, or for the entire ecosystem 768 g C/m^2/year.

Gross Productivity

To arrive at the elusive gross primary production (GPP) of a forest the vegetational respiration (R_a) must be added to NPP. At Oak Ridge, CO_2 uptake by leaves and their night respiration are monitored in the forest by infrared gas analysis, and pieces of excavated roots are subjected to O_2 sorption in respirometers. Total annual R_a is estimated (31) as 1452 g C/m^2, hence GPP is 2220 g C/m^2/year. The forest had a total standing crop of 8760 g C/m^2. Multiplication of grams of C by 2 approximates grams of dry biomass.

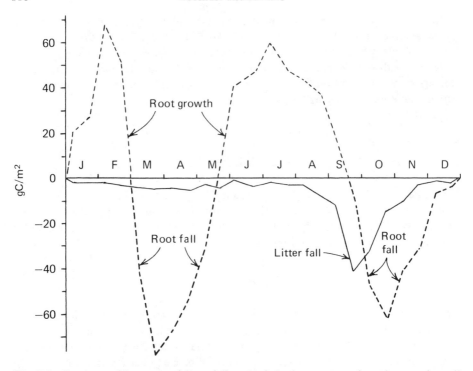

Fig. 5.3. Semi-monthly rates of litterfall and of the large rates of net-losses of small roots (rootfall), all plotted as negative quantities of carbon/m². Net productions of small roots are marked as positive rates. The roots were collected to 60 cm depth in a humid tulip poplar forest (23)

D. Nitrogen Fixation

In organisms nitrogen (N) ranks fourth in terms of atomic proportions, being preceded by O, C, and H. The all-important proteins contain 16-19% N.

Higher plants are unable to tap the plentiful reservoir of inert nitrogen gas (N_2) that makes up 78.0% of the volume of the atmosphere. Roots secure nitrogen from the soil as ammonium (NH_4^+) and nitrate (NO_3^-) ions that are delivered by microbes during oxidation of soil organic matter. While this explains nitrogen availability to plants, it does not answer the question of how nitrogen entered the ecosystem in the first place.

Abiotic Sources of Nitrogen

Nitrogen is a regular constituent of igneous rocks (69) but in too minute quantities to be a significant contributor to nitrogen in soils.

At 60 stations in the United States nitrogen in rains measured during 1 year amounted to 0.25 g/m² in the arid and 0.12 g/m², mostly NO_3, in the humid region (39). NH_4 ions in rain are believed to be soil-derived and their precursor, the ammonia gas (NH_3), tends to be expelled by alkaline soils and retained by acid soils. In industrial societies farmers fertilize their soils with as much as 30-40 g N/m²/year in the form of gaseous ammonia, ammonium sulfate, and Na- and Ca-nitrates.

Free-Living Nitrogen Fixers

The bacterium genus *Clostridium,* discovered by Winogradsky in the 1890s, tolerates wide ranges of pH and absence of oxygen. Its counterpart, the oxygen-requiring bacterium *Azotobacter chroococcum,* studied by Beijerinck in 1901, is common in the temperate zone and fixes nitrogen under laboratory conditions at pH values above 6. While absent in the acid soils of forests and pastures of high mountains, *Azotobacter* thrives in the Alps in manured garden plots under severe climatic stress. Several N_2-fixing *Beijerinckia* bacterial species that tolerate the wide pH range of 4-9 have been reported from tropical soils (20).

The overall reaction of N-fixation in gaseous form is (9):

$$N_2 + 3 H_2 \rightarrow 2 NH_3, \quad \Delta G^0 = -7950 \text{ cal} \tag{5.4}$$

or, in more detail,

$$N_2 + 6 e^- + 6 H^+ + n\text{ATP} \rightarrow 2 NH_3 + n\text{ADP} + n\text{Pi}$$

in which the nitrogen twins accept six electrons from the H atoms. The nitrogenase enzyme system that activates the triple bond of $N \equiv N$ can do the same with the triple C-bond in acetylene ($H - C \equiv C - H$), which has opened a new way for the pertinent enzyme assay in soils.

In *Clostridium* the N_2 reduction is supported by the enzymes nitrogenase, hydrogenase ($2 H^+ + 2 e^- \rightleftharpoons H_2$) and ferrodoxin. The required ATP is furnished by respiration and the organisms must have access to metabolic products such as sugars and fatty acids.

The bacteria congregate in the mucigel-rhizosphere zone near root surfaces, perhaps utilizing their excretions (57). The occasional high N gains observed under lysimeter and field conditions (up to 10 g/m^2/year) pertain to soils with prolific root systems, like those of domestic grass crops.

The photosynthetic purple bacteria found in Asian soils are self-sufficient and fix N_2 in daylight (57). Similarly, the *blue-green algae* (*Cyanophyceae*) are endowed with photosynthesis and they harness N_2 as well, as documented by Japanese and Indian scientists for rice paddy fields. In Arizona deserts, Cameron and Fuller (11) found up to 0.300% N in algal-mineral crusts and less than 0.055% in the soil material immediately below. In association with fungi in the form of lichen the algae are believed to fix N in tree crowns of conifers.

Bormann et al. (6) assign 1.42 g/m^2 of N gains in a northern deciduous forest to microbes associated with decaying wood.

Symbiotic Nitrogen Fixation

Though the Romans knew that the plowing-under of clover benefits the succeeding crop, over 2000 years passed until the mystery was pinpointed to nitrogen. In the 1880s Hellriegel and Wilfarth concluded that pea plants acquire N_2 from the air and that bacteria in their root nodules are responsible for it. Soon afterwards Beijerinck isolated a bacterial species belonging to the genus *Rhizobium.* Since then hundreds of workers have constructed an impressive edifice of knowledge concerning the intricate physiological relations involved. But, as Russell (60) cautions, it is geared to agriculture of temperate regions and very few of the 10,000 legume species that cover the earth as herbs, shrubs, and trees have been studied.

The bacteria that invade red clover through the root hairs do not excrete the critical enzymes pectinase and cellulase needed to break the bonds in macromolecules and fibrils of the cell wall, hence, the chemical mode of entry is unknown. Upon reaching the cytoplasm of root cells the bacterium multiplies and in turn the host undergoes rapid cell division leading to nodulation. The plant supplies carbohydrates to the bacteria—which is a drain on photosynthates—and the microbes reciprocate by donating amino acids to the tops.

The mutual contract is a classic case of *symbiosis*. The infection deviates into parasitism when the boron content of the soil is low (60), or long darkness prevails above ground. Successful symbiosis of a legume with the many species and strains of *Rhizobium* requires highly integrated behavior of both partners and depends on their genetic structures and on countless microenvironmental variables that affect host and invader individually as well as in their state of union. Thus, if nitrates are offered to the root, fixation is reduced or inoculation may be prevented by root exudates from adjacent nonlegumes.

The pathway of N_2 reduction in nodule bacteria is believed to be similar to that in *Clostridium*. Definitely, the presence of molybdenum (Mo) and cobalt (Co) is essential, the latter being a component of vitamin B_{12}. On Australian pastures response to Mo fertilization has been spectacular.

Actively N_2-fixing nodule systems display the red color of the iron-containing hemoglobin catalyst; absence of red color signifies absence of fixation. Under efficient agricultural practice on good soils, gains of 10-20 g N/m^2/year are not uncommon. In rare instances, these quantities may be doubled, even trebled (60). No free-living bacteria have ever been accredited with such high N fixation rates.

During World War I, thermodynamist Haber and engineer Bosch succeeded in imitating Eq. (5.4) by combining the gases N_2 and H_2 at high temperatures (400°C) and pressures (700 atm), in the presence of Fe and Mo catalysts. This extensively used process of nitrogen fertilizer manufacture helps to sustain the large human populations of Europe, the United States, and Japan, at high energy cost.

A few *nonleguminous plants*, mostly of woody type, have nodules containing an Actinomycete as symbiont. Species of the genus alder (*Alnus*) have large nodule structures and fixation rates up to 10 and 20 g/m² have been claimed. Other nodulating plants are the bog myrtle (*Myrica* sp.), the sea buckthorn (*Hippophae*) on sand dunes, the genera *Shepherdia, Coriaria, Dryas*, sagebrush, and others. For all of them Bond (5) documents N-fixation with nitrogen isotope (^{15}N) procedure. All species need the enzyme cofactors Mo and Co. For snowbrush, deerbrush, and other *Ceanothus* species of the North American Pacific Coast, Delwiche et al. (17) obtained ^{15}N fixation in excised nodule clumps. In practical pot and field tests proof relies on good growth of nodulated plant specimen in N-deficient soils.

Genes that synthesize nitrogenase have been transplanted successfully from a N-fixer (*Klebsiella*) to a non-N-fixing bacterium (9), which holds promise for expansion of world N supply.

E. Soil Organisms as Decomposers and Porositors

Green plants and a few soil bacteria are *producers* or *autotrophs* (self-nourishing) because they manufacture organic matter from inorganic materials; at the same time they respire, cleaving their own organic molecules. All other organisms, including man, are *heterotrophs* (hetero = other) who require prefabricated substances. They are decomposers, the most important of which are the bacteria and fungi and the small invertebrates, the animals without a spinal column. Together they macerate, digest, and oxidize fallen leaves and trunks, dead grass, offal and feces, defunct bodies, and some species devour one another. Many are makers of biopores. In the extreme, aerobic decomposers transform the organic to the inorganic realm:

$$\text{organic} \atop \text{matter} \left\{ \quad \xrightarrow[\substack{+O_2}]{\text{decomposers}} \quad \right\} \begin{array}{ll} \text{gases} & (CO_2, NH_3) \\ \text{water} & (H_2O) \\ \text{cations} & (\text{e.g., } K^+, Ca^{2+}, Fe^{2+}) \\ \text{anions} & (\text{e.g., } NO_3^-, SO_4^{2-}, PO_4^{3-}) \end{array}$$

Anaerobic decomposers additionally furnish the gases H_2 (hydrogen), CH_4 (methane), and H_2S (hydrogen sulfide). The chemical energy of the organic compounds is conveyed to the environment as work and entropy.

The Soil Microflora

The microbe populations of the soil space are better known than those of the vert space (57).

Bacteria. The tiny, unicellular, oval and rod-shaped soil bacteria are less than 1 μm wide and 1 or a few μm long. They lack differentiated nuclei. For good agricultural lands Clark and Paul (12) consider two billion (2×10^9) viable individuals in a gram of soil as representative.

Assuming a bacterial volume of 1 μm^3 and a density of 1.04 g/cm^3, the population of two billion weighs 2.08 mg. To 15 cm depth and a soil volume weight (d) of 1.3 g/cm^3 the biomass amounts to 406 g/m^2.

Stöckli's (70) estimated live-weights are over 1000 g/m^2; other workers (12) average 270 g/m^2, all to a depth of 15 cm. Bacteria contain about 10% C, 2% N, and 80% water. Since individuals divide and double in half a day (33), large seasonal fluctuations are common.

Most bacteria are *heterotrophs* and *aerobic,* requiring oxygen for respiration. The *facultative anaerobes* can live either with or without oxygen, whereas the *obligate anaerobes* cannot tolerate it. In a closed, moist soil pore the aerobes consume oxygen, give off CO_2, and eventually suffocate. Then, surviving obligate anaerobes proliferate. The important nitrogen-fixing bacteria, previously described, include both aerobic (*Azotobacter*) and anaerobic species.

The autotropic bacteria assimilate CO_2 gas to carbohydrates in the dark and secure the needed energy either by oxidizing ammonia (NH_3) to nitrate (NO_3^-), which characterizes the nitrifiers, or hydrogen sulfide (H_2S) and sulfur (S) to sulfate (SO_4) which is done by the sulfur-oxidizing bacteria. Like green plants they can synthesize all their carbohydrates, proteins, vitamins, and enzymes from inorganic substances.

Actinomycetes. These living networks of microscopic filaments are aligned with bacteria because their cellular architectures lack chromosomes and a true nucleus. They are aerobic and heterotrophic. Many endure elevated temperatures, partake in the fermentation of manure and compost piles, and occupy some of the root nodulations of nonleguminous plants. Several species are responsible for the earthy smell of newly plowed land.

In New Zealand's tussock grasslands (12) actinomycetes constitute 40-60% of the total microbe population, whereas in Canadian prairies the percentages are much smaller.

Species of the genus *Streptomyces* produce the antituberculosis drug streptomycin, the discovery of which brought the late soil microbiologist S. A. Waksman the Nobel prize in medicine. Like many extracellular enzymes the antibiotic metabolites may be adsorbed on the surface of clay and humus particles where they lose effectiveness.

Fungi. Popularly, fungi are known as mushrooms, molds, mildews, and yeasts. Many are plant parasites and pathogens that rot roots and stems. Fungi possess immense diversity in form and size, ranging from microscopic single cells to huge umbrella-type multistructures. In soils webs or mycelia consist of whitish cellular strands (filaments, hyphae) that may be extensive and millimeters thick. In oak forest litter in the Netherlands the summation of lengths of growth of mycelium caught in vertically inserted nylon gauze averages 2449 m/month/g of oven-dry litter. In the fresh L-layer 80% of it is alive; in the humified H-layer 95% is dead (49).

The quantity of fresh fungal tissue in a pasture soil is about 0.2% of the soil volume (34). For a light-weight soil (d = 1.0) the amount per square meter to a depth of 1 cm is about 20 g, which is of the same order of magnitude as the fresh biomass of bacteria quoted previously.

As decomposers of live and dead tissues *fungus physiology* is highly diversified and geared to a kaleidoscopic microenvironment of ions and macromolecules. Some species live on sugar and starch, some attack cellulose, and others are able to break down lignin, keratin, chitin, and humic substances by excreting hydrolase enzymes that split polymer molecules to smaller, readily diffusing units. All fungi are chlorophyll-free, need oxygen, and can synthesize their own types of proteins from NH_4, NO_3, and low-molecular organic nitrogen sources. Often they compete with plants for nitrates. On a given organic substrate a whole suite of species may take their turns in regular heterotrophic succession. Certain species eat other fungi and catch nematodes and amoebae. In turn, fungi are consumed by mites and other faunal species. Fungal C/N ratios are about 10.

Widely known genera include *Penicillium* of medical fame, *Fusarium,* the plant pathogen, *Gibberella,* the synthesizer of plant growth hormones, and *Neurospora,* the servant of modern molecular genetics. Nearly 700 fungal species have been isolated from soil. The black mold *Aspergillus niger* has high phosphate and potassium requirements and its growth of mycelium has been used routinely as a quantitative index of biologically available soil P and K.

Reproduction of fungi is by means of tiny spores familiar from the brown dust given off by puffballs.

A permanent structural association of fungi with living roots is named *mycorrhiza* (fungusroot). On pines and beeches the fungi live on the outer surface of the roots, form a sheath, cause stubbiness, and are named ectotrophic mycorrhiza. In orchids and in Ericaceae and other families the fungus invades the internal root tissue and is given the adjective endotrophic. Vesicular-arbuscular mycorrhiza are intermediate.

Mycorrhiza used to be described as feebly pathogenic but is now considered a symbiosis that aids roots in mineral uptake in infertile soils. External hyphae reach into the soil, effectively enlarging the absorbing root surface and augmenting the transfer of phosphorus and other ions (61). In turn, the symbiont draws soluble sugars from the host, estimated by Harley (61) as 25 g C/m^2/year.

Prominent are the intimate ties of fungi with algae that give rise to the distinct organism group of lichens consisting of some 17,000 species. Some decorate rocks, bare soil, tree barks, and decaying wood with vivid yellows, greens, and reds. Their enhancement of weathering of exposed rocks has been attributed by Schatz (65) to contact chelation of crystal-lattice Ca and Fe with physodic and lobaric lichen acids.

Soil Invertebrates

The small, spineless invertebrates act as mechanical blenders. They break up plant material, expose organic surface areas to microbes, move fragments and bacteria-rich excrements around and up and down, and function as homogenizers of soil strata.

Most soils contain up to 1000 species of animals that are locked in an intricate web of food chains and activities, often expounded as an equilibrium of soil organisms. As the subsurface creatures live in the dark, body coloring offers no advantage, and since nothing can be seen many species are blind; as compensation, they are endowed with highly developed organs of touch, taste, and smell, and probably hearing. Odor receptors detect exceedingly low concentrations of molecules, including CO_2. A moisture sense is said to be located in antennas.

Animal density declines rapidly with soil depth. Populations vary from soil to soil and climate to climate and quickly adjust to seasonal weather.

A sketchy enumeration of invertebrates is appended (63,75), and some of the species that are prominent soil formers beyond being decomposers are appraised in Chapter 13 on the Biotic Factor.

Protozoa. Protozoa are cosmopolitan, unicellular animals that are microscopic in size (< 100 μm) and live in films of water. Common are the flagellates, amoebae, and ciliates. They multiply by fission once or twice a day. Amoebae feed mainly on bacteria and yeasts by engulfing them bodily (pinocytosis).

Worms. Except perhaps for the protozoa, the worm-shaped creatures are the most widely distributed soil animals. They vary from microscopic dimensions to the Australian giants nearly 3 m in length.

The *Annelida* phylum embraces the tropical land leeches, the potworms, the thin mudworms, and the intensely studied true earthworms (Lumbricids) of which there are hundreds of species. Because of their mechanical activities as earth movers and porositors the latter are evaluated as a special biotic factor in Chapter 13.

Nematodes are usually small (< 2 mm), elongated, transparent cylinders of circular cross sections (round worms), but some forms may reach a length of 1 m. They are plant, microbial, and animal feeders and as destructive root parasites wriggle their way to great depth. Up to 10,000 individuals and 100 species are counted in 1 cm^3 of soil. Encased in cysts they remain at rest for years. Related round worms are the parasitic strongyles, the hookworms, and the lungworms. They infect wild and domestic mammals.

Rotifers or wheel animals are tiny, transparent, and live in moss. Upon drying they shrink and become resistant to great extremes of drought and temperature. They feed on nematodes and protozoa.

Turbellaria (flatworms) reach 1-2 mm in length (a few much longer). They occupy water-filled cavities and are carnivores.

Mollusks. Mollusks have soft, unsegmented bodies (locomotory "foot") that are covered in some species with external, frail shells. Prominent in the soil are the slugs and snails. They are herbivores and seek moist environments.

Arthropods. This phylum consists of articulated animals having jointed limbs.

Insects possess six legs and three clearly defined body regions. Many live in the soil continually, while others such as butterflies live in the soil periodically as eggs, larvae, nymphs, caterpillars, maggots, grubs, and pupae. Common insects are the beetles (scarabs, glow worms, weevils), ants, bees, wasps, flies (midges, gnats), bugs (cicadas, scales, aphids), thrips, lice, earwigs, grasshoppers, mole-crickets, termites, cockroaches, silverfish, and springtails.

Pollinating insects affect soil indirectly by modifying the proportions of species in the plant cover.

Termites, with cellulose-digesting protozoa in their hindguts, and the distantly related *ants* are ferocious decomposers and some of them acquire fame as social builders of mounds and hills (described as pedogenic factor in Chapter 13).

Springtails (*Collembolas*) are the most widely distributed soil insects. Their sizes and shapes are said to be adjusted to the pore sizes of soil horizons, and their droppings are humus precursors.

Myriapods include the vegetarian millipedes, the carnivorous centipedes, and the delicate white symphylas. They have numerous, similar body segments with legs, but not "ten thousand," not even in the foot-long tropical scolopenders. Some species are short (1 mm) and many roll into a ball. As litter eaters they prefer Ca-rich leaves. Myriapods are animals of the forest floor.

Crustaceans prefer wet spots. They include crabs, crawfish, slaters or sowbugs, sand hoppers, and curling-up woodlice (pill bugs). The isopod subgroup feeds mainly on dead vegetation.

Arachnids resemble insects superficially, but usually have eight legs. They are litter inhabitants and often predators. The class includes the very abundant mites, the parasitic ticks and gamasids, and also the subclass of spiders with the tarantulas and mygales, the "daddy longlegs," the scorpions and the microscopic moss-dwelling tardigrades.

Fauna and Soil Porosity

As a pond fills up with washed-in fine sediment the platy silt and clay particles settle in horizontal planes, thereby achieving the lowest potential energy of configurations in space. The resulting mud flat is an incipient soil, but it would not support much vegetation because the tightly packed gel retards oxygen diffusion, water circulation, and microbe and humus penetration.

It takes the lumbricoids and the four-, six-, eight-, and hundred-legged animal workers and their lifting, digging, and tunneling activities to improve the soil fabric by rearranging the platelets into open structures of low density (1.0 g/cm^3) that have channels, crevasses, and large pores padded with dragged-in humus molecules. The little, busy porositors do mechanical work that imparts potential energy to soil particles, enhances water percolation, and renders the substrate a more livable abode for microbes and fine roots. Soil animals work continually to keep their structural edifice from collapsing to its stabler state. While the energy devoted to faunal mechanical work is but a fraction of the total respirational calorie flux, its role as a feedback loop in maximizing photosynthesis at a site bestows on invertebrates status above mere decomposers.

Soil Biomass and Energy Flow

From an examination of estimates of biomass in the literature, representative means of fauna and microflora in a good soil are arranged in Table 5.1. The total stock of 1334 g/m^2 corresponds to 11,900 lb/acre, which equals the weight of 12 horses per acre. If the figures are at all realistic the soil organisms must be dormant or starving.

To cite recent data, in the *Liriodendron* forest at ORNL previously mentioned, heterotroph populations in forest floor and mineral soil are reported (31) in g C/m^2 as follows: invertebrates 6.92, fungi 37.7, and bacteria 20.3, for a total of 64.92 g C/m^2. Multiplication by 10 provides an approximation to live weights, as in Table 5.1.

Respiration. As a criterion of soil fertility, soil respiration has been determined in countless laboratory experiments which, however, are not suited to the present inquiry.

Table 5.1. Stationary Soil Biomass (Exclusive of Roots)[a]

	g/m^2		g/m^2
Bacteria	400		
Actinomycetes	100		
Fungi	460	Earthworms	200
Algae	14	Mollusks	50
Protozoa	30	Other invertebrates	80
Total microbes	1004	Total invertebrates	330
		Total biomass	1334

[a] Fresh weights, containing about 20% dry matter.

Special microrespirometers enable measurement of the O_2 consumption in basal metabolism of tiny animals in the laboratory. The findings are multiplied by the number of animals and their biomass estimated in the field. In this manner, O'Connor (52) computes the metabolism of a population of enchytraeid worms as 12.4 liters O_2/m^2/year. If $CO_2/O2$ is 1 as in Eq. (5.3), the gas volume also specifies CO_2 evolution. Because 1 mole of gas occupies 22.4 liters, each liter of O_2 consumed connotes 5015 cal of glucose oxidized in the body of the animal. For worm diet, O'Connor chooses 4775 cal/liter of O_2 and thereby brings their population metabolism to 59,210 cal/m^2/year.

For all invertebrates together, MacFadyen's (44) "grassland" has a metabolism of 1,138,300 cal/m^2/year. Dividing by 4775 and converting liters to grams the faunal CO_2 evolution is 468 g/m^2/year, or, multiplied by 0.2727 is 128 g C/m^2/year.

Hawkins' (32) elaborate field techniques with exceptionally good aeration produce for potato crops 13-26 g CO_2/m^2/day for soil and root respiration combined, and 9-21 g CO_2/m^2/day for soil alone, the roots contributing about one-third of the total. In the ORNL forest *annual* CO_2 evolution from forest floor and mineral soil *in situ* is 1040 g C/m^2. Root respiration *in vitro* is 392 g C/m^2 (31) and the difference, 648 g C/m^2, is assigned to heterotroph respiration of invertebrates and microflora.

Figure 5.4 depicts a vertical CO_2 profile measured by Lemon (43) at midmorning in a field of clover about 50 cm tall. CO_2 is diffusing into the vert space upward from soil and downward from the external atmosphere above the canopy. From a series of such profiles taken during a 24-hr period the computed production values in calories per square centimeter are as follows: gross photosynthesis 19, day respiration 7, night respiration 3.4, and net photosynthesis 8.6. With a conversion factor of 4000 cal/g of dry matter it corresponds to a diurnal net production of 21.5 g/m^2 of dry matter.

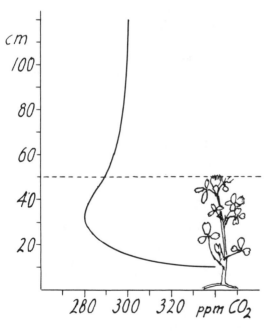

Fig. 5.4. Profile of CO_2 concentrations (volume ppm) at 10:40 AM in a field of clover 50 cm tall, after Lemon (43). High ppm near ground level relates to soil respiration; the bulge to the left to photosynthesis. Vertical axis represents height above ground level (Courtesy Academic Press, New York)

At Rothamsted, Monteith et al. (47) attribute 20% of the carbon content of field crops to soil respiration. Eiland (23) correlates CO_2 emission of soils with their ATP content.

Schlesinger's (66) survey of soil respiration (Rsp) in forests offers the latitude (Lat) dependency of CO_2 evolution: Rsp = 1721.5 – 24.2 Lat, in g C/m^2 per year. The correlation coefficient r is 0.78. At latitude 35°, which passes near Los Angeles, California, and Memphis, Tennessee, the average forest soil respiration is 875 g C/m^2/year. One-third is derived from litterfall and two-thirds from root respiration and humus oxidation. Emission of CO_2 is highest at the equator and again one-third is assigned to litterfall.

Energy Flow. Procedurally, plant parts, organisms, and their feces are oxidized in a bomb calorimeter which gives their calorific content or energy change ΔH. Treating each organism as a subsystem, a chain of energy flow—actually, organic matter · ΔH-flow—among heterotrophs is conceived as follows:

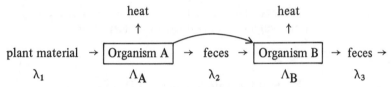

In this sequence, organism A is a herbivore, B a caprophage or a carnivore. The Greek lambdas, λ_1, λ_2, λ_3, denote energy consumed or excreted, expressed as cal/m^2/unit time (day or year), and Λ_A, Λ_B, are the calories in the standing stock (stationary biomass) of organisms, as cal/m^2. The differences $\lambda_2 - \lambda_1$ and $\lambda_3 - \lambda_2$ reflect growth of organisms and heat of respiration. Mechanical work done by organisms is considered negligible, or is classified as heat emitted. Often, steady state condition is assumed which facilitates the drawing of balance sheets of d-transfers because outflux is then equal to influx.

Instructive is MacFayden's approximation of energy flow in a grazed, English meadow ecosystem, presumably at steady state as far as organic matter is concerned. The calories in standing crop and stock, as cal/m^2 are 2 million for grass, 155,000 for cattle, and 11,900 for man. Gross photosynthesis is 16,700 cal/m^2/day. It is partitioned into grass respiration (2400), consumption by herbivores (4300), and production of grass litter (10,000). The animals respire (1540) and produce manure (2680), and beefeating man does the same. According to Figure 5.5, the sum of above-ground respiration (upper row) is 4014 cal/m^2/day, and the sum of calories transported from vert space to soil space (lower row) is 12,686 cal/m^2/day, which is 76% of the gross photosynthesis. It is soil influx I_s.

In the soil 91% of the standing stock of 4,819,400 cal/m^2 is attributable to microflora, the rest to invertebrates. The former respire 83% of the influx calories I_s, after part of them have passed through the guts of the invertebrates. The latter oxidize 12% of I_s, which leaves a slight imbalance of the energy budget. Multiplication of respired calories by 1.122×10^{-4} converts them to approximate g C/m^2.

In recent years interest in energy transfer is being supplemented with concerns of molecular flow. Modern society creates an abundance of practically nonbiodegradable molecules, like the chlorinated hydrocarbons (e.g., DDT), that become concentrated in segments of the food chain and may injure and eliminate soil organisms.

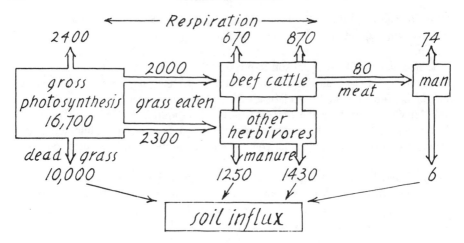

Fig. 5.5. Compartments of MacFadyen's (44) energy flow through a meadow eco-system. Values are in calories/square meter/day

F. Soil Organic Matter and Humus

Inside soil space any organic carbon assembly, large or small, dead or alive, is classified as *soil organic matter*. Some think that only dead bodies should be so considered, which stresses the abiotic concept of soil, but we do not yet know how to separate live and dead small organisms and rootlets *in situ* without destroying the soil fabric. Anyhow, a holistic stance is favored. A number of ecologists include humus in biomass.

Organic Layers and Their Importance

Dark colored organic tissues that are too small or too disorganized to be recognized with the naked eye as such are named *humus*. The word further denotes the dark-brown and black substances dissolved in soil-pore solutions, waterways, and ponds.

In natural grassland the soil is permeated by fine, fibrous roots and the amount of humus in horizons is an image of root distribution (Figure 7.15). Root weights of perennial native grasses often match or exceed the tops (Table 12.2). The A and B horizons of forest soils receive humus substances from rotting surficial litter layers. While the massive supporter roots keep growing year after year dieback of rootlets supplies a seasonal source of organic matter. Trees seldom invest much over a third of their permanent biomass in soil space (62).

When disjunct from the underlying mineral soil the forest floor is an acid, peat-like *mor humus*; when incorporated by invertebrate activities the layer is a crumbly *mull*.

Leaf molds cushion the destructive splash-impact of raindrops and reduce the speed of runoff waters. Organic layers serve as nutritious seedbeds for revegetation and afford seeds a chance to escape detection by foraging animals. Unequalled is the role of soil organic matter as the reservoir of nitrogen for plant growth. In quantity per unit area, to-tal nitrogen in soil exceeds by far the supply in the standing vegetation (e.g., Table 5.2).

Table 5.2. Narrowing of C/N Ratios (Percentage divided by Percentage) from Vert Space to Soil Space for a Pine Ecotessera (Fig. 1.5) Having a Circular Cross-section of 116.7 m^2 [a]

Components	% N	kg N	C/N
Needles	0.994	0.997	52.2
Branches	0.118	0.646	446.6
Trunk	0.078	1.802	668.9
Roots	0.176	1.613	284.7
Litterfall	0.651	0.160	80.1
Forest floor	0.666	3.551	70.9
Fine earth (fi), $<$ 2 mm, in mineral soil (0.2-0.5 g fi/cm^3 of soil space)			
0-30.5 cm	0.216	15.29	28.4
30.5-61.0 cm	0.065	8.44	24.0
61.0-91.4 cm	0.045	8.26	17.8
91.4-122.0 cm	0.034	6.24	17.4
122.0-152.4 cm	0.028	5.14	17.9

[a] Hobergs, California, Salminas stony loam. (see p. 138)

Before rain waters enter the mineral soil they pass through the organic surface layer where they acquire cations and anions and a miscellany of organic molecules and colloidal particles that steer processes in the soil below.

In a balanced, inorganic nutrient solution plants grow well *without humus.* Yield-stimulations induced by humus amendments to soil are customarily explained as indirect, improving soil structure and assisting the delivery of microelements. Flaig and co-workers (24) grow plants under sterile conditions in the presence of ^{14}C-labeled degradation products of humic acids (e.g., polyphenols) and attribute the stimulation of plant growth to a weak uncoupling of "oxidative phosphorylation" within the plant. Humus components are seen as bioregulators.

Designations of Organic Horizons

In Blume's (4) description of the upper portion of a soil developed in an open birch stand on a slate moraine in arctic Lapland, distance counting is upward from the surface of the mineral soil.

7 to 6 cm   Litter and fresh moss remains; slightly brownish leaves and broken stems of dwarf shrubs (*Vaccinium, Empetrum*) but hardly any birch leaves.

6 to 5 cm   Litter-humus; small (1-2 mm) leaf and stem remains enveloped by fungal hyphae; living and dark dead roots; brown, round organic aggregates (30-60 μm), probably insect and snail droppings.

5 to 3 cm   Litter-humus, coprogene (fecal) aggregates (50-200 μm) and blackened, minute litter remains ($<$ 200 μm); many fungal hyphae; a few mineral particles ($<$ 10 μm).

3 to 0 cm   Litter and pitch humus; brownish-black aggregates with fine mineral particles and tiny, blackish stem and root remains; open microstructure. All the above horizons have many live and dead roots, and they are very acid.

       0 cm   Boundary to mineral soil.
  0 to 4 cm   First mineral horizon, yellow brown, a few coarse mineral particles, rounded, brownish-black humus aggregates (250μm) containing mineral particles ($< 40 \mu$m), small, brown root remains embedded in mycelia, agglomerate structure; high acidity.

The lower A and B horizons are not reproduced here.

Naturalists lump fresh litter and humus layers together and speak of mulch, tangle, litter, and ground litter. Ecologists call everything detritus or debris. Foresters distinguish silviculturally significant strata as mor humus and mull humus, already mentioned, and as moder, felt, dry peat, duff, and raw humus, the last expression being applied by Blume to the humus type of the aforementioned Lapland profile. Under magnification numerous humus microstructures or "fabrics" are discernible.

The widely used Swedish triad L, F, H assigns L to fresh, unaltered litter, F (fermentation layer) to the decomposing plant parts below L, and H (humus) to the dark, heavily humified stratum resting on the mineral soil; F + H together are identified with mor. Soil Taxonomy of the United States separates organic from mineral horizon by the proportion of clay and offers two organic subdivisions: 01, essentially unaltered vegetative matter, mixed with soil fauna and their egesta and fungal hyphae; 02, the original tissues can no longer be recognized with the naked eye.

When the boundary to mineral soil is fairly distinct, 01 and 02 together are the forest floor. The 01 includes L and some of F; 02 relates to the remainder of F and to H. Other correspondences refer 01 to Aoo and 02 to Ao.

Determination of Soil Organic Matter and Humus in Bulk

In fine-textured mineral soil horizons the content of organic matter is lower than the loss of ignition (volatile matter) because clay particles give off crystal lattice water which masks the humus contribution. For accurate routine work organic carbon (C) is determined by the dry combustion procedure in which C is oxidized to CO_2 and weighed as such.

Conventionally, ash-free organic matter (OM) in the < 2mm soil fraction contains 58% carbon and has an OM/C conversion factor of 1.724, established by van Bemmelan over a century ago. For Belgian beech forests conversion factors are 2.00 for Al and lower mineral horizons, 1.73 for Ao-horizon, and 1.70 for fresh litter (16). Aleksandrova (1) reports ratios of 1.7-1.85 for humic acids and 2.0-2.2 for fulvic acids.

The old Grandeau humus probe acidifies the soil to eliminate carbonates, leaches it with ammonium hydroxide, and then dries, weighs, and ignites the black extract. In tropical Colombia (37) humus determined in this manner correlates well ($r = 0.95$) with total soil nitrogen of the Kjeldahl method.

Ranges

Over the Earth as a whole organic matter contents range from practically nothing in very young soils and in deserts to 80% and more in mucks. Absolute inventories are arranged in Tables 5.2, 5.3, 9.1, 13.1, and 13.2. Kononova (46) lists a rich Russian grassland soil (Chernozem) as containing 3580 g N/m^2 to a depth of 100 cm. In the writer's collection high values in *well-drained profiles* originate in the equatorial regions of Colombia, S.A. In fine earth ($<$ 2 mm) of the mineral soil the Chinchiná profile contains to a depth of 127 cm 3486 g N/m^2 and 44.40 kg C/m^2, corresponding to 77 kg/m^2 of organic matter (Table 5.3). A páramo soil at high elevation has 2.66% N and 41.62% C in the loamy Al horizon, and to a depth of 158 cm harbors 143 kg C/m^2 (37) all figures exclusive of forest floor and coarse roots.

Nature of Humus Substances

Chemists subject soil horizons, peat layers, and rotting plant materials to diverse identification procedures that began with Sprengel and Berzelius early last century (74).

Extractions. In dilute (2%) NaOH extraction the undissolved organic residue is called *humin* and is considered recalcitrant to biodegradation. When the soluble portion, the black humus sol, is acidified with HCl, a light-brown humus fraction stays in solution and is named *fulvic acid;* the precipitated part is black and is *humic acid.* These definitions are strictly operational.

The separation scheme is summarized as follows:

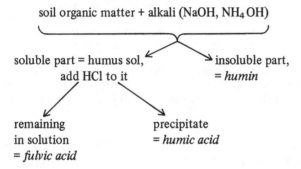

soil organic matter + alkali (NaOH, NH_4 OH)

soluble part = humus sol,   insoluble part,
   add HCl to it   = *humin*

remaining   precipitate
in solution   = *humic acid*
= *fulvic acid*

Fulvic acids are higher in oxygen and lower in nitrogen and carbon than humic acids, but each fraction is composed of diverse substances, some of them still unidentified. For classifying organic soils (Histosols) in the new Soil Taxonomy the extraction agent is sodium pyrophosphate (which mobilizes Al- and Fe-humus).

Table 5.3. Decomposition Coefficients k' of Assumed Steady-State Litterfalls and Forest Floors, Ash-Free (Volatile Matter), and Humus Inventory of Mineral Soils[a,b]

Locality	maT	Annual litterfall		Forest floor		k'	Humus (C·1.724) in mineral soil	
	(°C)	g/m^2	%C	g/m^2	%C	(%)	cm	g/m^2
Calima, Colombia	25.6	730	53.5	432	52.0	62.8	0-76	62,850
Chinchiná, Colombia	22.0	935	54.5	1,455	54.6	39.1	0-127	76,547
Shaver Lake, CA								
Oak (*Q. Kelloggii*)	10	149	50.1	2,517	46.6	5.6	0-127	17,899
Pine (*P. ponderosa*)	10	305	53.7	12,635	51.2	2.4	0-127	18,712

[a] From Jenny (36). (see p. 143)
[b] C/N in columns from top to bottom; for litterfall 37.6, 32.5, 58.7, 106.4; for forest floor: 25.5, 22.4, 39.4, 55.3; for mineral soil: 14.6, 12.7, 17.3, 20.4.

It is unknown to what extent the various extracts may be identified with the actual humus inside the soil. Yet, many of the organic particles in a light-yellow-colored, natural leachate of a Canadian Podzol soil had the characteristics of fulvic acid (67). With gentle means Burns (8) isolated a humus fraction that contained the active enzyme urease.

Along a north-south transect in Russia the amount of humic acid rises, reaches a maximum in the central Chernozem belt, and then declines, always paralleling total carbon and nitrogen, as reported by Kononova (41).

Colloidality of Humus Substances. Waters emerging from humus-rich, strongly acid (pH < 4) soils of the Podzol or Spodosol type are coffee brown and swarm with colloidal particles of H-humus. The color darkens as pH rises (27). In arid regions water extracts of alkali soils (pH 9-10) are colored blackish from dissolved Na-humus. Between these pH extremes the soil solutions are nearly colorless because humus is precipitated as Ca-Mg-Fe-Al varieties.

Colloidality of humus is easily recognized by the Tyndall beam. Viewed in an ultramicroscope, the solution appears as a dark sky studded with twinkling little stars which are the light-diffracting humus particles in perpetual Brownian motion.

Students of clays characterize colloidality by particle size, saying "clay of 0.2-2 μm diameter." Humus chemists prefer "molecular weights" which are the gram weights of Avogadro's number or of 6.06×10^{23} particles of a given size and composition of humus. The weights may be as low as 700 for fulvic acid and may exceed a million (68). Assuming a density of 1, a spherical humus particle of molecular weight 100,000 would have a diameter of 68 Å and would be much smaller than many clays.

The particles are globular in shape and unlike clays fail to yield detailed X-ray diffraction patterns. In their natural state they are negatively charged but may become neutral at very low pH. Humus substances strongly sorb water, act as colloidal weak acids, participate in cation exchange, fix polyvalent metal ions, and interact with clays.

Functional Groups and Models. Today's researchers emphasize isolating specific molecules and groups, an approach pioneered by Schreiner and Shorey (74) early in this century. On the formidable lists of identified compounds one recognizes hydrocarbons, phenols, aliphatic and aromatic acids and alcohols, quinones, vanillin, heterocyclic pyridin and purine, fats, waxes and phosphatides, pentoses, uronic acids and polysaccharides, amino acids, proteins, vitamins, enzymes, and antibiotics.

On surfaces and in the pores of humus structures the hydrogen atoms attached to C as $-CH_3$ or $-CH_2$ groups do not readily dissociate but those linked to oxygen bridges swiftly react with bases. Prominent are the carboxyls ($-COOH$), previously depicted in Figure 3.14, and the hydroxyls ($-COH$) as aliphatic alcohols and reactive phenolic OH-groups (Fig. 5.8), followed by the carbonyl ($=C = 0$) and methoxyl ($-OCH_3$) groups. Per gram of weight, fulvic acids exceed humic acids in numbers of carboxyls.

The acid carboxyl group is neutralized as shown in Chapter 3. The alcohol radical titrates with NaOH at high pH to become $R-CO^-Na^+$ where R denotes the remainder of the molecule. Titration endpoints are often reached slowly because the acquired Na^+ induces swelling which distorts the humus framework and exposes previously inaccessible sites to surface reactivity. The neutralizations of $-COOH$ and $-OH$ correspond to pH-dependent exchange capacities which achieve magnitudes of 300-600 me/100 g and which are highest for fulvic acids.

The small monomers (e.g., phenoles) are joined together—are polymerized—by fungal enzymes, e.g., oxidases, explored in detail by Flaig and his co-workers (25).

Models of humus are still tentative. Schnitzer (68) proposed the structure shown in Figure 5.6 for fulvic acid. It is a lace of phenolic carboxylic acids, joined by H-bonds, and permeated by voids and holes that permit access to interior functional groups.

Metal Complexes and Chelates. As said, humus participates in cation exchange, the sites of actions being the functional $-COOH$ and $-OH$ groups. For K^+, Na^+, Ca^{2+}, Mg^{2+}, and often Al^{3+} the replacements are reversible. Some humus components bind polyvalent ions of Fe, Al, Mn, Zn, and others with great firmness, rendering them nonexchangeable, which reduces CEC. The combinations are *metal-organic complexes,* termed *chelates* in case the cation becomes part of an organic ring structure. As dispersed or dissolved humus complexes the polyvalent elements stay in solution over much of the soil pH range instead of precipitating as near-insoluble oxides and hydroxides.

Consider the titration or neutralization curve of *citric acid*, $(CH_2)_2 COH(COOH)_3$ (H_4Cit), which is a plant and microbial product. The addition of the base NaOH diminishes the acidity, pH rises, and at pH 7 the three carboxyl groups are neutralized to sodium citrate, Na_3HCit.

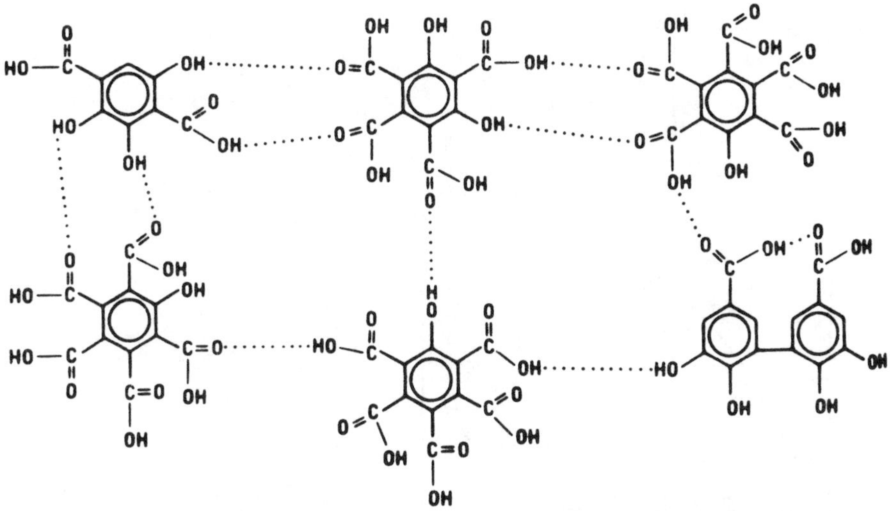

Fig. 5.6. Schnitzer's (68) structure of fulvic acid, an open network of phenole rings and accessible $-COOH$ groups, and connecting H-bonds (dotted lines) (Marcel Dekker, Inc., New York)

When iron nitrate, $Fe(NO_3)_3$, having a pH of about 3, is introduced in equimolar amounts pH is lowered immediately because Fe^{3+} liberates protons (H^+) by ion exchange and near pH 2 joins the molecule as tightly bound HFeCit, which is a soluble chelate, observable optically. Three H^+ have been exchanged:

$$
\begin{array}{ccc}
\text{CH}_2-\text{COOH} & & \text{CH}_2-\text{COOH} \\
| \quad\ \text{COOH} & & | \quad\ \text{COO} \\
\text{C} & & \text{C} \\
| \quad\ \text{OH} + Fe(NO_3)_3 & \rightleftharpoons & | \quad\ \text{O--Fe} + 3HNO_3 \\
\text{CH}_2-\text{COOH} & & \text{CH}_2-\text{COO} \\
\text{citric acid} & & \text{chelated iron}
\end{array}
$$

Adding more NaOH ionizes the fourth proton at about pH 3 and iron citrate becomes the chelate anion $FeCit^-$ which is balanced by Na^+. It remains soluble at higher pH. In absence of citrate most of Fe^{3+} would be precipitated (Fig. 4.9).

The cursory description neglects the tendency of Fe^{3+} to surround itself octahedrally with H_2O and OH^- ions and polymerize to colloidal $Fe(OH)_3$. Thus, at pH 2.4 a reddish color appears, ascribed to a ferric citrate hydroxyl complex that consumes extra base. The tendency toward $Fe(OH)_3$ is held in check by adding more citric acid, a 5-fold excess at pH 6.5. Iron may then join two or more citrate molecules. Depending on the proportions of Fe to citrate, colloidal iron hydroxide either stays in suspension stabilized by adsorbed citrate anions, or settles out, dragging the citrate along, as observed by Muir et al. (48). Further, Na-citrate in moderate amounts flocculates a colloidal suspension of $Fe(OH)_3$ and disperses the precipitate again at high concentrations.

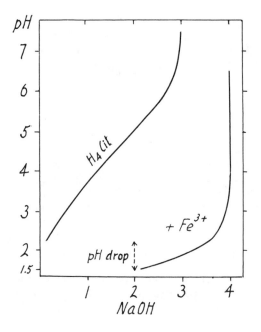

Fig. 5.7. Titration curve of citric acid (H_4Cit) alone and in the presence of equimolar iron nitrate ($Fe(NO_3)_3$), after Warner and Weber (76)

The actions of citric acid serve as models of humus performance. Titration curves similar to Figure 5.7 have been obtained for fulvic and humic acids, proving their complexing power, though the degree of chelation is still controversial.

The pH drop (ΔpH) at salt concentrations of $S = 1$ has been used as a measure of affinity of cations and humus particles. Van Dijk (72) finds

$$K^+ < Ca^{2+} < Mn^{2+} < Fe^{2+} < Zn^{2+} < Pb^{2+} < Fe^{3+}$$

ΔpH increases from left to right, but is confounded with metal-hydroxyl complexions.

Canopy drippings in forests and water extracts of leaves and needles of pines, larch, aspen, ash, and oaks are organic colloidal suspensions with intense complexing ability (3).

Beyond the pedogenic elements humus complexes ions of pollution (e.g., lead, mercury, arsenic) and, by different mechanisms, acts as solubilizing carriers for herbicides and pesticides.

Biogenesis of Humus Substances

In summer-dry climates mature grass culms 60-120 cm tall often deteriorate in standing position over a period of 1-2 years, meanwhile being engulfed by new shoots. When forced to ground, folds and mulches of grass persist. In midwestern grasslands mats of 6-15 cm thickness and average weight of 400 g/m^2 have been observed.

Prior to autumnal shedding, the maturing leaves of trees translocate ions to younger tissue and to leaf wash of rain and they suffer infections by fungi which sorb the oozing sugars, amino, malic, and citric acids. In a day, fresh litters of ash, alder, and birch lose 12-20% of their dry weight in water, but only 1-2% in the case of pine and spruce needles (51).

On the ground, new armies of fungi invade the fallen leaves and their interiors are hydrolyzed and consumed without altering the exterior shape. Within days, the microbial attacks subdue and faunal species such as mites and *Enchytraeidae* ("white worms") take over. The hordes of tiny wreckers leave millions of heaps and trails of black dung pellets that invite new invasions of fungi and bacteria to break the covalent bonds neatly put together in the vert space. Babel's (2) colored microphotographs of thin sections trace the alterations in great detail.

In Broadbent's laboratory study (7) of rotting of cereal straw (Fig. 5.8) the bulk of loss of organic matter as CO_2 takes place during the first 6 months. Since proteins, fats, sugars, hemicelluloses, and cellulose are rapidly consumed by microbes, lignin and —COOH accumulate in the residue and CEC rises in proportion. In advanced states of decomposition the straw remainder shares properties with humus, and this general trend leads many investigators to view lignin as the ultimate source of humus.

Nature impregnates woody tissues with lignin to the extent of 20-30% by weight, even to 50% in some tropical species. Lignin is linked by chemical bonds to hemicellulose and cellulose in the cell wall (25). The large, highly branched molecule is built up from alcohols of the coniferyl (a), sinapyl (b), and coumaryl (c) varieties, the first one having the monomer configuration shown in Figure 5.9. Preferentially, coniferous trees fabricate molecule (a), broadleaf trees (a) and (b), and grasses all three. Some 20 monomers from which H atoms have been stripped (dehydrogenated) are polymerized to lignin in the network outlined in Figure 5.9.

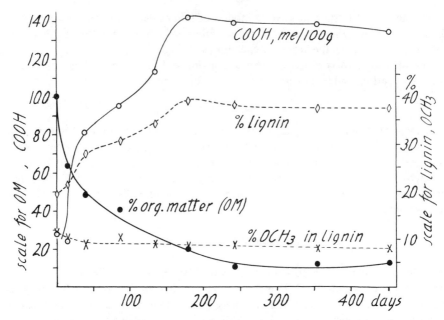

Fig. 5.8. Decomposition of cereal straw in the laboratory, from Broadbent's experiment (7)

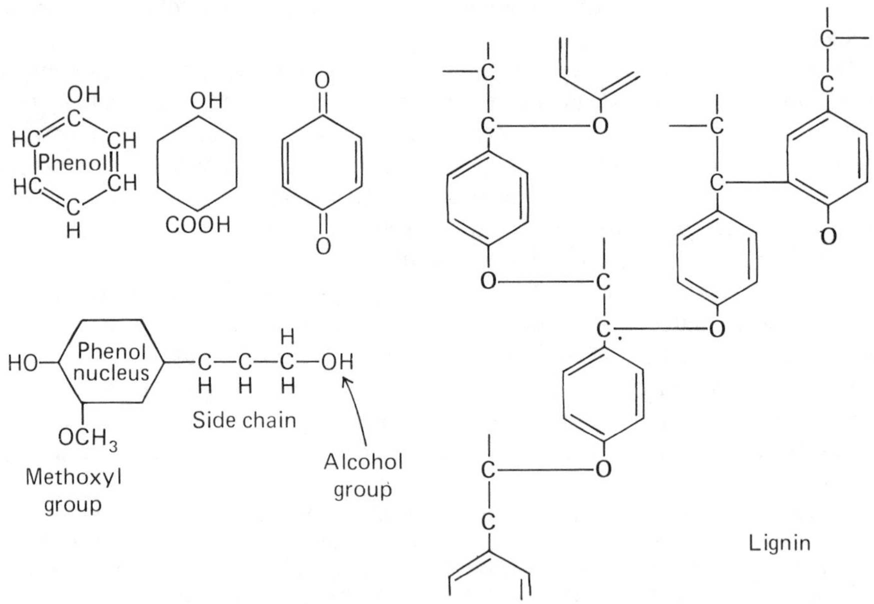

Fig. 5.9. Molecular components of humus chemistry: phenol, phenol with carboxyl group (most of the C and H not lettered), quinone, coniferyl alcohol and its polymerization grid in lignin, as visualized by Freudenberg (25)

During humification the side chain is shortened and the methoxyl group eliminated, making the compound more hydrophilic ("water loving"). The benzene ring is cleaved by oxidase enzymes which shift the Cs to carboxyls. Simultaneously nitrogen of amino acids and proteins is incorporated in numerous degradation products. These finally condense to dark-colored humus substances, but by what path is not yet known.

Haider, Martin and associates (29) are scrutinizing humus precursors metabolized inside microbes. The fungus *Epicoccum nigrum* transforms glucose and other carbohydrates to aromatic phenols which polymerize with nitrogenous and other organic radicals inside and outside of the cell to dark-colored, humic-type pigments. Rates are accelerated by clay particles which stimulate CO_2 evolution, biomass production, and polymer genesis.

Age of Humus Substances

In the atmosphere, cosmic radiation converts nitrogen gas to carbon-14 (^{14}C) which has a half-life of 5568 years. Like ordinary CO_2, $^{14}CO_2$ participates in photosynthesis and eventually is incorporated into humus fractions. When these are buried ^{14}C keeps decaying and its analysis dates the burial and provides an estimate of the age of humus. Owing to "dilution" with younger roots and later humus infiltrations the ages obtained are minimal and are designated as "apparent mean residence times" (AMRT). In surface soils values of centuries and a few millennia are common. In lower horizons infiltration of new humus is scant and AMRT of 10,000-20,000 years have been encountered (64).

G. Nitrogen in Humus and Its Availability

Dhar (19) estimates that for world food and fodder production some 350 million tons of available nitrogen are needed annually. To this amount precipitation contributes 7-10 million tons, N-fixation by legumes and N in farm yard manure 5 million tons each, and world fertilizer manufacturing 20-40 million tons. The difference, nearly 300 million tons, must come from mineralization of N in humus which exists there in quantities of 0.5-5%. It has been claimed that in nonindustrialized countries, like India and China of yesteryear, the population size is eventually limited by the plant protein supply derived from annual mineralized soil nitrogen.

Molecular Linkages of Nitrogen

Because mineralization of organic nitrogen to ammonia proceeds more slowly with humus than with plant and animal proteins (gluten, milk, flesh), Waksman (74) advanced the idea that soil nitrogen exists as a ligno-protein, or as a ligno-peptide in the words of more recent workers. Mayaudon (46) isolates biologically stable humic proteins. Flaig et al. (25) and others suggest models of nitrogenous monomers (Fig. 5.10) either with an amino acid attached to phenol (left-side formula), or with N as an integral part of a pyridine ring (middle formula), or with N in a larger, heterocyclic unit (right-side formula).

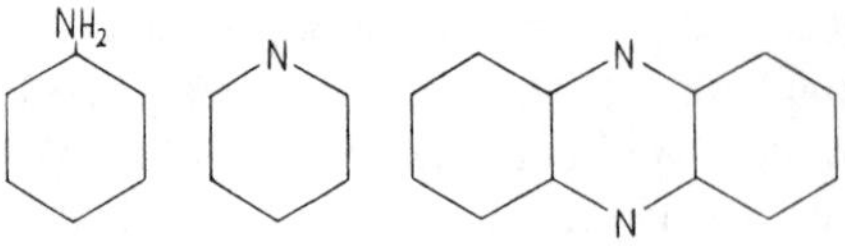

Fig. 5.10. Positions of nitrogen atoms in some of the molecular components of humus, after Flaig et al. (25). Left to right: aromatic amine, pyridine ring, heterocyclic segment

Carbon-Nitrogen Ratio

Oak leaves decaying in bottles for 448 days lose carbon but no nitrogen, whereas grass and oat hay export some N in gaseous form (42). In all decompositions the carbon-nitrogen ratio, % C/% N, narrows. In any forest the ratio attenuates conspicuously from vert space to soil space (Table 5.2, see p. 129). Conversion to agricultural land further reduces C/N and values of 10-12 predominate, which is advantageous for nitrogen mineralization. In C horizons C/N may diminish to 5-6, which is partly caused by trapping of NH_4 ions in clay crystals.

Mineralization of Organic Nitrogen

Amino acids are at the high end of the spectrum of biodegradability and millennia-old humic acids at the low one. Action is slow when enzymes become inactivated by humic polyphenols or by adsorption on clay particles, or when nitrogen atoms are physically unreachable. Mineralization requires a diverse population of organisms which is assured by the global dispersion of microbes by winds.

Ammonification. In the breakdown of proteins, amino acids, and nucleic acids by microbes the volatile ammonia gas (NH_3) is a waste product.

The molecule reacts with dry, acid clay particles to form exchangeable NH_4^+ cations by proton (H^+) acceptance. The discovery initiated the nowadays enormous consumption of anhydrous ammonia gas in fertilizer practice on large farms.

In soil water, ammonia becomes ammonium hydroxide and neutralizes carboxyl groups and acid clay

$$\boxed{\text{Humus}}\ \text{CO}\bar{\text{O}}\ \text{H}^+ +\ \ NH_4^+\ OH^-\ \rightleftharpoons\ \boxed{\text{Humus}}\ \text{CO}\bar{\text{O}}\ NH_4^+\ +\ \text{HOH}$$

$$\boxed{\text{Clay}}\ \begin{matrix}Al^{3+}\\H^+\end{matrix}\ +\ 4NH_4^+\ OH^-\ \rightleftharpoons\ \boxed{\text{Clay}}\ (NH_4^+)_4\ +\ \begin{matrix}Al(OH)_3\\HOH\end{matrix}$$

The attached NH_4^+ ions are exchangeable and are available to roots and microbes by displacement with metabolically produced H^+ ions. Exchangeable NH_4^+ is not readily leached out of the soil because release from particles occurs by hydrolysis, which is the right-to-left trend in the above neutralizations and which is inefficient.

Nitrification. The bacteriological oxidation of NH_3 and NH_4^+ to nitrate, discovered by Winogradsky in 1890, proceeds in two steps:

ammonia		nitrite		nitrate
(NH_3)	$\xrightarrow{\text{first oxidation}}$	(NO_2^-)	$\xrightarrow{\text{second oxidation}}$	(NO_3^-)

Autotrophic *Nitrosomonas* bacteria oxidize NH_3 to nitrous acid (9):

$$NH_3 + \frac{3}{2} O_2 \rightarrow HNO_2 + H_2O \quad \Delta G^0 = -65,000 \text{ cal}$$

ammonia   oxygen   nitrous   water
acid

and they use the release of free energy $-\Delta G$ to convert CO_2 to cell carbohydrates. Presence of copper is essential. Nitrous acid and its salts, the nitrites, e.g., KNO_2, dissociate into cations H^+, K^+, and NO_2^- anions, the latter being toxic to many plants.

In the second step *Nitrobacter* bacteria oxidize nitrous acid to nitric acid (9):

$$HNO_2 + \frac{1}{2} O_2 \rightarrow HNO_3 \quad \Delta G^0 = -18,200 \text{ cal}$$

nitrous   oxygen   nitric
acid   acid

The acid and its salts, the nitrates, e.g., KNO_3, dissociate into cations and anions (H^+, K^+, NO_3^-), the latter being a main N source for plants. At high concentrations nitrate in plant tissues is suspected of causing animal and human health problems.

In most soils *Nitrosomonas* and *Nitrobacter* operate in temporal sequence, but above pH 7.5 the chain may be broken and enough NO_2^- may accumulate to damage domestic plants severely. Many pesticides sharply inhibit nitrification.

In the overall oxidation N donates eight electrons to four oxygen atoms, and the weak base NH_3 is transformed into a strong acid (HNO_3). For the oxidation of ammonium sulfate [$(NH_4)_2SO_4$], which is a prominent fertilizer, the conversion burdens the soil with 3 moles of strong mineral acids (2 HNO_3, H_2SO_4). Plant growth has been harmed by high, localized acidity and the ensuing release of toxic quantities of Al and Mn ions from clay.

Since silicate clay and humus particles do not adsorb NO_3^- the anion and its cation partner tend to be washed into the country drainage. Nitrate leaching acidifies and debases mineral soils.

The conventional thought that all NH_4^+ is nitrified and that most plants prefer NO_3^- to NH_4^+, even though in leaves and roots the anion must be reduced to function in amino acid metabolism, may have to be reexamined. In the 1920s Hesselman found only limited quantities of NO_3^- in acid Swedish forest soils, and a preponderance of NH_4^+ over NO_3^- in acidic woodland soils has since been found repeatedly. In a vegetation succession on abandoned agricultural fields to a climax oak-pine forest, Rice and Pancholy (58) observed upward shifts of NH_4^+ from 1-3 ppm under weeds to 3-7 ppm under trees, and, simultaneously, a drop of NO_3^- from 2-4 ppm down to 0.1-1.8 ppm, all to a soil depth of 15 cm. In line with others, the authors contend that in forests nitrifying bacteria are being inhibited by tannins.

Scarcity of nitrifiers in tropical savannas and grasslands and in arctic tundra (26) is well documented, but the purported low NO_3^- level in tropical evergreen forest soils is not supported. The seemingly conflicting behavior of NH_4^+ and NO_3^- in virgin versus cultivated soils (59) has ramifications in balancing cation and anion uptake by roots of native plants and in intensity of nutrient leaching by nitric acid.

Denitrification. When soils in pots and lysimeters are monitored for changes in total nitrogen and for nitrate efflux in leachates the balance is sometimes negative, meaning that the element has been volatilized. With the reliable ^{15}N tracer techniques losses as high as 10-30% of N added as fertilizer have been registered in poorly drained soils.

When soils are waterlogged and deprived of oxygen, bacteria of the genera *Pseudomonas, Bacillus, Micrococcus,* and others switch from oxygen respiration to "nitrate respiration" in which NO_3^- rather than O_2 accepts the e^-s from (H) of organic matter and becomes reduced to nitrogen gas: $2HNO_3 + 10e^- + 10H^+ \rightarrowtail N_2 + 6H_2O$. Several intermediate steps are identified (54):

$$NO_3^- \rightarrowtail NO_2^- \rightarrowtail NO \rightarrowtail N_2O \rightarrowtail N_2$$

| nitrate anion | nitrite anion | nitric oxide gas | nitrous oxide gas (laughing gas) | nitrogen gas |

In the process organic matter is being oxidized (dehydrogenated). The reaction is energy yielding, generates ATP and is catalyzed by reductase enzymes. Mo ions mediate the e^- transport. Humus also may induce denitrification.

Rates of transformations are slow below pH 5 and rapid at pH 7-8. Raising temperatures decrease the population of denitrifiers but increase their efficiency (21). In a well-drained soil a clod may have an aerobic environment on its outside and an anaerobic one inside because of slow O_2 diffusion into tortuous microchannels that house bacteria. Denitrification is the main avenue of returning soil nitrogen to the atmosphere and balancing nitrogen fixation (Fig. 5.11).

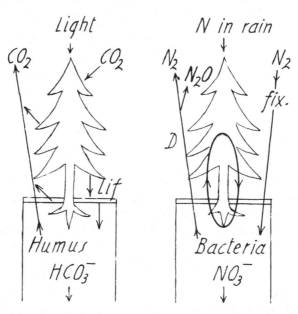

Fig. 5.11. Major traffic lanes of carbon (left) and nitrogen (right). Lif, Litterfall; R, respiration; fix., fixation; D, denitrification. *Within* the ecosystem C cycling (uptake of HCO_3^-) is physiologically less important than N cycling (uptake of NO_3^-)

Nitrous oxide gas that escapes into the soil environment, estimated as $0.1 \text{ g/m}^2/$ year and less (55) reduces ozone (O_3) to oxygen (O_2). Since deforestation and agricultural ammonia fertilization augment nitrification and denitrification, fear has been voiced that N_2O may harm the ozone shield in the upper atmosphere, therewith increasing ultraviolet radiation and incidence of skin cancer.

Correlating Plant Biomass and Soil Nitrogen

So crucial is soil nitrogen for biomass production that agriculturists have searched for a century for tests and equations that would link the two. Legions of soil analyses and field and greenhouse trials unfold endless complexities, and a myriad of climatic, plant, and soil variables is involved. Daily or weekly monitoring of nitrates in the field has not been found practicable. Instead, soil samples are taken to the laboratory and are incubated for weeks. Good accordances between crop yields in the field and nitrate production in test tubes have been amassed by agricultural workers for field crops and by Zöttl (78) for Austrian forests.

Alignments of *biomass and total soil nitrogen* have been observed with corn yields in the Middle West (35) and with site index of firs and pines (73,77). The correlations do not possess generality because of the diversity of nitrogen forms in humus and the variety of microenvironments for microbe action. However, when the individuals of a population of soils are lined up along state factors they become homologous sequences, and conditions for correlations are more favorable (Fig. 10.2).

Overview of C and N Pathways

In summary fashion the routes of major gains and losses of carbon and nitrogen in natural ecosystems are sketched in Figure 5.11. Carbon additions are mediated by CO_2 fixation in photosynthesis; nitrogen gains by microbial N-fixation. Neither can proceed without the other and the two graphs should be fused together. When inputs and outputs are matched the organic matter is at steady state, to be discussed in the next section.

H. Steady-State Formulations

When a soil experiences equality of gains and losses of C, N, and accessory humus elements a *steady state* is achieved. The organic matter content (OM) of the soil remains time-invariant, or, in the case of rhythmic patterns, cycle-invariant, written in either case $\Delta OM/\Delta t = 0$. Steady state differs from ordinary chemical equilibrium because the system is open and experiences equality in amount and composition of material inputs and outputs, or gains and losses, or influxes and effluxes.

Forest Floor

Steady state is visibly suggested in a redwood forest preserve. Each year the long-living trees shed some 300 g/m^2 of needles and twigs that cover the ground about 1 cm deep. In absence of decomposition a 2-3 thousand-year-old tree would be standing in a litter mass 10-20 m thick. Actually, the accumulated forest floor is roughly 10 cm thick and fluctuates little over decades.

In a model of a *deciduous forest* at steady state the decomposing floor F consists in late summer of humified plant remains of the mor- or mull-humus type. Few fresh leaves are on the ground. When the annual litterfall A, in grams/square meter, descends suddenly in toto on October 1, it is assumed, the debris on the ground amounts to $F + A = \mathcal{F}$. Because of decomposing during the ensuing year (Fig. 5.12) F loses $k'_F \cdot F$ grams (dashed curve) and A loses $k'_A A$ grams, and by year's end the F-loss is replenished by the humified remainder A_H of the fresh litter, which is $A - k'_A A$ or $A(1 - k'_A)$. When k'_F and k'_A are unknown, an overall loss coefficient k' is determined from $A = k'\mathcal{F}$, that is (38),

$$k' = \frac{A}{\mathcal{F}} = \frac{A}{F+A} \tag{5.5}$$

The coefficient characterizes decomposer activities in natural settings.

A 1-year litterfall (ash-free basis, Table 5.3) in grams/square meter is 149 for oak and 305 for pine, and the loss coefficients are 5.6% for oak and 2.4% for pine. At other sites oak values are 10.8 and 11.7% and adjacent pine values are 1.5 and 2.9%. The k' discrimination of broadleaf and needle trees is confirmed by Manil (45) in Belgian forests. In cool climates coniferous crops may experience declining growth rates because N is being tied up in the thickening layers of forest floor which may hold as much N as the standing vegetation (Fig. 1.5).

In a model of *evergreen forest* litterfall occurs every day, minute, and second and the steady-state forest floor keeps losing the same organic matter at every instant. Combination of the two actions leads again to Eq. (5.5).

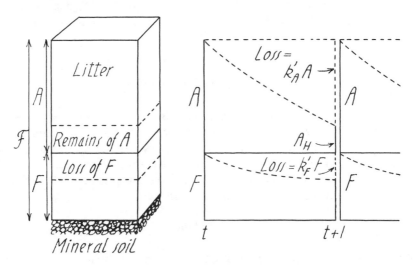

Fig. 5.12. Model of a steady-state, deciduous forest floor. Conceptual analysis begins on October 1, labeled t, and ends a year later, at $t + 1$. Subsequent, identical dates are $t + 2, t + 3$, etc.

The instantaneous rate of litterfall Q is dQ/dt which is taken as constant and which is equal to the yearly rate A. Forest floor F loses organic matter (Lo) at the constant rate dLo/dt which matches the annual rate $k'_F F$. The loss is compensated by the humified portion of litterfall, dQ_H/dt, in which $Q_H = Q - k_Q Q$. The instant rate equals the annual rate $A - k'_A A$. Therefore, $dQ_H/dt = dLo/dt$ which is $A - k'_A A = k'_F F$, which becomes Eq. (5.5).

For Nakane's (50) collection of leaf litterfall (237 g C/m^2/year) and floor respiration in Japanese evergreen oak forest, k' is 29-30% and the loss is accounted for by CO_2 emission. Carbon storage in three mineral soils varies from 8500 to 10,900 g C/m^2/60 cm.

In the broad-leaf rain forest of Calima, Colombia, at 30 m altitude a year's litterfall is 730 g/m^2 of volatile matter (Table 5.3, see p. 131). The forest floor F is a low 432 g/m^2 and its decomposition coefficient is 62.8%. At cooler Chinchiná (1630 m) k' is 39.1%. In the Yangambi forest (Zaire) Belgians secured k' values of 68-76%. Compared to the much lower figures from the temperate oak sites the high k' values in tropical regions place in view the climatic quickening of decomposer activities.

For evergreen forests Greenland and Nye (28) and Olson (53) prefer to write $k = A/F$. It elevates all k' coefficients, the one of Calima from 62.8 to 169%.

Grassland Mineral Soil

On a near-pristine prairie on a Putnam clay pan soil (Mollic Albaqualf) in central Missouri, Dahlman and Kucera (14) collected tesseras at 4-month intervals during a period of over 2 years. The set provides annual *net* organic productions of 508 g/m^2 in tops and 510 g/m^2 in roots. It is noteworthy that 64% of the standing biomass is underground root mass, a proportion which is the opposite of tree specimen.

Root biomass R exhibits seasonal maxima and minima, similar to Figure 5.3, and the difference ΔR between them is root dieback or, assuming steady state, yearly root production. The ΔR values in grams/square meter are 429 in A1, 40 in A2, and 41 in B horizons. A turnover value is calculated by dividing maximum root biomass R into ΔR. It may be recognized as being identical with the k' value of Eq. (5.5), the maximum corresponding to $\mathcal{F}$, and ΔR to $\mathcal{F} - F$ which is A. For the three horizons the k's are 26.5, 25.5, and 26.6%. It means that every year about one-quarter of the root mass is being replaced by new growth. A similar turnover is found for above-ground biomass.

Dahlman and Kucera (14) covered 10 large vegetation quadrats with plastic hoods, exposed the grass underneath for 6 hr to radioactive $^{14}CO_2$, and then monitored the pathways of ^{14}C during a 2-year period. Up to 80% of the assimilated ^{14}C is transferred to the root system to all depths. In the densely rooted A1 horizon root radioactivity decreases linearly with time. Extrapolation to zero activity yields a span of 52 months, which agrees with the 4-year turnover value calculated from root biomass production.

Carbon-14 appears in soil humus but does not stay there long, possibly because of respiration losses as $^{14}CO_2$. Annual decay rates are estimated as 2% for young organic soil fractions and 0.4% for the resistant humus portions, which agrees with findings of others.

N Balances

Nitrogen in stabilized ecosystems is in balance with *gains* from microbial N_2-fixation that culminates in proteins, and with *losses* of NH_3 gas, NO_3^- leaching, and denitrification. The latter conversion returns N_2 to the atmosphere (Fig. 5.11). It is probably the preeminent pathway in stabilized soils of arid regions.

In the case of the Missouri prairie, calculations of steady state N-inventories (15) in g N/m^2 yield for the entire soil humus reservoir 600, for live shoots and leaves 3.6, for the abundant, standing litter in absence of grazing 4.2, for the roots 6-12. Annually, 6 g N/m^2 move from the soil pool into living biomass. For an undisturbed northern hardwood forest a detailed N-budget with transfer rates was secured by Bormann et al. (6).

The uppermost question of how steady states of entire N and C profiles are recognized in the field is taken up in Chapter 9 dealing with time functions.

I. Review of Chapter

Soil organisms are distantly linked to green chloroplasts where sunlight splits water molecules into oxygen gas (O_2) and where CO_2 is reduced to energy-rich carbohydrates. Aided by water and nutrients from the soil, the plants generate the biomass that nourishes directly or indirectly all living specimens. Among the millions of soil organisms a few fix nitrogen molecules (N_2) from the air and endow and perpetuate the nitrogen economy of ecosystems.

As respiring decomposers, the soil bacteria, fungi, actinomycetes, and invertebrates consume O_2 and revert the organic molecules to CO_2 and H_2O and transfer chemical energy to environmental entropy. Through mechanical work fauna creates soil porosity that helps stimulate the life-supporting photosynthesis and energy flow.

In the soil dead and live plant and animal tissues are being physically disintegrated and chemically degraded and black humus substances (fulvic and humic acids) are synthesized metabolically. Excluding peats and mucks, well-drained soil profiles vary enormously in contents of organic matter, from practically nothing in deserts to 247 kg humus (C · 1.724) per m^2 in equatorial mountains.

Humus molecules are frameworks of various phenols and organic accessories. They sorb water, act as colloidal, weak acids, participate in cation exchange, fix polyvalent metals, and interact with clays. Humus performs as a critical reservoir of slowly available N sources and through microbial action supplies plant roots with NH_4^+ and NO_3^-. In the process of denitrification N returns to the environment as N_2 gas.

Since organic synthesis and decay are more dynamic than rates of weathering and clay formation, the soil organic matter regime is often viewed in short perspective as a steady-state system with inputs of C and N balancing their egress.

References

1. Aleksandrova, L. N. 1960. *Soviet Soil Sci.* 2: 190-197.
2. Babel, U. 1975. In *Soil Components,* J. E. Gieseking, ed., Vol. 1, pp. 369-473. Springer-Verlag, New York.
3. Bloomfield, C. 1965. In *Experimental Pedology,* E. G. Hallsworth and D. V. Crawford, eds., pp. 257-266. Butterworths, London.

4. Blume, H. P. 1965. *Z. Pfl. Düng. Bod.* 111: 95-114.
5. Bond, G. 1976. In *Symbiotic Nitrogen Fixation,* P. S. Nutman, ed., pp. 443-474. Cambridge Univ. Press, Cambridge.
6. Bormann, F. H., G. E. Likens, and J. M. Melillo. 1977. *Science* 196: 981-983.
7. Broadbent, F. E. 1954. *Soil Sci. Soc. Am. Proc.* 18: 165-169.
8. Burns, R. G. (ed.). 1978. *Soil Enzymes.* Academic Press, New York.
9. Burns, R. C., and R. W. F. Hardy. 1975. *Nitrogen Fixation in Bacteria and Higher Plants.* Springer-Verlag, New York.
10. Cameron, R. E., and G. B. Blank. 1966. *Desert Algae.* Jet. Prop. Lab. T. Rep. 32-971, Pasadena, CA.
11. Cameron, R. E., and W. A. Fuller. 1960. *Soil Sci. Soc. Am. Proc.* 24: 353-356.
12. Clark, F. E., and E. A. Paul. 1970. *Adv. Agron.* 22: 375-435.
13. Cooper, J. P. (ed.). 1975. *Photosynthesis and Productivity in Different Environments.* Cambridge Univ. Press, New York.
14. Dahlman, R. C., and C. L. Kucera. 1965. *Ecology* 46: 84-89.
15. Dahlman, R. C., J. S. Olson, and K. Doxtader. 1969. In *Biology and Ecology of Nitrogen,* C. C. Delwiche, ed., pp. 54-82. Nat. Acad. Sci. (Davis Conf.), Washington, D.C.
16. Delacour, F., and A. El Attar. 1964. *Pédologie* 14: 55-63.
17. Delwiche, C. C., P. J. Zinke, and C. M. Johnson. 1965. *Plant Physiol.* 40: 1045-1047.
18. Denison, W. C. 1973. *Sci. Am.* 228(6): 75-80.
19. Dhar, N. R. 1968. In *Organic Matter and Soil Fertility,* P. Salviucci, ed., pp. 244-360. Pont. Acad. Scripta varia, No. 32, Vatican.
20. Döbereiner, J. 1977. In *Recent Developments in Nitrogen Fixation,* W. Newton, J. R. Postgate, and C. Rodriguez-Barrueco, eds., pp. 513-522. Academic Press, New York.
21. Doner, H. E., and A. D. McLaren. 1978. In *Environmental Biogeochemistry and Geomicrobiology,* W. E. Krumbein, ed., pp. 573-582. Ann Arbor Science, Ann Arbor, MI.
22. Edwards, N. T., and W. F. Harris. 1977. *Ecology* 58: 431-437.
23. Eiland, F. 1979. *Soil Biol. Biochem.* 11: 31-35.
24. Flaig, W. 1968. In *Organic Matter and Soil Fertility,* P. Salviucci, ed., pp. 723-770. Pont. Acad. Sci. Scripta varia, No. 32, Vatican.
25. Flaig, W., H. Beutelsbacher, and E. Rietz. 1975. In *Soil Components,* J. E. Giese-king, ed., Vol. 1, pp. 1-211. Springer-Verlag, New York.
26. Flint, P. S., and P. L. Gersper. 1974. In *Soil Organisms and Decomposition in Tundra,* A. J. Holding, O. W. Neal, S. F. MacLean, Jr., and P. W. Flanagan, eds., pp. 375-387. Tundra Biome Steer. Comm., Stockholm.
27. Gjessing, E. T. 1976. *Physical and Chemical Characteristics of Aquatic Humus.* Ann Arbor Science, Ann Arbor, MI.
28. Greenland, D. J., and P. H. Nye. 1959. *J. Soil Sci.* 10: 284-299.
29. Haider, K., J. P. Martin, Z. Filip, and E. Fustec-Mathen. 1972. In *Humic Substances,* D. Povoledo and H. L. Golterman, eds., pp. 71-85. Pudoc, Wageningen.
30. Hall, D. O., and K. K. Rao. 1977. *Photosynthesis,* 2nd ed. E. Arnold, London.
31. Harris, W. F., P. Sollins, N. T. Edwards, B. E. Dinger, and H. H. Shugart. 1975. In *Productivity of World Ecosystems,* D. E. Reichle, J. F. Franklin, and D. W. Goodall, eds., pp. 116-122. Nat. Acad. Sci., Washington, D.C.
32. Hawkins, J. C. 1962. *J. Sci. Food Agr.* 13: 386-391.
33. Hissett, R., and T. R. G. Gray. 1976. In *The Role of Terrestrial and Aquatic Organisms in Decomposition Processes,* J. M. Anderson and A. MacFadyen, eds., pp. 23-39. Blackwell, Oxford.
34. Jackson, R. M., and F. Raw. 1966. *Life in the Soil. Studies in Biology,* No. 2. St. Martin's Press, New York.
35. Jenny, H. 1941. *Factors of Soil Formation.* McGraw-Hill, New York.
36. Jenny, H. 1950. *Soil Sci.* 69: 63-69.

37. Jenny, H., F. Bingham, and B. Padilla-Saravia. 1948. *Soil Sci.* 66: 173-186.
38. Jenny, H., S. P. Gessel, and F. T. Bingham. 1949. *Soil Sci.* 68: 419-432.
39. Junge, C. E. 1958. *Trans. Am. Geoph. U.* 39: 241-248.
40. Kononova, M. M. 1968. In *Organic Matter and Soil Fertility,* P. Salviucci, ed., pp. 361-379. Pont. Acad. Sci. Scripta varia, No. 32, Vatican.
41. Kononova, M. M. 1975. In *Soil Components,* J. E. Gieseking, ed., Vol. 1, pp. 475-526. Springer-Verlag, New York.
42. Kuo, M. H., and W. V. Bartholomew. 1966. In *The Use of Isotopes in Soil Organic Matter Studies,* pp. 329-335. Rept. FAO/IAEA, Völkenrode, 1963.
43. Lemon, E. R. 1967. In *Harvesting the Sun,* A. San Pietro, F. A. Greer, and T. J. Army, eds., pp. 263-290. Academic Press, New York.
44. MacFadyen, A. 1961. *Ann. Appl. Biol.* 49: 216-219.
45. Manil, G. 1971. *Bull. Soc. Roy. Forêts Belg.* 53: 217-250.
46. Mayaudon, J. 1968. In International Atomic Energy Agency, *Isotopes and Radiation in Soil Organic Matter,* pp. 117-188. Vienna.
47. Monteith, J. L., G. Szeicz, and K. Yakuku. 1964. *J. Appl. Ecol.* 1: 321-337.
48. Muir, J. W., R. I. Morrison, C. J. Brown, and J. Logan. 1964. *J. Soil Sci.* 15: 220-237.
49. Nagel de Boois, H. M., and E. Jansen. 1971. *Rev. Ecol. Biol. Sol.* 8: 509-520.
50. Nakane, K. 1975. *Jap. J. Ecol.* 25: 204-216.
51. Nykvist, N. 1963. *Stud. Forest Suecica* 3: 1-31.
52. O'Conner, F. B. 1963. In *Soil Organisms,* J. Doeksen and J. van der Drift, eds., pp. 32-48. North Holland, Amsterdam.
53. Olson, J. S. 1963. *Ecology* 44: 322-331.
54. Payne, W. J. 1973. *Bact. Rev.* 37: 409-452.
55. Pratt, P. F. 1977. *Climatic Change* 1: 109-135.
56. Prigogine, I., and R. Defay. 1954. *Chemical Thermodynamics.* Longmans, London.
57. Quispel, A. (ed.). 1974. *The Biology of Nitrogen Fixation.* North Holland, Amsterdam.
58. Rice, E. L., and S. K. Pancholy. 1972. *Am. J. Bot.* 59: 1033-1040.
59. Richards, B. N. 1974. *Introduction to the Soil Ecosystem.* Longman, New York.
60. Russell, E. W. 1973. *Soil Conditions and Plant Growth.* Longmans, London.
61. Sanders, F. F., B. Mosse, and P. B. Tinker. 1975. *Endomycorrhizas.* Academic Press, New York.
62. Santantonio, D., R. K. Hermann, and W. S. Overton. 1977. *Pedobiologia* 17: 1-31.
63. Schaller, F. 1968. *Soil Animals.* Univ. of Michigan Press, Ann Arbor, MI.
64. Scharpenseel, H. W. 1972. In *Humic Substances,* D. Povoledo and H. L. Golterman, eds., pp. 281-292. Pudoc, Wageningen.
65. Schatz, A. 1963. *Agr. Food Chem.* 11: 112-118.
66. Schlesinger, W. H. 1977. *Ann. Rev. Ecol. Syst.* 8: 51-81.
67. Schnitzer, M., and J. S. Desjardins. 1969. *Can. J. Soil Sci.* 49: 151-158.
68. Schnitzer, M., and S. U. Khan. 1972. *Humic Substances in the Environment.* Marcel Dekker, New York.
69. Stevensen, F. J. 1965. *Adv. Agron.* 10: 1-42.
70. Stöckli, A. 1946. *Nat. Ges. Zürich* 91: 1-18.
71. Terry, N., and A. Ulrich. 1973. *Plant Physiol.* 51: 43-47.
72. Van Dijk, H. 1971. *Geoderma* 5: 53-67.
73. Viro, P. J. 1961. *Unasylva* 15: 2.
74. Waksman, S. A. 1938. *Humus,* 2nd ed. Williams & Wilkins, Baltimore.
75. Wallwork, J. A. 1970. *Ecology of Soil Animals.* McGraw-Hill, New York.
76. Warner, R. C., and I. Weber. 1953. *J. Am. Chem. Soc.* 75: 5086-5094.
77. Zinke, P. J. 1960. *Trans. 7th Int. Congr. Soil Sci.* 3: 411-418.
78. Zöttl, H. 1960. *Plant Soil* 13: 183-206.

6. Soil Colloidal Interactions and Hierarchy of Structures

In many parent materials the sorting of individual grains appears in haphazard manner. As soil formation commences the particles, the voids between them, and the accessory life forms become organized to structures or architectures that stamp each soil as an individual.

A. Flocculation and Deflocculation

Concentrations of salts, acids, and bases govern the stability of sols and gels.

Stability of Clay Suspensions

In muddy waters the clay particles may be minute single platelets, as is typical of Na-clay, or thicker packages and "books" made up of several sheets held together by shared K^+ and Ca^{2+} ions and H-bonds, or still bulkier edge-to-face constructs (34). The suspensions, called sols when the particles are of colloidal size, may stay turbid for months and years (Table 1.1) and for the light humus macromolecules the settling times are even longer.

Small suspended particles undergo *Brownian movement* which is seen in an ultra-microscope as a violent zig-zag motion. It is caused by the thermal, irregular bombardment by water molecules. When two clay particles are propelled toward one another their swarms of exchangeable ions initiate repulsion and guarantee sol stability. Absence of double layers invites collision and the particles adhere to each other by means of short range cohesion and adhesion forces. Lumps and flocs settle out. The overall interaction energy (16) is a function of the distance between particles and exhibits an energy barrier B (Fig. 6.1). To the right of it repulsion prevails, to the left attraction. The barrier is higher the more diffuse the ion swarm (Fig. 3.11). Its thickness runs parallel to the exchangeability of adsorbed ions and to the electrophoretic migration velocity expressed as ζ-potential. In Table 6.1 sols with monovalent exchangeable ions are more stable than those with divalent, trivalent, and tetravalent (Th) ions.

Eventually, clay particles settle out and form a gel. The platelets still tend to repel one another and they orient themselves snuggly in positions of lowest potential energy

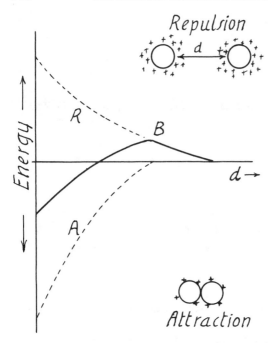

Fig. 6.1. Energy profile (solid curve) of two colloidal particles as a function of the distance d between them. When the curve of repulsion energy R of the long-range repulsion forces of the swarm of exchangeable cations (+ signs) is superposed on the curve of attraction energy A of short-range, strong attraction forces, the solid curve results. B is a stability hump. After Hamaker (16)

(Fig. 6.2). Stiff clay gels that liquify temporarily when pressured, stirred, or shaken are "thixotropic."

Flocculation

Addition of NaCl to a sol of Na-clay represses the ion swarm and lowers barrier height B. Brownian movement pushes particles over the lowered energy hump, which leads to attraction and settling out. The salty flocs are open-structured and occupy a larger space than the salt-free sedimentation volumes (Fig. 6.2). Coagulation or flocculation occurs on a grand scale at the seashore where muddy rivers flow into salt-rich sea water and the settling clays create mud flats and underwater benthic soils.

The salt concentration that induces flocculation under a given set of conditions is the flocculation value Fl. For various Putnam clays (beidellites) the Fls, expressed as S-concentrations (S), are listed in Table 6.1, last column. Sodium clay is the most stable of the sols because it requires the highest amount of salt for flocculation. Acid Al, H-clay sol is least stable. For all clays the flocculation values correlate positively with ζ-potential and cation exchange.

Soil solutions are mixtures of ions and the flocculation series in Table 6.1 is diversified by simultaneous ion exchange reactions. Whereas Fl is 14.8 for Na-clay + NaCl, for Na-clay + $CaCl_2$ it is only 1.4. In a practical valency series of flocculation efficiency the cations line up as: $Na^+ < Ca^{2+} < La^{3+}$, Al^{3+}, $Fe^{3+} < Th^{4+}$. A concentrated clay sol in a test tube will flocculate with salt to a rigid jelly without settling of particles. Powders of $CaSO_4$ and $CaCO_3$ are effective flocculators.

Table 6.1. Stability of Suspensions (Sols) Related to Exchangeable Cations[a]

Suspensions (sols)	ζ-potential (mV)	Cation exchange with NH_4^+, K^+, or H^+ at $S=1$ (%)	Flocculation value for common ion (S-concentrations[b])
Na^+-clay	57.6	66.5 (NH_4Cl)	14.8 (NaCl)
K^+-clay	56.4	48.7 (NH_4Cl)	7.8 (KCl)
H^+-clay[c]	—	—	—
Mg^{2+}-clay	53.9	31.3 (KCl)	0.69 ($MgCl_2$)
Ca^{2+}-clay	52.6	28.8 (KCl)	0.55 ($CaCl_2$)
Al^{3+}, H^+-clay (electrodialized clay)	48.4	14.5 (KCl)	0.36 (HCl)
			0.5 S of KCl[d]
Na^+-humus	24.1	79.1 (HCl)	0%
Ca^{2+}-humus	11.2	73.9 (HCl)	63.4%
H^+-humus	17.6	13.3 (KCl)	50.9%

[a] From Jenny, H., and R. F. Reitemeier (19), and Baver, L. D., and N. S. Hall (4).
[b] Clay sol, 0.109%; CEC = 60 me/100 g clay. Humus sol, 0.28%; CEC = 403 me/100 g of α-humus.
[c] Probable position of columnated H-clay.
[d] Percentage of humus that settles upon addition of 0.5 S KCl.

Extracting peat, Baver and Hall (4) convert H-humus to sols of Na- and Ca-humus which are negatively charged. In the lower part of Table 6.1 flocculation is measured as the percentage of humus that settles out in the presence of ½ S of KCl. In addition to ionic double layers humus sols are further stabilized by strong particle hydration (4,6). H-humus is more stable than Ca-humus which has implications for profile differentiations in acid soils. Na-humus is readily flocculated by $AlCl_3$ and $FeCl_3$.

The effectiveness of metal ions in flocculating fulvic acid extracted from a Canadian Chernozem is expressed by Khan (21) as the minimal amounts of cations (in milligrams) needed per gram of fulvic acid. The flocculating power of elements lines up as: Mn^{2+} (1428 mg) < Zn^{2+} (654 mg) < Cu^{2+} (254 mg) < Fe^{3+} (65 mg) < Al^{3+} (27 mg). These values far exceed those needed for clay flocculation, which implies that humus sols are much more stable than clay sols.

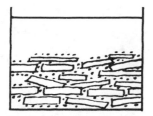

Fig. 6.2. Attracted by gravity, clay platelets settle out and form gels. Left: with repelling, diffuse ion swarms (dots) the particles orient themselves. Right: with ion swarms reduced or absent particles adhere to each other and form open structures. In either gel many pores have colloidal dimensions

In principle, *exchange flocculation* underlies Hilgard's famous reclamation of hard, unproductive alkali soil of arid lands by the incorporation of gypsum (solid calcium sulfate), formulated as:

$$\boxed{\text{clay}}\begin{matrix}^-\\^-\end{matrix}\begin{matrix}Na^+\\Na^+\end{matrix} + n(Ca^{2+} + SO_4^{2-}) \rightarrow \boxed{\text{clay}}\begin{matrix}^-\\^-\end{matrix}Ca^{2+} + \begin{matrix}2Na_2^+ + SO_4^{2-}\\(n-1)(Ca^{2+} + SO_4^{2-})\end{matrix}$$

| alkali clay | dissolved | ion exchange | in soil solution |
| (pH 8-10) | gypsum | and flocculation | of pH 7 |

The reaction renders the soil friable and soft and the enlarged micropores facilitate leaching of sodium sulfate by rain and irrigation water.

Visibly instructive is Chang's (9) experiment with montmorillonitic, salt-free Stockton clay. Into large pans he places dry soil, either natural or with salt additions. The samples are soaked with water and then let to dry out slowly. As seen in Fig. 6.3, the natural clay shrinks to massive, stone-hard angular polygons up to 12 cm in diameter and separated by wide cracks. The clod corners are sharp and the faces planar. Contrastingly, the salt-enriched, flocculated soil develops fine fissure cracks and small, subangular fragments. The larger lumps crumble under light pressure.

Salt-free, reddish clay soils that are dominated by iron oxides dry to polygons with serrated edges and irregular clod faces. Shrinkage is minimal and cracks are narrow.

What has been said about mechanism pertains to *positive suspensions* as well, except that the valencies of anions come to the fore. Deb's (11) Fe_2O_3 sol is flocculated

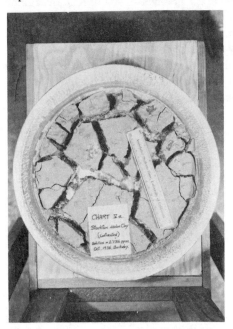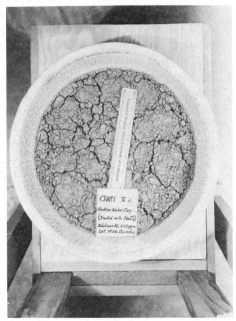

Fig. 6.3. Drying of wetted Stockton clay soil (9). Left: natural soil; right: after addition of excess NaCl

by KCl and $CaCl_2$ at equal concentrations but requires only 1/40 as much of K_2SO_4 because the SO_4^{2-} anion is better adsorbed than monovalent Cl^-. Phosphate (PO_4^{3-}) anions are effective coagulators of positive sols.

Deflocculation or Dispersion

A Na-clay sol runs through filter paper, but the same sol flocculated with NaCl has its clusters retained. Leaching the mass of flocs on the filter with distilled water removes the excess salt and the particles separate to a sol that runs through the paper again. The process is deflocculation, dispersion, or peptization.

> To demonstrate deterioration of soil structure by Na^+, fill a cylinder which has an opening at the bottom with loam or clay loam soil. Percolate a strong solution of NaCl (e.g., sea water) through it. Flow rate is good because the clay remains flocculated by high salt concentration even though it is being converted by ion exchange towards a Na-soil. Now, discontinue the salt percolation and leach the column with distilled water. As excess salt is being displaced downward, Na-clay above begins to disperse to a sol and clogs up the fine soil pores. Eventually outflow of solution ceases, yet water still stands on top of the column.

Soil bathed in sea water has its exchangeable Na fraction augmented to nearly 40% (22); consequently, clay dispersion has to be faced wherever land is being reclaimed from the ocean.

B. Mutual Interactions of Unlike Particles

The dry clay clods of Chang withstand strong pressures, but in water they slake to paste and mire. *Water-stable* aggregates are fabricated by interactions of silicate clays with humus particles and sesquioxide clays.

Silicate Clays and Iron Oxides

Sand grains in water are negatively charged; the water is positive and moves *en masse* toward a negative platinum electrode.

> Passage of a positive iron oxide sol through a column of sand, followed by water wash, alters the inherently negative sand skeleton to positive charge; subsequent percolation of negative Na-clay sol and water wash makes the matrix negative again. Repetition of the sequential treatments drastically cuts down the rate of flow of water because coprecipitation builds up sandwich-type coats of oxide/clay/oxide/clay/ . . . that narrow the pores (19). The layering has been verified microscopically (31).

The interaction of negative clay and positive hydrous oxide particles is mutual flocculation or coprecipitation, portrayed schematically as:

$$
\begin{array}{c} Na^+ \\ \\ Na^+ \end{array}
\begin{bmatrix} - & - \\ \multicolumn{2}{c}{\text{Smectite}} \\ - & - \end{bmatrix}
\begin{array}{c} Na^+ \\ + \\ Na^+ \end{array}
\begin{array}{c} Cl^- \\ \\ Cl^- \end{array}
\begin{bmatrix} + & + \\ \multicolumn{2}{c}{\text{Fe-oxide}} \\ + & + \end{bmatrix}
\begin{array}{c} Cl^- \\ \\ Cl^- \end{array}
\rightarrow
\begin{array}{c} Na^+ \\ \\ Na^+ \end{array}
\begin{bmatrix} - & - \\ \multicolumn{2}{c}{\text{Smectite}} \\ - & - \end{bmatrix}
\begin{bmatrix} + & + \\ \multicolumn{2}{c}{\text{Fe-oxide}} \\ + & + \end{bmatrix}
\begin{array}{c} Cl^- \\ \\ Cl^- \end{array}
+ \; 2NaCl
$$

The bonds between the two particles might be of the types $-O-Fe\begin{smallmatrix} O- \\ OH \end{smallmatrix}$.

The floccule is water stable, amphoteric, has an isoelectric point, and may display combinations of cation and anion exchange, as previously outlined.

Water-stable granulation by sesquioxides is widespread in reddish tropical soils. They drain well and resist erosion in spite of high clay contents. For kaolinitic B horizons of red soils of North Carolina (2) the amount of iron oxide extracted chemically and the quantity of silicate clay thereby freed of iron are related to stability of medium-sized crumbs (< 0.25 mm diameter) as follows:

$$\text{crumb stability} = -2.01 + 0.96\% \text{ clay} + 3.33\% \text{ iron oxide}$$

The percentages refer to total soil weight. Comparing coefficients (0.96, 3.33%), the larger value of iron oxide points to its dominant importance in aggregation.

Though H-clay is a tenuous species, its action upon sesquioxides:

$$
\begin{bmatrix} & H^+ & \\ - & - & H^+ \\ - & & H^+ \\ - & & H^+ \\ - & - & H^+ \\ & H^+ & \end{bmatrix}
+ \begin{array}{c} \boxed{Fe(OH)_3} \\ + \\ \boxed{Al(OH)_3} \end{array}
\begin{array}{c} \rightleftharpoons \\ \rightleftharpoons \end{array}
\begin{bmatrix} - & - & Fe^{3+} \\ - & \\ - & - & Al^{3+} \end{bmatrix}
+ \quad 6H_2O
$$

is a crucial, yet poorly evaluated, pedogenic reaction. To what extent will an H-clay dissolve sesquioxide particles? Is the reaction proceeding more to the right or to the left? The ferrolysis soil process (Chapter 7) assumes the latter trend. During 2 months of shaking of H^+-montmorillonite with iron oxide the clay acquired 26 me/100 g of exchangeable Fe, calculated as trivalent, according to Pugh (29). Similarly Chernov and Kislitsyna (10) obtained Al-clay but the conversion is complicated by the internal Al and H replacement discussed in Chapter 3.

Particle Bondage with Humus

In an extension of his structure experiments Chang incorporates 10 g of ground alfalfa into 100 g of salt-free Stockton clay and after moist incubation for 4.5 months the drying soil crumbles into a medley of mellow, coarse-granular and medium subangular pieces and the quantity of water-stable aggregates rises from 80 to 500 g/kg of soil.

Biologically induced aggregation, often explained as a physical entanglement of particles by organic exudates, polysaccharides (sugars), and fungal hyphae, is prominent in grasslands and pastures with abundant roots. Possibly, the active microorganisms are themselves structurally integrated in the crumbs. Since the binder molecules are consumed by microanimals, biogranulation is dynamic and lasts only as long as the metabolites are being regenerated.

Freshly precipitated Al- and Fe-hydroxide gels attract fulvic and humic acids (2). On clay surfaces the adsorption of molecules having carboxyl groups is strong in the presence of Al, either as exchangeable ion or as lattice constituent exposed at the edges of the platelets (26). For exchangeable Al on a planar surface a charge-balanced equation reads as follows

$$\left[\begin{array}{c}- \\ - \\ -\end{array}\right]\!\!>\!\!Al^{3+} \quad + \quad H^+ \ \overline{O}OC - R \quad \rightleftharpoons \quad \left[\begin{array}{c}- \\ - \\ -\end{array}\right]\begin{array}{l}>\!Al^{3+}\!\!-\!\!-\overline{O}OC\!-\!R \\ -\!\!H^+\end{array}$$

in which R is the remainder of the organic chain. A humus macromolecule studded with —COOH radicals could engage a number of platelets and construct an aggregate. Calcium also has been proposed for "cation bridges" between humus and clay, as in the triplet *clay-Ca-humus-Ca-clay*.

Humus in a water extract from rotting clover leaves (18) is strongly adsorbed by aluminum-rich allophane clays (SiO_2/Al_2O_3 = 1.32), but is barely adsorbed by montmorillonite. The darker, larger humus particles of molecular weights up to 10,000 are preferentially retained and are rendered stable toward leaching and microbial degradation, which implies large bonding areas with many points of contact within the aggregate. Humus may enhance the CEC of clays and silts and accelerates their electrophoretic mobility (12).

Cationic organic molecules (e.g., amino acids) and large positive proteins (17) readily occupy interlayer spaces in montmorillonite, but the negative fulvic acid globule tends to be repelled unless polyvalent cations on the clay take hold of it (24). Neutral organic molecules may become protonated by acid clays, accepting H^+ ions, in consequence of which they become adsorbable cations. Adsorption of organic molecules is not confined to humus constituents. Clay particles interact with herbicides, pesticides, enzymes, mucigel, and bacteria (14,25,33).

Humus as Protecting and Sensitizing Agent

Peat extract and tannic acid protect kaolinite clay from flocculation by salt, as do leaf leachates (5). Neutral humus given to a clay sol of the beidellite type prevents flocculation with NaCl. Hence, humus initiates "protective action."

When gently shaken, coarse sediment of iron-clay-coprecipitate disperses to a sol upon addition of humus. Deb (11) notes significant *dispersion* of iron oxide by water-soluble peat humus in amounts of 10 parts of humus to 100 parts of Fe_2O_3. An iron sol protected by humus is not absorbed by soil (11) and a humus-enriched silicate clay fails to undergo mutual flocculation with an iron sol (20).

Contrastingly, Aarnio's (1) dark-brown water extracts of peats *flocculate* iron hydroxide sols. In Schnitzer and Kodama's (30) titration of fulvic acid in the presence of $FeCl_3$ an iron-humus complex is formed. It breaks up at higher pH and $Fe(OH)_3$ separates out. A similar trend with citrate has been mentioned in Chapter 5.

A partial integration of these diverse reponses is attempted in Figure 6.4. To positive iron sol Winters (37) adds negative humus sol extracted with water from decayed leaves of a deciduous hardwood forest and determines the flocculation value Fl with KCl. Small additions of humus sensitize the positive iron sol, lower its stability, and

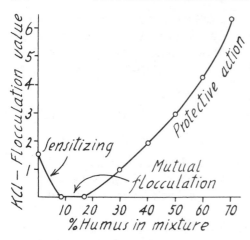

Fig. 6.4. Interaction of positive Fe(OH)$_3$ sol (0.1%) with negative humus sol, as revealed by flocculation values (in moles/liter). After Winters (37)

reduce Fl. Higher amounts cause instant coprecipitation and Fl is zero. Above 20% of humus the mixture becomes stabilized as a negative sol and humus exerts protective action.

C. Peds, Crusts, and Voids

Ultrafine particles assemble to visible pedogenic granules, crumbs, plates, clods, and columns (Fig. 6.5), all of them spoken of as *peds*.

Skins or Cutans (7)

Best seen with a hand lens, many peds in B and C horizons are covered with wave-like, shiny cutans of gray, brown, and black coloring. They are up to 1 mm thick and the thin foils consist of clay, humus, and iron and manganese oxides. Clay cutans are explained as downward translocation of clay in pores and fissures, followed by flocculation (8,23). Absence of clay skins in B and C horizons is interpreted as destructive squeezing by swelling pressures.

Crusts

Raindrops falling on bare soil and water flowing in rills and furrows cause orientation of silt and clay particles to thin *crusts* that reduce gas exchange and impede seed germination. The structures, sometimes called Wollny layers, are not lasting.

Pouring a positive Fe(OH)$_3$ sol over a layer of sand, which is negative, initiates attraction. When dried the mixture resembles a fragment of iron hardpan (Ortstein). It dissolves in strong acid. Or, drying iron oxide-coated quartz grains in the presence of colloidal silica imitates properties of an iron-silica duripan. Alternating extractions with strong HCl and NaOH will break it up. In the field the hardpans form slowly and intermittently over long periods of time, which promotes layering and stratification.

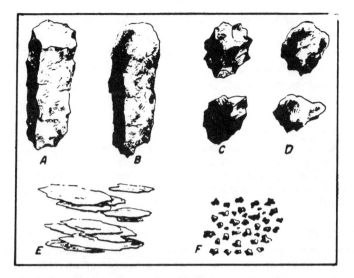

Fig. 6.5. Types of soil structures, from Soil Taxonomy (32). (A) Prismatic; (B) columnar; (C) angular blocky; (D) subangular blocky; (E) platy; (F) granular

Voids

The patterns of voids are as important as those of the solids (7). Networks of tunnels
and passageways of temporary nature are left by rotting roots and are generated daily
by fossorial rodents and invertebrates. Surveyors in deserts encounter conspicuous
vesicular structures that support in surface soils dense webs of isolated, bubble-like, air-
filled vesicles up to 5 mm in diameter. When water-soaked and stirred the structures
collapse, but form again on drying.

The sum of all voids, Psp, in a given volume of soil is calculated from bulk density
(Bd) and particle density (Pd), the latter often assumed to be 2.65, by the formula:
$Psp = 1 - (Bd/Pd)$. Total pore space or porosity averages about 50% for mineral soils,
but the biologically critical distribution of shapes and sizes must be put together from
visual and microscopic observations. Wiersum (35) concludes that higher order roots
of many species cannot enter a rigid channel opening of less than 0.2 mm in diameter;
thence, the number of these *biopores* in unit soil volume is crucial for crop growth.

Recovery of a biogenic soil structure is strikingly demonstrated on experimental
plots on reddish Ramona loam (Haploxeralfs), a near-virgin arid soil that possesses
excellent rates of infiltration of irrigation water. When disked in dry state continually for hours the granular soil structure is effectively destroyed. Clouds of dust
arise, porosity is lowered, and the rate of water infiltration is catastrophically reduced. Mustard (*Brassica nigra*) practically ceases to grow. When the land is left to
mustard and volunteer weeds and is given an annual summer irrigation, plant
growth and soil structure gradually improve and after 8 years of *noncultivation* the
porosity and water infiltration become restored to the original, optimal conditions
(27).

D. Swelling, Shrinking, and Their Turbations

In alternating, prolonged wet and dry periods some clayey soils expand and shrink so much that plants suffer root damage, stones move, fence posts are loosened, road pavements crack, and foundations tilt. On slopes slipping and sliding are induced.

Swelling Process

As soon as water molecules encounter a dry gel of montmorillonite the exchangeable ions hydrate, dissociate to an ion swarm, and initiate platelet repulsion, which is swelling. The swollen volume is always less than the volume sum of dry gel and water imbibed because hydrating ions disrupt the normal configuration of water molecules (volume contraction or electrostriction) in their neighborhood.

Among the swelling tests reported in Table 6.2, Al,H-bentonite of montmorillonitic nature more than doubles its volume whereas sesquioxide-rich kaolinitic Cecil clay barely expands. Water uptake parallels cation exchange capacity and both appear collinear with the ratio $SiO_2/(Al_2O_3 + Fe_2O_3)$, hence the table is akin to Figure 4.5. For Putnam clay the authors document the superiority of exchangeable Na^+ over K^+ and Ca^{2+} in causing the dry gel to swell. Colony montmorillonite in H^+-form outswells all other ionic garnitures.

Each degree of swelling is associated with a characteristic vapor pressure and the potential of swelling contains the familiar term $RT\ln(p/p_0)$ of Chapter 2.

The *heat of swelling* (q, $-\Delta H$) often follows the hyperbolic curve in Figure 6.6. The slopes of the curve are the differential heats that indicate the calories developed when 1 g of water is imbibed by a large quantity of soil or clay of moisture content i. The slopes are greatest near the dry end of the curve where tremendous swelling pressures develop and heats amount to hundreds of calories per gram of H_2O (380 cal in Fig. 6.6). The curvatures provide elucidation of the process of particle hydration (3).

When vapor pressures of the partially swollen clays are measured the free energy change, ΔG, and, therefore, the entropy reduction $-\Delta S$ ($T\Delta S = \Delta H - \Delta G$) may be estimated. It reflects the lowered degree of randomness of the water molecules on the clay particles.

Table 6.2. Swelling and Nature of Soil Clays[a]

Soil series names and dominant clay minerals[b]	Swelling = uptake of cm³ H_2O/1 g clay	SiO_2/R_2O_3	CEC (me/100 g)
Cecil (halloysite)	0.070	1.3	13
Susquehanna (beidellite-halloysite)	0.560	2.3	47
Putnam (beidellite)	0.960	3.2	65
Lufkin (beidellite-montmorillonite)	1.160	3.8	82
Bentonite-montmorillonite	2.320	5.0	95

[a] After Winterkorn and Baver (36).
[b] Al, H-forms, electrodialyzed.

For field evaluation the "linear extensibility" is calculated from the difference in bulk densities of clods, once when moistened at 1/3 atm. tension and again when oven dry. For a clay soil (Marias) of 234 cm depth the vertical linear expansion reaches 19 cm. It indicates the rise of the land surface from dry to moist soil (15).

Swell-turbation

All soils swell, more or less, depending on their clay species and electrolytes. In line with Table 6.2, montmorillonite, beidellite, and vermiculite swell strongly, mica, illite, kaolinite, and halloysite swell weakly, and sesquioxides swell hardly at all. The *Vertisols* (from verto, turn) expand and contract so spectacularly that they have been accorded the rank of a soil order. They develop in climates with prolonged wet and hot-dry periods, and on shales, marine and lacustrine clays, and argillic limestones from which they inherit high clay contents (40-60%) of montmorillonite quality. Volcanic ash and basalt weathering also may produce Vertisols.

During the long dry season cracks a meter deep and up to 25 cm wide criss-cross the landscape. Surface soil granules slough into the crevasses. With the onset of the rainy season, swelling closes the cracks, more rapidly in the upper horizons than farther down. Trapped material eventually expands inside the soil body causing distortion, convulsion, and churning of the mass (turbation). Side pressures induce lentilar, dense peds with faces smoothed and grooved by plastic flow (slickensides), and the soil surface buckles to a wavy microrelief called gilgai, an Australian name. Sparse stands of drought-tolerant eucalyptus, acacia, and mesquite (*Prosopis*) endure the stresses and strains. Vertisols are grassland and steppe favorites.

Turbation in Vertisols interferes with horizon formation which requires soil bodies at rest. Parsons et al. (28), mapping Bashaw soils (Pelloxererts) on alluvial surfaces of various ages, find little profile variation, whether the deposits are 550 or 6000 years old or of middle Pleistocene age.

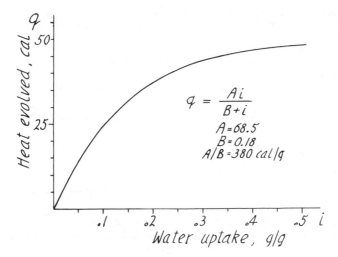

Fig. 6.6. Heat q evolved when 1 g of dry allophane clay takes up i g of water; calculated from Wada in (13). In thermodynamic language (Chapter 1) the heat released in a reaction is listed as negative, $q = -\Delta H$

E. Review of Chapter

The submicroscopic particles of clay and humus that belong to the colloidal domain ($< 2 \times 10^{-5}$ to $> 10^{-7}$ cm) command high proportions of surface area to mass that magnify the reactions of ion exchange, hydration, and adsorption of molecules.

Ionic double layers initiate repulsions between clay particles that encourage dispersion of wetted soil to muds and turbid sols and foster hydration of clayey soils to swollen Vertisols. Salts reduce the actions of the exchangeable ions and cause flocculation of individual particles to aggregates that assemble to water-permeable clay gels.

Negative humus molecules adhere to negative clay particles provided polyvalent cations (e.g., Ca^{2+}, Al^{3+}) on the silicate surfaces enter bondage with the carboxyl (–COOH) and –C–OH radicals.

Soil permanence is bolstered by negative clay surfaces reacting with positive sites of hydrous oxide clays to form water-stable aggregates that are common in red, erosion-resistant soils of lower latitudes.

References

1. Aarnio, B. 1913. *Int. Mitt. Bodenk.* 3: 131-140.
2. Arca, M. N., and S. B. Weed. 1966. *Soil Sci.* 101: 164-170.
3. Barshad, I. 1960. *Eighth Nat. Conf. Clay, Clay Min.* 84-101.
4. Baver, L. D., and N. S. Hall. 1937. *Colloidal Properties of Soil Organic Matter.* Missouri Agr. Exp. Sta., Res. Bull. 267.
5. Bloomfield, C. 1963. *Rep. Rothamst. Exp. Sta.* 226-239.
6. Boratynski, K., and S. Mattson. 1938. *Ann. Agr. Coll. Sweden* 7: 63-119.
7. Brewer, R. 1964. *Fabric and Mineral Analysis of Soils.* Wiley, New York. (Latest def. in *Geoderma* 8:81.)
8. Buol, S. W., F. D. Hole, and R. J. McCracken. 1973. *Soil Genesis and Classification.* Iowa State Univ. Press, Ames, Iowa.
9. Chang, C. W. 1941. *Soil Sci.* 52: 213-227.
10. Chernov, V. A., and L. P. Kislitsyna. 1955. *Pochvo.* 3: 7-16.
11. Deb, C. 1949. *J. Soil Sci.* 1: 112-122.
12. Dixit, S. P. 1978. *Agrochimica* 22: 25-31.
13. Dixon, J. B., and S. B. Weed (eds.). 1977. *Minerals in Soil Environments.* Soil Sci. Soc. Am., Madison, Wisc.
14. Dommergues, Y., and F. Mangenot. 1970. *Écologie microbienne du sol.* Masson et Cie, Paris.
15. Grossman, R. B., B. R. Brosher, D. P. Franzmeier, and J. L. Walker. 1968. *Soil Sci. Soc. Am. Proc.* 32: 570-573.
16. Hamaker, H. C. 1937. *Rec. Trav. Chim. Pays-Bas* 56: 727-747.
17. Harter, R. D. 1977. In *Minerals in Soil Environments*, J. B. Dixon and S. B. Weed, eds., pp. 709-739. Soil Sci. Soc. Am., Madison, Wisc.
18. Inove, T., and K. Wada. 1968. *Trans. 9th Int. Congr. Soil Sci.* 3: 289-298.
19. Jenny, H., and R. F. Reitemeier. 1935. *J. Phys. Chem.* 39: 593-604.
20. Jenny, H., and G. D. Smith. 1935. *Soil Sci.* 39: 377-389.
21. Khan, S. U. 1969. *Soil Sci. Soc. Am. Proc.* 33: 851-854.
22. Kelley, W. P., and G. F. Liebig, Jr. 1934. *Bull. Am. Assoc. Petrol. Geol.* 18: 358-367.
23. Khalifa, E. M., and S. W. Buol. 1968. *Soil Sci. Soc. Am. Proc.* 32: 857-861.
24. Kodama, H., and M. Schnitzer. 1968. *Soil Sci.* 106: 73-74.

25. McLaren, A. D., G. H. Peterson, and I. Barshad. 1958. *Soil Sci. Soc. Am. Proc.* 22: 239-244.
26. Mehta, N. C. 1956. Thesis, Univ. of California, Berkeley.
27. Parker, E. R., and H. Jenny. 1945. *Soil Sci.* 60: 353-376.
28. Parsons, R. B., L. Moncharoan, and E. G. Knox. 1973. *Soil Sci. Soc. Am. Proc.* 37: 924-927.
29. Pugh (Page), A. L. 1960. Ph.D. Thesis, Univ. California, Davis, Calif.
30. Schnitzer, M., and H. Kodama. 1977. In *Minerals in Soil Environments,* J. B. Dixon and S. B. Weed, eds., pp. 741-770. Soil Sci. Soc. Am., Madison, Wisc.
31. Soileau, J. M., W. A. Jackson, and R. J. McCracken. 1964. *J. Soil Sci.* 15: 117-123.
32. Staff (Soil Survey). 1975. *Soil Taxonomy.* U.S.D.A., S.C.S., Agr. Handb. 436.
33. Tschapek, M., and A. J. Garbosky. 1950. Adsorción del Azotobacter y su importancia agronomica. Inst. Suelos y Agrot. Publ. 14, Buenos Aires.
34. van Olphen, H. 1963. *Clay Colloid Chemistry.* Interscience, New York.
35. Wiersum, L. K. 1957. *Plant Soil* 9: 75-85.
36. Winterkorn, H., and L. D. Baver. 1934. *Soil Sci.* 38: 291-298.
37. Winters, E. 1940. *The Migration of Iron and Manganese in Colloidal Systems.* Ill. Agr. Exp. Sta. Bull. 472.

7. Pedogenesis of Horizons and Profiles

When uniform rock material transforms to soil, *horizons* appear at various depths, recognizable as dark humus layers, gray and reddish color bands, zones of clay accumulations, carbonate strata, and iron and silica hardpans.

A. Transport Considerations

To assess pedogenic displacements of soil constituents, horizons are compared analytically with an assumed, initial state of the soil. With a lucky choice the balance sheet tells what happened during millennia of genesis, but does not reveal what processes are operating now.

New directions in technology of moisture assay permit collecting soil solutions from large pores of undisturbed horizons in place to find out what is going on today. How far backward or forward the momentary insight may be projected varies from profile to profile.

During passage of solutes the soil matrix acts as an exchanger and chromatography column, and release of Mg, Ca, Na, and K and dissolution of carbonates are promoted by the biologically generated carbonic, sulfuric, and nitric acids. The last one excels but is elusive because NO_3^- is taken up by roots in exchange for HCO_3^-. Transition elements, Cu, Zn, and many others, are effectively mobilized by humus extracts.

In the long run, the rate of flow of water through the soil may not be the controlling process of weathering and debasing. To occupy sites accessible to leaching, ions in crystal lattices must migrate through weathering rinds, which is a slow process. This rate-limiting step causes questioning of the notion that a doubling of rainfall is equivalent to halving the time of soil formation.

B. Clay Migration

On the southern Piedmont Plateau of the Atlantic Coast whitish, sandy-loam A horizons sit on red, clay-loam B horizons. The wide geographic extent of these Ultisols (Yellow-Red Podzolic soils) intimates a genetic horizon bondage: clay migrates from A to B. To north and west, in Alfisols (Gray-Brown Podzolic soils) and Mollisols

(Prairie soils), the A/B texture contrast is not as colorful but may be equally strong (Fig. 7.1).

Argillic Horizons

In loamy soils, clay-enriched "textural B horizon" becomes a *Bt horizon* when the ratio of percentage clay ($< 2 \mu m$) in B to that in A exceeds 1.2. In clay pan soils the ratio may reach 3.0. The proportion becomes glaring, 7.7, for the mobile fine clay ($< 0.2 \mu m$) in the Arago Podzol of Quebec.

Migration is not the sole cause of a clay bulge in B. Higher moisture in the subsoil likely enhances clay formation. In alluvial parent materials B-clay may be inherited from unconformity in sedimentation of sand, silt, and clay, as happened in the perplexing Dayton profile in Oregon (5,50). For Lapeer loam in Wisconsin, investigators (14) infer ancient influx of windblown clay dust that washed downward into B21.

The deep Puerto Rican Oxisol of Flach et al. (26) contains 83% clay in the B and only 25% in the C horizon which is a saprolite or weathered rock. Yet, both horizons have equal CEC, equal kaolinite contents in sand, silt, and clay, and the same internal surface area of horizon mass. Mainly, the clay in B has become readily dispersible in conventional agents.

Ecological Features

Massive Bt horizons confine roots to A horizons because Bt pores are small and expansion and contraction of clays break the fine rootlets of native herbs. Under alternating wetting and prolonged desiccation argillic horizons diversify to prismatic and columnar structures (Fig. 11.6), especially in the presence of Na^+ ions. On loess soils in semiarid Idaho annual, virgin herbage production is 38 g/m^2 on a columnar Natrargid and 76 g/m^2 on an adjacent silty Mollisol (51). The strong woody roots of sage shrubs (*Artemisia* spp.) push through 40-cm-long Bt fissures but remain unbranched and are

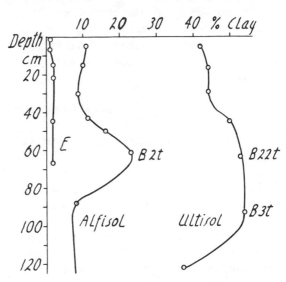

Fig. 7.1. Example of clay-depth functions. E, Entisol on sand, Alfisol on glacial till, Ultisol on basalt, all from Soil Survey Staff collections

flattened and deformed. On Cornell University grounds young spruce roots zigzag conformably with structure channels and are compressed.

Experimentation

Continued slow percolation of 8000 cm of distilled water through a column of red, kaolinitic Aiken soil having an effective mean pore diameter of 15.4 μm moves 3.6% clay ($<$ 2 μm) from the upper portions into the lower ones (12). In montmorillonitic Yolo soil of pore diameter 8.0 μm, the "rainfall" of 1300 cm transports almost the same amount (3.5%). Flow is laminar and no clay appears in the effluent.

A year of leaching, interrupted by seven drying periods (11), induces clay translocation and clay coatings that, however, are severely restricted in soils having high Ca concentrations ($>$ 20 ppm) in their solution. In the presence of $CaCO_3$ no clay moves in the columns of Wurman et al. (69), yet Bt horizons in calcareous profiles exist. Hallsworth's (34) sand-clay mixtures (montmorillonite) crack on drying and more clay moves downward in the fissures than between them.

Mechanism

Most pedologists accept the ubiquity of clay skins on peds in lower horizons as qualitative proof of clay transfer from above, though migration of clay precursors and subsequent synthesis is not to be ruled out.

In colloid terminology, migration of clay begins with *dispersion* of aggregates in the A horizon by changing the particle's ion swarm, e.g., $Ca^{2+} \rightarrow H^+$ or Na^+, by anchoring humus molecules (clay cutans are high in humus) and by mechanical disruption of peds by shear forces and faunal action. Small particles disperse more readily than large ones.

Transport

Transport is accomplished as percolating sol or as slurry-creep along ped faces. *Deposition* is preceded by lowering of ζ-potential (H-clay $\rightarrow$ Al-clay, Na-clay $\rightarrow$ Ca-clay) followed by flocculation by Ca and Mg of subsoil carbonates (4), by rise in pH with soil depth, and by mutual interactions with oxides. Downward flow of water and clay may be arrested by texture discontinuities and by subsoil desiccation inherent in evapotranspiration.

Gain and Loss Calculations

Many attempts have been made to differentiate quantitatively genesis and migration of clay particles.

The difference in weight of the whole soil and of its clay portion is denoted by Barshad (5) as *nonclay* fraction. It is understood that the proportion of clay to nonclay depends on the efficacy of dispersion and on the upper particle size assigned to clay.

Uniformity of Initial State. Unless the initial properties of the present-day profile are known, material balances cannot be set up. In practice a reference horizon, most often the C horizon, is selected and it is assumed that every horizon originally possessed this same make-up. The hypothesis is amenable to verification by assay of weathering-resistant *index* minerals, advocated by C. E. Marshall and discussed in Chapter 10.

Calculating the Initial Profile. Once the degree of uniformity of the initial state is assessed, any chemically or mineralogically stable criterion may be chosen to calculate the proto-profile.

The proportion of any soil property of the C horizon to its index criterion is projected onto every horizon above, based on its own quantity of the index. For 1 cm² cross section of soil tessera area and for any horizon h,

$$\frac{J_h}{J_c} = \frac{W_h{}^i}{W_c}, \text{ or } W_h{}^i = W_c \cdot \frac{J_h}{J_c}$$

where, $W_h{}^i$ is initial weight of horizon h, W_c is weight of the C horizon, J_h is grams of index mineral in horizon h, and J_c is grams of index mineral in the C horizon.

For these calculations the volume weight of each horizon must be known.

Balance Sheet. Gain (+) or loss (−) of *soil mass* in any horizon h is the difference D_h of initial (i) and present-day (t) horizon weights, or

$$D_h = W_h{}^i - W_h{}^t$$

A Case Study. Total chemical analyses of horizons and their isolated clay ($< 1 \mu m$) fractions in a Gray Forest soil from Bulgaria (maP 54 cm, maT 9.4°C) were made available by A. Hajiyanakiev, but he is not responsible for any conclusions drawn, the more so as a few minor chemical irregularities were adjusted by interpolation with adjacent horizons. Emphasis is on procedure. The soil is derived from an older loess and has a calcareous C horizon. Size variations among the resistant coarser quartz particles and their ratios appear randomized and do not point to significant depositional horizon discontinuities.

All properties of the soil, freed of $CaCO_3$, are recalculated on ignition (600°C) basis. Subtraction of ignited clay from ignited soil yields ignited nonclay.

In the condensed summary (Table 7.1) of present-day profile features, percentage TiO_2 in *nonclay* is chosen as the index oxide. In the C horizon 1 g TiO_2 is associated with 100/0.814 = 122.85 g nonclay, and 1 g of nonclay relates to 0.3843 g

Table 7.1. Features of a Bulgarian Gray Forest Soil for a Tessera of 1 cm² Cross Section and 200 cm Depth[a]

Horizon	Depth (cm)	Bulk density ovendry (g/cm³)	Ignited clay (g)	Ignited nonclay (g)	TiO_2 in nonclay (%)
Ah	0-6	1.49	1.62	6.55	0.895
AB	6-25	1.53	6.54	20.55	0.969
B1	25-60	1.88	25.52	37.33	0.977
B2	60-120	1.87	39.01	68.79	0.920
B3	120-145	1.79	14.82	28.46	0.867
C	145-200	(1.70)	(19.40)	(50.47)	0.814

[a] Note that clay + nonclay = total ignited soil.

of clay. These proportions fix the initial state of each horizon in accordance with its present-day TiO_2 content in nonclay. Thus, the Ah horizon contained originally 7.20 g nonclay and, by multiplication with 0.3843, 2.77 g clay. The sum is 9.97 g of soil on ignition basis. The balance sheet is constructed in Table 7.2.

To begin with *total soil mass*, the solum tessera of 1 cm^2 area, 145 cm depth, and 255.44 g initial weight lost 6.25 g of soil. Most of the reduction is confined to the subsurface horizon AB, which speaks against sheet erosion as having been the cause.

Nonclay disappeared from every horizon, the sum of 22.84 g being much greater than the total loss of soil mass (6.25 g); hence, for the most part the weathering products derived from nonclay remained in the profile. The percentage loss of nonclay in a horizon is a measure of intensity of its transformation. It is largest in B1 (16.7%) and AB (16.0%), less in B2 (11.5%) and Ah (9.0%), and least in B3 (6.1%), a trend that has been repeatedly observed by Barshad. The low Ah value does not support the belief that weathering is highest in the surface soil.

Clay inherited from the parent material left the A horizons to the extent of 4.01 g. The three B horizons gained 20.60 g of clay, and the difference, 16.59 g, reflects clay genesis in the entire profile.

The *crucial question* of how much clay was translocated from A to B and how much was synthesized in B is therefore not answered because nonclay in Ah and AB weathered to the extent of 4.56 g and part of it also migrated downward as clay, thereby exaggerating clay formation in B. The answer hinges on knowing to what extent the *disappearance* of nonclay in a given horizon may be identified with the *appearance* of clay in it. In general, loss of nonclay means conversion to clay by weathering and decementation and outright leaching of components. The latter amounts to 6.25 g for the tessera in question.

Table 7.2. Balances in Grams, Ignited Basis, for Bulgarian (Gray Forest Soil of Tessera Dimensions of 1 cm^2 Cross Section and 145 cm Depth)

Horizon	Initial soil	Present soil	Gain or loss (−)
Ah	9.97	8.17	−1.80
AB	33.86	27.09	−6.77
B1	62.02	62.85	0.83
B2	107.63	107.80	0.17
B3	41.96	43.28	1.32
	255.44	249.19	−6.25
	Initial nonclay	**Present nonclay**	**Gain or loss (−)**
Ah	7.20	6.55	−0.65
AB	24.46	20.55	−3.91
B1	44.80	37.33	−7.47
B2	77.75	68.79	−8.96
B3	30.31	28.46	−1.85
	184.52	161.68	−22.84
	Initial clay	**Present clay**	**Gain or loss (−)**
Ah	2.77	1.62	−1.15
AB	9.40	6.54	−2.86
B1	17.22	25.52	8.30
B2	29.88	39.01	9.13
B3	11.65	14.82	3.17
	70.92	87.51	16.59

By resorting to *chemical analyses of total soil* (not included here) and calculating a balance for each of the 11 elements determined in each horizon, or 66 balances in all, it turns out that the leachate of 6.25 g is composed foremost of silica, 5.2 g, followed by sodium, 0.51 g, then sulfate and calcium. Neither Al_2O_3 nor Fe_2O_3 left the profile; curiously, their overall balances are slightly positive instead of zero, likely caused by soil variability, nutrient cycling by trees, dust influx, experimental errors, and invalid assumptions.

The source of the leachate lies in the transformation of nonclay and in dissolution of inherited clay. Assuming chemical stability of initial clay and attributing all loss to alteration of nonclay, the amount of clay produced in the entire profile becomes 22.84 − 6.25 = 16.59 g, which is a lower limit, the true value lying somewhere between 16.59 and 22.84. To partition the clay formed among the five horizons a good choice is setting it proportional to disappearance of the nonclay in Table 7.2. For Ah the procedure gives 0.47 g of clay created. For the other horizons clay genesis amounts to 2.84 g for AB, 5.43 g for B1, 6.51 g for B2, and 1.34 g for B3.

In this light, *clay migration* from the two A horizons to the three B horizons comprises 4.01 g of initial clay and 3.31 g of new clay or 7.32 g total. Clay formed in B adds up to 13.28 g. Hence, the 20.60 g of clay gain in the textural B horizons is composed of 35.5% illuviation and 64.5% of synthesis *in situ*, which is an answer to the pedogenic information wanted.

Combinations of chemical analyses of soil and of its clay fraction allow calculation of a third analytical table of 66 entries: "percentage of oxides in the nonclay fraction" of each horizon. Since differences of determinations are involved and differences of differences are subsequently calculated, errors are compounded and statistical reliability is lowered.

Referral of A and B compositions to those of C divulges the chemistry of products of pedogenesis, the clays generated, and the ions leached out.

Thus, the ratio SiO_2/Al_2O_3 is 3.74 for the clay in the C horizon and 4.07 for the average clay synthesized. The corresponding ratios of $(MgO + K_2O)/Al_2O_3$ are 0.38 and 0.47. Original and acquired clays are similar in chemical composition, but Na is much lower and Ca higher in the synthesized than in the parent material clay, which is to be expected from ion exchange principles.

Quantitative balances have been published by numerous authors (e.g., 11,28,33).

C. Mottling, Gleyzation, and Concretions

Paraphrasing Hilgard (1860) (Chapter 1), when soil is kept wet its rust-brown color from iron oxides turns to blue-gray and greenish tints caused by the influence of fermenting vegetable matter at points where air has little or no access.

Reduction

Rainwater may be saturated with oxygen from the air and contain about 10 ppm of the gas. Because of microbial respiration soil air is low in oxygen and 0.2 ppm in soil solutions is cited as indicative of reducing conditions. Freshly planed wood stakes quickly blacken. Provided that readily oxidizable organic matter is present and air influx is curtailed, the reducing reaction of Chapter 3 is

$$Fe(OH)_3 \quad + \quad e^- + H^+ \quad \rightarrow \quad Fe(OH)_2 \quad + \quad H_2O$$

| insoluble ferric hydroxide | H of dehydro- genation of organic matter | fairly soluble ferrous hydroxide |

Iron reduction is paralleled by CO_2 production of oxidizing humus.

Accompanied by anions (e.g., HCO_3^-) ferrous iron leaches vertically or laterally in massflow, or diffuses in solution and by itself on clay surfaces in any direction from higher to lower concentrations (activities) as might be provided by local variations in pH, *Eh*, and by loci of iron-oxidizing Mn^{3+} compounds.

Mottling

Upon reoxidation, blanched, grayish, horizon domains are seen alternating with smaller, brownish Fe and Mn accumulation flecks and blotches, from millimeter to penny size, which are called *mottles.* They may be scarce or numerous, faint or prominent, spherical or irregular.

In fine-textured soils the mottling design simulates a network of threads and streaks, the reticulate pattern, as if Fe had followed desiccation cracks, and it imitates root contours because of oxygen excreted by roots. The blurred and flaming shapes remind workers abroad of stained marble and they designate the redox color palette of reddish-yellows, olive grays, blues, and blacks (with sulfur compounds) as *gley* and attach the symbol g to a gley horizon. The reduction zone is G_r, that of oxidation G_o.

Gley Regime

In English, the redox soil processes in hydromorphic soils are called *gleying* or *gleyzation.* For its chemical characterization the total iron content of a horizon is insensitive; instead, amorphous, microcrystalline, and organically linked "free iron" is being dissolved by mild extractants such as oxalates, dithionite-citrates, and Na-pyrophosphate.

In Table 7.3 Kha and Duchaufour (39) compare the iron status of a permanently waterlogged *gley* in a colluvial depression to a *seasonally* waterlogged Pseudogley on silt deposits. Gley supports wet meadow flora, mostly *Carex maxima* and diverse species of *Juncus* (rushes); pseudogley belongs to a glade of beech (*Fagus sylvatica*), birch (*Betula verrucosa*), and aspen (*Populus tremula*) trees.

In the alternating wetting and drying regime of Pseudogley the original minerals have liberated nearly all of their lattice iron (high alteration index) and a relatively small proportion, 7-15%, is retained as free Fe^{2+}, the remainder having been leached laterally from the very acid horizons. Gley, on the other hand, is still high in total Fe and preserves a high proportion of Fe^{2+} in the reduction horizons (G_r) because the nearly neutral soil water is barely moving.

Table 7.3. Iron Regimen in Seasonally Wet (Pseudogley) and Permanently Wet (Gley) Soils in France[a]

Horizons	Depth (cm)	pH	Clay (%)	Alteration index			Reduction index	
				Free Fe ($^o/_{oo}$)	Total Fe ($^o/_{oo}$)	Ratio (%)	Free Fe^{2+} ($^o/_{oo}$)	Free Fe^{2+}/Free Fe (%)
Pseudogley								
Al	0-15	4.2	15	8.5	10.7	79	1.3	15
g	15-35	4.4	17	20.8	21.0	99	2.3	11
Bg	35+	4.8	27	19.2	24.6	78	1.4	7
Gley								
Al	0-10	6.4	16	22.9	41.0	56	8.0	35
Gr'	10-25	6.2	16	22.5	41.0	55	10.1	45
Gr	25-70	5.3	20	41.5	58.0	72	19.1	46
Gr/C	70+	6.7	36	32.0	62.0	52	7.6	24

[a] Kha and Duchaufour (39).

Ferrolysis

The Fe^{2+} and Mn^{2+} ions generated in the soil solution participate in ion exchange with clay surfaces. If the replaced Ca, Mg, K, and Na ions are washed away during rains and an oxidizing regime resumes, the adsorbed Fe^{2+} and Mn^{2+} become Fe^{3+} and Mn^{3+} and may revert to hydroxide precipitates leaving H^+ ions on the clay, as Brinkman (15) points out. Schematically,

$$\begin{bmatrix} Fe^{2+} \\ Mn^{2+} \\ Ca^{2+} \end{bmatrix} + \begin{matrix} \tfrac{1}{2} O_2 \\ 5 H_2O \end{matrix} \rightarrow \begin{bmatrix} Fe^{3+} \\ Mn^{3+} \end{bmatrix} + \begin{matrix} 4 H_2O \\ Ca(OH)_2 \end{matrix} \rightarrow \begin{bmatrix} H^+ \\ H^+ \\ H^+ \\ H^+ \\ Ca^{2+} \end{bmatrix} + \begin{matrix} Fe(OH)_3 \\ Mn(OH)_3 \end{matrix}$$

The clay-acidification causes clay destruction and Al-interlayering.

Break-up of iron-clay aggregates during iron reduction and leaching of Fe and Si during ferrolysis is conducive to development of albic horizons low in clay and to silica cementation in lower strata.

Concretion Profiles

In medium-textured soils the mottles seemingly shrink to soft pellets and after many cycles of drying and wetting harden to brownish-black "buckshot" *concretions* up to a few millimeters in diameter and imbedded in a gray or white sand or silt matrix. Extreme ion selectivity produces nodules high in Fe, Al, Ti, and Mn (Table 7.4). Iron oxide pellets may contain colloidal illite, goethite, and hematite in concentric arrangement.

Table 7.4. Composition of Soil Concretions and Surrounding Soil Material

Types	SiO_2 (%)	Al_2O_3 (%)	Fe_2O_3 (%)	TiO_2 (%)	Mn-oxides (%)		Source
Soil material	79.6	8.7	3.0	–	0.14	(Mn_3O_4)	Illinois[a]
Fe, Mn-con-cretion	50.7	11.3	14.0	–	11.3	(Mn_3O_4)	Illinois[a]
Soil material	2.34	28.68	45.37	9.43	–		Hawaii[b]
Al-concretion	1.54	61.72	5.64	0.80	–		Hawaii[b]
Fe, Ti-con-cretion	2.59	6.90	73.63	18.50	–		Hawaii[c]
Mn-concretion	13.06	15.29	8.12	1.88	44.19	(MnO_2)	Hawaii[d]

[a] Winters (68).
[b] Sherman and Ikawa (57).
[c] Sherman and Kanehiro (58).
[d] Sherman et al. (59).

Because many plants transport oxygen from leaves to roots the ions Fe^{2+} and Mn^{2+} become oxidized near the root periphery and precipitate as hydroxides and persist as casts and tubes long after the roots have died.

On glacial till of Wisconsin age in Indiana, well-drained *Miami* silt loam (Alfisol or Gray-Brown Podzolic) is located on a gentle convex slope in a virgin forest of sugar maple (*Acer saccharum*), beech (*Fagus grandifolia*), and oaks (*Quercus* sp.). Nearby, the finer textured Bethel silt loam occupies beech- and maple-covered flat uplands of the same drift, has slow external and internal drainage, and suffers waterlogging during wet periods.

In the profiles exposed by Brown and Thorp (16) the weight percentages of Fe, Mn nodules in the fine gravel and sand fractions of the wetter Bethel soil exceed those of the drier Miami soil even though their A horizons have identical contents of total Fe_2O_3, namely, 2.9-3.3% inclusive of the pellets (Fig. 7.2). A source of the moisture divergence lies in the Bethel bulge of "silt plus clay" at 35-50 cm depth that causes temporary water stagnation during and after rains and reduction of iron in the A horizons, accompanied by horizon graying and slight acidification.

Reduction enhances the availability of essential Fe and Mn to organisms but excesses become toxic. At solution concentrations of a few ppm Mn, leaf rims of domestic crops turn pale and yellow and yields are reduced. In Hawaii manganiferous soils cause pineapple chlorosis; in Quebec regeneration of forests of jack pine (*Pinus banksiana*) is harmed by high Mn^{2+} concentrations that induce iron deficiency.

D. Podzolization

In the original Danish edition (1878) of P. E. Müller's (49) work on humus forms the author compared sandy soils under oaks with those under heather and thereby helped launch the latter profile to international, pedologic prominence.

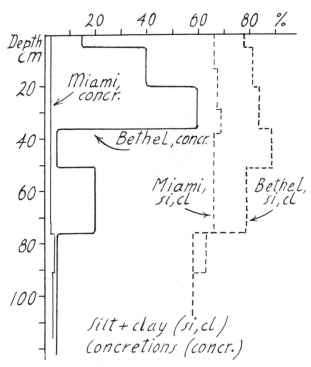

Fig. 7.2. Depth functions of iron-manganese concretions and of silt plus clay contents
of well-drained Miami soil and poorly drained Bethel soil (16)

Bleicherde-Ortstein

Three of Müller's color renditions of soil profiles (his word) using his own "layer"
designations a, a′, a″, a‴, b, c, which are the historical forerunners of the modern A,
Al, A2, B, C horizon nomenclature, are sketched in Figure 7.3.

Oaks stand on an a-c profile. The a-layer is a mull, a gray-brown, humus-rich, loosely structured, crumbly mass, with many roots and fauna-excrements, fungus mycelia,
earthworms, insects, spiders, millipedes, and scolopenders. It merges gradually into the
ochre-colored, sandy substrate c.

In the *heath profile* a 2- to 3-cm-thick crust of peaty humus, dense and tough, rests
on the mineral layer a′, which is webbed by roots and impregnated with humus, but is
low in microfauna and lacks earthworms; below this is bleached earth a″ (Bleicherde),
a whitish (albic) A2 horizon that is underlain by a‴, a blackish humus band with numerous heather roots, designated in the United States as Bh horizon. Directly underneath is hardened, dark-brown "ortstein" (localized stone) composed of sand cemented
with iron, aluminum, and humus substances. It is the American Bmir horizon.

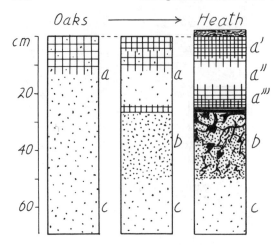

Fig. 7.3. Forester Müller's rendition (1878) of transition from soil under oaks to Bleicherde-Ortstein profile under heath vegetation in Denmark. Compare Fig. 1.4

Nomenclature

When Dokuchaev in Russia designated soils with grayish surface horizons as Podzols (pod = under, zola = ash) the Bleicherde-Ortstein profiles soon became *Podzols*. For the same horizon sequence—if well developed—Soil Taxonomy coins the term *Spodosol* (spodos = wood ash) and elevates it to a soil order. The master diagnostic criterion is the spodic horizon which is the illuvial B horizon rather than the etymologically expected eluvial A2 horizon. Precise chemical limits are assigned to it. Both terms are employed here, but the reader should be aware that all Spodosols are Podzols but not all Podzols are Spodosols (they may be Inceptisols). For the genetic processes the word *podzolization* is retained because of the long history of research and the extensive international literature.

Spodosol and Podzol Variations

In a transect excavation through Blacklock soil (Spodosol) on old, intensely weathered beach-sand deposits (Chapter 9) 21 profiles 10 m apart unfold the spatial variations sketched in Figure 7.4. While all A2 horizons are white, some are decorated with chocolate-colored humus tongues which are remnants of roots. Bh is present in some specimens and absent in others. The spodic horizon is an indurated hardpan that varies in thickness from 40 to 80 cm. In places it is weakly cemented, even absent.

Near Bogotá in equatorial Colombia, at an altitude of 2460 m and with maT 17°C and maP 200 cm the Sabaneta soil is a giant iron podzol (6). Under 46 cm of Al the gray and white A2 (2.5 Y 8/3, dry) extends to a depth of 142 cm and rests on a 10-cm-thick, yellowish red spodic horizon containing 8.7% humus and a high 40.6% of amorphous Fe_2O_3.

Many lowland tropical Podzols (40) in America and Asia conform to the syndrome of "white sand and black water," the latter being a dilute humus sol. In Surinam, Dutch pedologists encountered a provocative 230-cm-thick sandy A2 horizon laying on a humus hardpan (Humuspodzol).

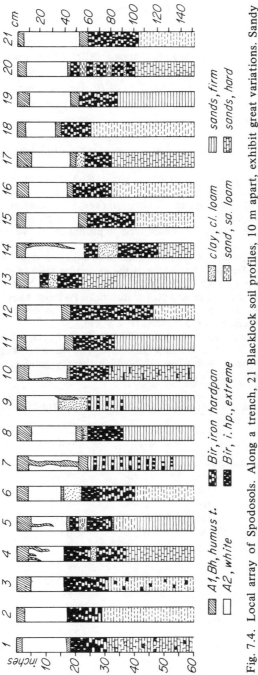

Fig. 7.4. Local array of Spodosols. Along a trench, 21 Blacklock soil profiles, 10 m apart, exhibit great variations. Sandy parent materials are of Pleistocene age (see Chapter 9)

Legend:
- A1, Bh, humus t.
- A2, white
- Bir, iron hardpan
- Bir, i. hp., extreme
- clay, cl. loam
- sand, sa. loam
- sands, firm
- sands, hard

Composition

In the albic horizon the quartz grains are shiny, having been "cleaned" by humic acids. In spodic horizons innumerable dark, fine organic pellets (20-50 μm) appear, and under magnification the quartz grains are bridged with amorphous gels of humus and sesquioxides. From the fusion analyses (Table 7.5) of *Becket fine sandy loam* derived from glacial till in a virgin conifer-deciduous forest in Massachusetts the following generalizations may be drawn.

1. Clay content ($<$ 2 μm) is low throughout the profile, which is an earmark of many Podzols.
2. The low percentages of the oxides in the dark A1 horizon, which here includes the lower litter layer, are consequences of dilution by organic matter. Recalculated on ignited basis, A1 resembles the C horizon, except for small declines in Fe and K. The B1 and B2 horizons also have high ignition losses, in part from humus migration and in part from decayed roots which tend to proliferate above the enriched spodic layer.
3. The bases K, Na, and Ca have undergone little change relative to Al, as evidenced from the near-constancy of ba, which is the molecular ratio $(K_2O + Na_2O + CaO):Al_2O_3$. Weathering of rock minerals has been minimal, which is characteristic of most of the North American and North European Podzols. They are young soils of postglacial age. In old Pleistocene Spodosols, like the aforementioned Blacklock soil, mineral weathering is far advanced (Fig. 9.21).
4. The molecular ratio saf, which is $SiO_2:(Al_2O_3 + Fe_2O_3)$, is augmented in A2 and reduced in B1 and B2, in comparison to C. The shifts are interpreted as preferential removal of sesquioxides from A2, which thereby gains in quartz percentage, and deposition of Al and Fe in B.

Table 7.5. Chemical Composition of a Spodosol (Becket Fine Sandy Loam)[a]

Horizons	A1	A2	B1	B2	C
Depth (inches)	0-6	6-11	11-13	13-24	24-36
Depth (cm)	0-15	15-28	28-33	33-61	61-91
SiO_2 %	52.95	83.32	69.60	72.67	77.86
Al_2O_3 %	7.04	6.73	9.61	10.32	10.00
Fe_2O_3 %	1.08	1.69	3.99	3.58	3.15
MnO %	0.01	0.01	0.01	0.02	0.03
CaO %	0.90	0.54	0.65	0.62	0.54
MgO %	0.15	0.18	0.33	0.41	0.48
K_2O %	2.06	2.89	3.41	3.45	3.79
Na_2O %	0.40	0.46	0.46	0.67	0.55
TiO_2 %	0.66	0.90	0.79	0.70	0.53
P_2O_5 %	0.13	0.04	0.08	0.08	0.08
SO_3 %	0.36	0.13	0.20	0.14	0.13
Ignition loss %	34.40	2.75	11.25	7.27	2.54
Total %	100.14	99.64	100.38	99.93	99.68
N %	1.04	0.05	0.14	0.09	0.02
Clay ($<$ 5 μm) %	3.8	7.0	10.6	9.5	8.9
pH	3.8	3.7	3.9	4.1	4.5
sa	12.8	21.0	12.3	11.9	13.2
ba	0.64	0.72	0.59	0.58	0.60

[a] Collected by W. J. Latimer, analyzed by G. H. Hough and G. E. Edgington (37).

These saf trends for total soil are not peculiar to Podzols, for mere downward migration of clay will produce a similar chemical pattern. The coincidence induced earlier pedologists to designate many soil profiles as podzolic, exemplified by the former Red-Yellow Podzolic soils, now Ultisols (37).

Analysis of the extracted *clay fraction* proves more discriminating than chemical digestion of the entire horizon. In Figure 7.5 sa's of Spodosol fine fractions are plotted relative to the C horizon, that is, sa of clay of every horizon is divided by sa of clay in C. The latter's absolute value is recorded on the unity line. Much more Al than Si moves out of A, increasing its sa, and more of the sesquioxides stay in B, decreasing its ratio. Mere clay migration would not generate these depth functions.

Podzols are low in exchangeable bases and high in exchangeable aluminum. In the Colombian Sabaneta profile which has a pH range of 5.3 to 5.7 the exchange acidity occupies over 80% of CEC at pH 7, followed by Ca and Mg. The French values (33) exceed 90% in all horizons (pH 3.8-4.2).

Thermodynamically, kaolinite is expected to be synthesized in an acid, leaching environment. Instead, amorphous gels, poorly crystallized vermiculites, and some montmorillonites (wide sa values) register on the X-ray films, mostly in A2, less so in B. For New York State Spodosols, Coen and Arnold (18) perceive eolian influx of these smectite species, whereas Gjems (31) sees the Swedish instances as neogenic,

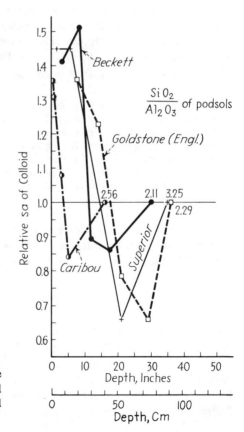

Fig. 7.5. Depth functions of relative SiO_2/Al_2O_3 (sa) of clays in Podzol profiles. The sa of C horizons is marked on the unity axis (sa= 1)

having been formed in 300 years. Marshall (44) points to immobilization of aluminium by humus as fostering in the soil solution an elevated silica $(Si(OH)_4)$ regime that favors montmorillonite synthesis.

Mechanisms

Podzolization is here defined as migration of sesquioxides from A to B horizons, regardless of whether or not a full-fledged Podzol or Spodosol materializes.

Mobilization. Reduction processes favor the solubilization of Fe and Mn but since Al^{3+} is not reducible, yet moves prominently, electron transfer is not deemed crucial. Most pedologists invoke organic matter as the sponsor of transportation. Older theories suggest colloidal sesquioxide particles being dispersed by humus fractions, as outlined in Chapter 6. Newer ideas rely on solubilization of Fe and Al as individual atoms attached to chelating hydroxy acids and polyphenols, as recounted in Chapter 5 on humus. Bloomfield (9) and others emphasize organic ligands of leaf extracts as mobilizers, rather than fulvic and humic acids. In view of the ubiquity and abundance of complexing organic molecules the geographic scarcity of Spodosols is unexpected.

Deposition. Why the mobile humus compounds of Fe and Al precipitate or flocculate in B rather than continue into nonsoil below has baffled pedologists.

In artificial trials percolation of a limited quantity of hydroxy acid through a column of quartz sand coated with iron oxide dissolves Fe in the top portions, leaving behind bleached sand as "A2," and precipitates Fe in the lower portion as a "B horizon"; however, Crawford (20) distrusts such testimonials.

> In Singer and Ugolini's (62) field experiment with porous, ceramic plates at 0.1 bar suction inserted in Podzol (Cryandept) horizons of volcanic ash, "soluble" Al (0.68-0.75 ppm) and Fe (0.03-0.04 ppm) transfer from 02 and from sandy clay loam A2 into sandy loam B, but barely transfer into the more clayey horizons beyond, whereas Si (3.4-4.4 ppm) leaves the profile. In the leachates from A2 scanning electron micrographs reveal fulvic acids of 1 μm size carrying minor amounts of Al and Fe. The B horizon effluent abounds in mineral particles, including silt-size siliceous spheres (66).

The organic Al, Fe complexes may break up as pH rises with soil depth, or when overloaded with Fe. Destruction of organic ligands in Bir by microbes could not be proved (22), the more so as polyphenols exert bactericidal action.

Podzol Vegetation

The soils harbor floras that are tolerant of soil acidity and low nutrient status. Legumes are scarce or absent. In Europe, the shallow-rooted shrubs *Calluna vulgaris* and *Erica carnea* reside on Iron Podzols and *Erica tetralix* favors the humus type. In North America the heathers are replaced by manzanita (*Arctostaphylos* sp.) subshrubs. On both continents pines, blueberries (*Vaccinium* sp.), and rhododendrons are common Podzol companions. The richer flora of North America adds hemlocks (*Tsuga* sp.) and cypresses (*Cupressus* sp.) to the list. On extreme sites all growth forms are dwarfed. The fertility-demanding redwoods (*Sequoia* sp.) "avoid" Spodosols.

Pines never crossed the equator naturally. South of the equator the needle-leaved podocarps (*Podocarpaceae*) encourage podzolization. In New Zealand the Kauri pine (*Agathis australis*) is famous as a Podzol indicator. Tropical lowland Podzols of the Americas maintain grass-shrub and savanna vegetation and very poor forest as edaphic climaxes with sharp soil and vegetation boundaries (19).

E. Laterization

In the red earth country of India's Malabar Coast, F. Buchanan, M.D., in 1800, observed Indians cutting slabs out of soft red and ivory mottled clay strata, air-drying them to a hard rock, and using them as building stones; this was photographed over a century later by Fox (27) and by Pendleton in Asia (52). Buchanan called brick and hardened soil mantles *laterite,* from the Latin word "later" meaning brick. Soil Taxonomy uses the synonym plinthite for the soft iron-rich clay strata from which bricks are still being made today.

The sketch of a classic laterite profile in Figure 7.6 extends to a depth of 30 m and appalls the pedologist of the temperate region. Agriculturally oriented soil taxonomists restrict "soil" to the 1 to 2-m-thick red loams and clays above the massive hardpan. Here, a broader and more genetic view is taken.

It should be emphasized that deep laterite profiles with crusts are end products of long-time soil genesis. They are restricted to high-level, ancient terraces and peneplains and do not necessarily typify tropical landscapes.

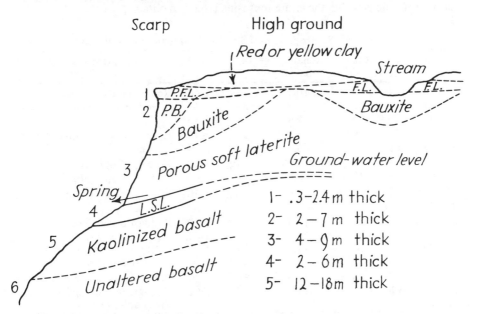

Fig. 7.6. Fox's (27) classical laterite "profile" from India. P.F.L., Pisolitic ferruginous material, often cemented to a hard crust (Panzer, cuirasse); F.L., ferruginous lateritic material; P.B., pisolitic bauxite; L.S.L., laminated siliceous lithomarge

A Subtropical Red Earth

On a 1.2% slope near Salisbury, Zimbabwe (Rhodesia) (maP 84 cm, maT 18.3°C), Maufe (45) and Ellis (25) analyzed a 6-m-deep red soil underlain by gray-green basic igneous rock (dolerite). The original open savannah is now in pasture.

Color, Texture, and Clay Content. Red and brown colors of iron oxides are accented by red-yellow mottling and by blackish Fe-Mn pellets (Table 7.4). Horizontal differentiation is feeble and no albic band is discernible.

Whereas Podzols are coarse-textured, the red earths acquire the attributes of finer loam and clay. In many tropical countries "Red Loams" are accorded high-level taxonomic status. The crumbs of iron-rich clay barely disperse in water and they render the red soils permeable and resistant to erosion.

Clay content in Table 7.6 increases from bottom to top, except for the top surface layer. The substantial percentages are minimal because the aggregates of silicate-sesquioxide clays are difficult to disperse. To check the clay content, Soil Taxonomy measures soil moisture at 15 bar suction and multiplies it by 2.5.

Soil Reaction and Exchangeable Bases. At depth of rock decay pH is neutral; above it pH is moderately acid. The exchangeable bases are scarce because the red clay materials have limited adsorption capacities (CEC). The surface soil holds a fair reserve of humus (0.16% N) which augments the meager base supply. The current belief is that in tropical forests the mineral nutrients are mostly in the living vert space, not in the soil, but quantitative ecosystem analysis inclusive of exchangeable and slowly available cations and anions within the entire root zone is hard to obtain.

Table 7.6. Subtropical Red Soil from Zimbabwe (Rhodesia)[a]

Depth (cm)	pH	Clay (%)	Exchangeable bases[b]	sa[c]	saf[d]	Description (excerpts)
0-15	6.1	48	10.9	2.06	1.46	Red and reddish brown friable
15-38	5.8	62	6.7	2.00	1.44	soil, many roots, no stains
76-107	6.0	53	5.4	1.99	1.45	
137-145	5.3	48	5.0	2.00	1.45	Hard, partly cemented, Fe, Mn
145-183	5.6	49	5.3	2.01	1.44	stains
213-244	5.6	47	5.6	2.05	1.47	Brown, loamy, friable, pebbly
244-274	5.8	47	6.1	–	–	concretions (Mn,Fe)
274-305	5.7	45	6.6	2.15	1.50	Red-yellow mottling
305-335	5.7	39	8.6	–	–	
335-366	6.4	32	11.7	2.34	1.59	Reddish brown clay
396-427	5.9	29	11.7	2.32	1.64	Light brown clay
427-457	6.2	16	11.9	2.60	1.79	Gritty, last stages of rock
457-488	6.4	10	11.3	3.35	2.26	decomposition
518-549	7.2	10	13.6	3.27	2.20	Brown-green decomposed rock
579-610	6.8	9	14.1	4.02	2.60	Gray-green decomposed rock
610+						Fresh dolerite rock

[a] After Ellis (25).
[b] Exchangeable bases, me/100 g of soil.
[c] sa of clay, SiO_2/Al_2O_3.
[d] saf of clay, SiO_2/R_2O_3.

Total Soil Fusion Analysis (Table 7.7). Dolerite rock is well supplied with Al, Ca, Mg, and Fe. During soil genesis mono- and divalent cations suffer severe losses in percentages and more so relative to Al, given by *ba*, which is moles of $(CaO + K_2O + Na_2O)$ divided by moles of Al_2O_3. The departure of ions signifies deep-seated weathering of the primary minerals, which is corroborated by the rise of ignition loss, made up of OH ions of clay gels and their crystals. These products retain Al and Fe without much change in their proportion, as judged by the trend of *al/fe* which is the molecular ratio Al_2O_3/Fe_2O_3.

Relative to the sequioxides, silica experiences a loss, documented by lowering of *sa* (SiO_2/Al_2O_3) and *saf* (SiO_2/R_2O_3). It is most noticeable in the zone of decomposing rock, a feature that is corroborated for other low-latitude sites (1, 13).

Clay Fusion Analysis. In the *clays* isolated (Table 7.6) the mole ratios *sa* and *saf* diminish systematically from the clay pockets between disintegrating rock pieces to the biologically active surface layer where the sa of 2 coincides with that of kaolinite.

Similar depth functions of clay desilication are obtained for many red soils of the Ultisol and Oxisol kind, as seen in Figure 7.7. Here, sa of clay in any horizon is divided by sa of the chosen C horizon. The latter's sa is recorded on the right-hand segment of the unity axis.

Clay Mineralogy. Crystal structure of clays is expected to parallel the depth trend of silica, as observed in Schellmann's serpentine soil on Borneo (Fig. 4.4). Toward the soil surface gibbsite, hydrous iron oxides (e.g., goethite), and kaolinite dominate the crys-

Table 7.7. Subtropical Red Soil from Zimbabwe (Rhodesia). Fusion Analysis of Total Soil[a]

Oxide	Fresh dolerite rock (%)	Decomposed rock[b] (%)	Red subsoil[c] (%)	Red surface soil[d] (%)
SiO_2	47.66	43.82	40.94	42.50
Al_2O_3	15.28	23.28	27.23	25.22
Fe_2O_3	1.02	16.65	17.92	16.91
FeO	8.01	–	–	–
TiO_2	0.67	1.41	1.63	1.87
MnO	0.14	0.20	0.18	0.27
MgO	9.62	2.94	0.50	0.93
CaO	13.74	2.70	0.23	0.18
K_2O	0.55	0.21	0.17	0.19
Na_2O	1.86	0.76	0.10	0.09
P_2O_5	0.12	0.06	0.05	0.03
Ignition loss[e]	1.34	8.43	11.89	11.98
Total	100.01	100.46	100.84	100.17
sa	5.29	3.20	2.55	2.86
saf	3.74	2.20	1.80	2.00
al/fe	2.42	2.19	2.38	2.33
ba	1.87	0.27	0.03	0.02

[a] From Maufe (45).
[b] Depth, 526-734 cm.
[c] Depth, 48-526 cm.
[d] Depth, 0-48 cm.
[e] Ignition loss

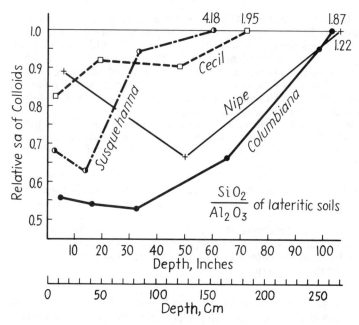

Fig. 7.7. Depth functions of relative SiO_2/Al_2O_3 (sa) of clays from laterized soils of Oxisols, Ultisols, and Alfisols (37). The sa of C horizons is marked on the unity axis
$(sa=1)$

tal suite. The silicate mineral may owe its persistence to continuous influx of $Si(OH)_4$ from dissolving plant opals.

Profiles with a Laterite Crust

Alexander and Cady's (1) profile shown in Figure 7.8 is located in a granite quarry near Samaru, Nigeria (maT 27°C, maP about 130 cm). The cut exposes a quartz vein that passes from bedrock to the soil surface and thereby establishes normalcy of horizon sequence.

Laterite Crust (0-80 cm Depth). The laterite crust is a hard, porous laterite stone or indurated plinthite, with conspicuous dark-red blackish, globular, pea-sized concretions or pisolites. Cavities and tubular channels, emptied of their clay fillings, are up to and over 1 cm in diameter and placed in random orientation. Coarse quartz grains are imbedded and fine goethite crystals dominate the clay mineral assembly, with hematite, gibbsite, and boehmite trailing. In thin sections hardened flow structures composed of colloidal particles are strongly double refractive, meaning that the submicroscopic crystals are well oriented.

Soft Laterite or Plinthite (80-160 cm Depth). Red (10 R 4/6, dry), yellowish-red, brown, and purple (10 R 5/4, dry) mottles merge and form the continuous color phase

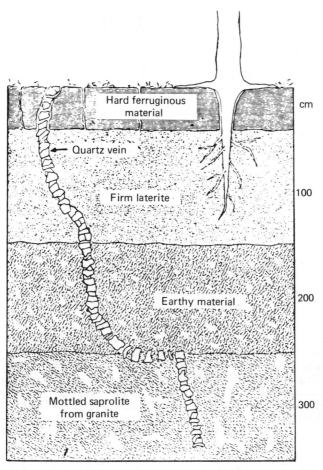

Fig. 7.8. Nigerian laterite profile to a depth of 3 m, described by Alexander and Cady (1) and d'Hoore (21). A quartz vein extends from bedrock to crust surface

with isolated whitish, grayish spots, dots, and smears of clays as dispersed phases. The showy, decorative design reminds French workers of gingerbread architecture. Crystalline white gibbsite, dark red (10 R 3/4, dry) iron concretions, and shiny quartz grains impart a slight hardness. The horizon is suitable for brick making. Plinthite genesis has been linked to gleysation (46).

Saprolite (160-300 cm Depth). Saprolite means decayed bedrock. The altered granite is soft, pale-red with dark and light mottles. The feldspars and nearly all of the micas have weathered, leaving their shapes intact as pseudomorphs that are filled with kaolinite "books" stained with iron oxide. Quartz remains largely intact.

Chemical Composition. The chemical alterations (Table 7.8) appear as a continuation or exaggeration of those in the subtropical Rhodesian profile. From rock to

Table 7.8. Chemical Composition of a Nigerian Laterite Profile[a]

Formula	Granite rock (%)	Saprolite (%)	Soft laterite (plinthite) (%)	Ironstone crust (%)
SiO_2	74.50	67.31	55.15	35.67
Al_2O_3	13.34	18.37	21.41	18.84
Fe_2O_3	2.73	6.27	13.70	33.94
TiO_2	0.19	0.89	1.11	1.39
CaO	0.76	v.s.[b]	v.s.	v.s.
MgO	0.20	v.s.	v.s.	v.s.
K_2O	5.58	v.s.	v.s.	v.s.
Na_2O	3.20	v.s.	v.s.	v.s.
Ignition loss	0.35	0.63	1.16	1.46
Total	100.85	99.89	100.29	99.85
sa	9.47	6.22	4.37	3.21
saf	8.38	5.11	3.10	1.50
al/fe	7.71	4.62	2.44	0.87

[a] After Alexander and Cady (1).
[b] v.s., Very small amounts.
[c] Ignition loss.

crust the *silica* content is halved, the remainder probably being resistant quartz. In Lacroix's (37) laterite crust derived from quartz-free diabase rock in Guinea, a reduction of SiO_2 from 51.27 to 1.30% has taken place.

Mobilized silica descends in the profile, locally resilicating $Al(OH)_3$ to kaolinite, impregnating clays (siliceous lithomarge in Fig. 7.6), crystallizing to centimeter-size neogenic quartz crystals (27), and precipitating as hard, head-sized chunks of microcrystalline chalcedony (35) that are still seen at fossil, Tertiary sites in central Europe. Laterite spring waters at Gimbi, Zaire (67), bear 1-24 ppm of SiO_2 at pH 4.8-6.8, < 1 ppm of ferrous iron, and no Al. In Australia the silica effluents cement gravels and sands by SiO_2 polymerization to "silcrete" rock (65) or duripans.

The *bases* Ca, Mg, K, and Na practically disappear, which is corroborated by Lacroix's Guinea site (37) at which relative bases (ba) drop from 1.90 in fresh rock, which is high, to a vanishing 0.0051.

The efflux of silica and bases entails a *residual accumulation of Fe, Al, and Ti* as hydrous oxides and oxides. As the soil ages, kaolinite of the granite saprolite desilicifies to gibbsite and boehmite. In red soils of Cuba (8) Fe_2O_3 climbs to 72% and in Hawaii TiO_2 rises to 21% (61). In Lacroix's aforementioned hardpan it is Al_2O_3 that experiences the highest accretion, rising from 12.36% in the rock to 60.19% at the surface. The layer is bauxitic laterite which contrasts with the more common ferruginous type of Nigeria. In Hawaii (56) sheets of white gibbsite clay with 65% Al_2O_3 develop from basalt under impeded drainage that allows exit of iron in reduced form. Both ferruginous and bauxitic soils are mined commercially for Fe and Al recovery.

Crust Formation. To create an iron stone Fe must accumulate and the horizon must harden. *Enrichment of Fe* is called relative (21) when Si and bases disappear. The proportions of Fe and Al remain practically unchanged, as in the Rhodesian profile. In the Nigerian case concentration of Fe from saprolite to crust overshadows the relative gains of Al and Ti, and al/fe in Table 7.8 narrows decisively. In all likelihood an absolute gain (21) of Fe by influx is operating. As in podzolization, one is confronted with the question: how does Fe move?

To join the surface crust Fe^{2+} might ascend by capillary rise from a seasonal water table located at bedrock or from the pedogenic one in Figure 7.6, though a vertical climb is limited to 2-3 m at best and the source of electrons to be donated to Fe^{3+} remains hidden. Lateral Fe flow within the crust horizon is questionable in view of the high level position of the site. Crusts many meters in thickness are linked by Waegemans (67) to slow tectonic uplifting of the landscape and concomitant lowering of water tables. For the buried type of crust, downward Fe-chelate migration has been invoked (42), but the present writer has not yet seen an albic or Bh horizon in red earth or laterite landscapes of India, Colombia, Kenya, Hawaii, or California.

Hardening. Hardening of ferruginous soil entails stripping kaolinite of its Si tetrahedra and is accompanied by heightened crystallinity of sesquioxides, as seen in color micrographs (1). As silica is being leached hematite spherules merge (55) and become knitted to a rigid framework that coats channels, pore walls, and nodule exteriors with a glossy, reddish-brown glaze.

Hardening to a 2-cm-thick, slaglike ply has been observed during 15 years of drying and wetting (1). Monsoon climates with wet-dry periods encourage subsoil induration at 1-2 m depth (42). In rain forest soils teeming with pea-sized iron concretions induration to rock sheets is believed to require the presence of a permanent water table (ground water laterites). In Hawaii (60) removal of vegetation and exposure of soil to sun oxidizes humus, shrinks peds, intensifies acidity from pH 4.6 to 3.8, and slashes CEC from 41 to an infertile 4 me/100 g. Amorphous iron gels are converted to hematite crystals and concretions become strongly magnetic. Soil bulk density mounts from an optimal 0.91 to 2.13 g/cm^3 of the hardened crust.

Vegetation of Laterized Soils

The slender, tall hardwood trees of the broadleaf *rain forests* receive copious quantities of precipitation, 200-1300 cm/year, and prosper on the deep, leached, red soils that have feebly developed texture horizons because Al- and Fe-humus are not dispersers. Worms are few and pH may be < 4. Understory is scant and ground game is rarely seen. At comparable sites annual litterfall can be four times higher than in the temperate region and rates of decay are fast (Chapter 5). The rapid turnover keeps the sparse nutrients circulating and permits the crowns to rise higher and higher.

Under lesser rainfall and in wet-dry climates savanna grasslands with scattered trees and open groves hold sway. When present in humid, laterized landscapes savannas are attributed to deforestation and recurrent burning by humans, and they are termed "fire climaxes." It seems logical to attribute laterite induration itself to desiccation induced by the activities of aborigines, but this anthropogenetic thesis cannot claim generality, for buried Tertiary crusts (35) antedate *Homo sapiens;* moreover, the Hawaiian Islands were populated in historical times, yet massive laterite crusts adorn Kokee State Park on Kauai.

An edaphic lateritic component of savanna origin is inferred by French (3), British (47, 48), and American (7) workers. The generalized drawing of savanna ecosystems by Morison et al. (47), shown in Figure 7.9, links grass, brush, and tree distribution in southern Sudan (maT 26°C, maP 110-140 cm) to soil depth above the iron crust.

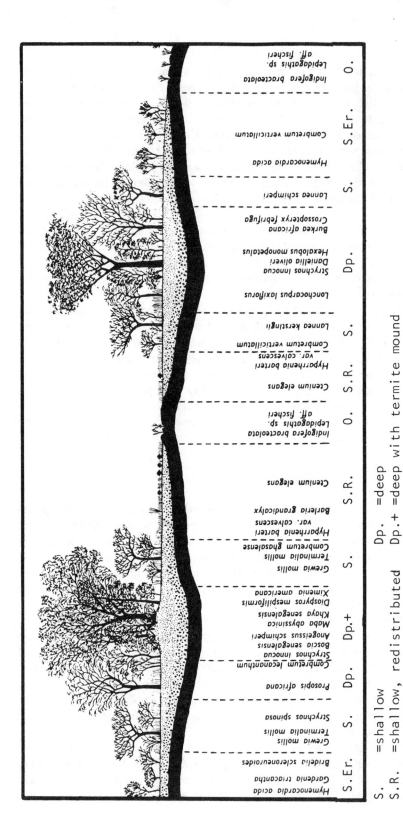

S. = shallow
S.R. = shallow, redistributed
S.Er. = shallow, eroded

Dp. = deep
Dp.+ = deep with termite mound
O = outcrop

Fig. 7.9. Savanna woodland, scrub, and short grass geared to depths of A horizon, pea-iron subsoil (dotted) and ironstone crust (black band). Sudan after Morison et al. (47). Trees reach 18 m in height, soil depth to hardpan < 40 cm. (Courtesy Blackwell Scientific Publications, Ltd., Oxford, England)

Where the inhospitable iron crusts are exposed vegetation is nil, except for a few grasses and shrubs in cracks and crevasses. These "bovals" (3), photographed by Pendleton (52) and Maignien (42), extend from a fraction of a hectare to many square kilometers.

The Californian fossil laterite crusts near Ione are very acid and harbor low-shrub manzanita vegetation, with pure stands of bronze-leaved, endemic *Arctostaphylos myrtifolia* on shallow phases. On plinthite slopes the rare, delicate wild buckwheat *Eriogonum apricum* survives.

The Meaning of Laterization

In summary, it would seem presumptuous to try to gather all the red soils of the world under a single, conceptual umbrella. Hence, in a humid tropical climate a tessera on a level plateau or gentle, upper slope is visualized. Down-slope and foot-slope positions with significant lateral influxes of materials, including Fe^{2+} solutions (42), are best assigned to the toposequences of Chapter 11. The parent material is to be an igneous or sedimentary rock well supplied with feldspars, micas, and ferromagnesian minerals, and soil formation time is long, up to a million years and more.

Under these constraints silica and total bases are leached severely, which is desilication and debasing, and inherited aluminum and iron accrue as kaolinite and as hydrated amorphous and crystalline hydrous oxides and oxides. Reddish colors are conspicuous, clay content is high, and reaction is acid. The clay fraction has low values of sa and saf in upper horizons and higher ones at greater depth. These criteria characterize *laterization* and the soils are laterized. Humus impregnation, clay migration, horizon induration, gleysation, and even podzolization may be superimposed. The process is imprinted on many members of Oxisols, Ultisols, and Inceptisols and on Red Earths and ferralitic soils of other classifications. The "laterite profile," as used here, has a laterite crust either at the surface or below it.

The custom of assigning the red, laterized soil enclaves of the temperate regions to *previous* tropical climates and calling the bodies *paleosols*, relicts, and fossil soils is still popular. It is not obvious, however, why year-round warmness should favor red soils. Parent materials rich in sesquioxides—the basic igneous rocks—and longtime weathering and soil genesis seem chemically more propitious. Importantly, soil depth and profile permanence are regulated by rate of erosion at the soil surface and by rate of rock weathering at the base. Since temperature elevation accelerates chemical reaction rates but not necessarily the physical rate of erosion, owing to enhanced evapotranspiration with less runoff, one anticipates soil deepening and profile stability from pole to equator, other factors being equal and exploitative man absent. From this vantage point a measure of laterization is to be expected in the humid temperate region provided land surfaces are old and parent materials are suitable. The red oxidic and ferritic soils on ultramafic rocks along the Pacific Coast (e.g., Red Mountain near Leggett, California, and Nickel Mountain, Oregon) endorse the thesis. Clayey Cornutt soil on olivin-rich peridotite rock carries in the red surface soil 76% $Fe_2O_3 \cdot H_2O$ as goethite with some hematite, and 97% in the yellowish subsoil, according to Barshad's unpublished analyses.

F. The Pedocal Process

Many Aridisols are endowed with "lime horizons" that comprise gravelly, nodular concretions and mycelium-like threads and webs of Ca- and Mg-carbonates (Fig. 7.10). In ordinary determinations acid (HCl) is added to soil and the escaping "fizz" is analyzed for CO_2. The result is expressed as "calcium carbonate equivalent," abbreviated $CaCO_3$. In soils $CaCO_3$ crystallizes most commonly as calcite (23).

Earlier Accounts

In Hilgard's (1892) opinion the "sheets" of lime-cemented subsoil result from downward movement of $CaCO_3$ dissolved in penetrating rainwater which reascends—without the carbonate—in evaporation and transpiration. The salt gradually accumulates in the subsoil, "chiefly at the lowest point usually reached by moisture." Marbut (43) coined the word *pedocal* for soils having in their profile a layer of carbonates that contains more $CaCO_3$ and $MgCO_3$ than the horizons above or below. In the depth function in Figure 7.10 (left) the depleted area to the left and the bulge to the right of the

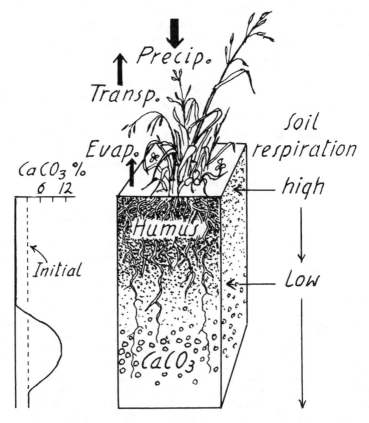

Fig. 7.10. A pedocal with an horizon of carbonate concretions. On the left is the corresponding $CaCO_3$ depth function

dashed vertical line, which represents $CaCO_3$ of the parent material, are about equal. Rocks free of $CaCO_3$ but high in calcium (e.g., basalt) also may develop pedocals. Influx of $CaCO_3$ dust brought by wind accentuates the horizon.

Microscopically, the soil fabric is dominated by carbonates that coat and engulf pebbles, sand, and silt grains as an essentially continuous medium (30). When cemented with silica the carbonate horizon is called petrocalcic.

Ca-Bicarbonate Equilibria

To evaluate Hilgard's picture and put it on a quantitative basis, solubilities of $CaCO_3$ in the presence of carbon dioxide (CO_2) must be known (Fig. 7.11). In the equations that follow the constants are taken from Langmuir (41).

Carbonic Acid in Water. Carbon dioxide gas (CO_2) dissolves in water proportionally to its content in the air in which it is recorded as "partial pressure" P in atmospheres. The proportionality factor α is 0.0339 moles of CO_2 in a liter of water at $25°C$ and at one atmosphere of pressure. At lower temperatures α more than doubles. Dissolved CO_2 reacts with water molecules to form carbonic acid (H_2CO_3) which dissociates into H^+ ions and bicarbonate (HCO_3^-) anions, having the dissociation constant $k_1 = [H^+][HCO_3^-]/\alpha P$ which is 4.34×10^{-7} at $25°C$. Bicarbonate dissociates further, $HCO_3^- \rightleftharpoons H^+ + CO_3^{2-}$. Its dissociation constant is $k_2 = [H^+][CO_3^{2-}]/[HCO_3^-]$ and has the value of 4.67×10^{-11}. The brackets denote activities. The pH of water acidulated with CO_2 is calculated as

$$pH = 3.916 - \frac{1}{2} \log P \qquad (7.1)$$

where the constant equals $- 1/2 (\log \alpha + \log k_1)$. Values for selected CO_2 pressures in ecosystems are given in Table 7.9. At 1 atmosphere, pH is 3.92.

The $CaCO_3$-H_2O-CO_2 Equilibria. Solid $CaCO_3$ added to a liter of distilled water solubilizes to the extent of a few milligrams and most of the molecules dissociate into Ca^{2+} and CO_3^{2-}. The two enter the solubility product as $k_s = [Ca^{2+}][CO_3^{2-}]$ which is 3.98×10^{-9} at $25°C$. The anion reacts immediately with a water molecule to become a bicarbonate anion:

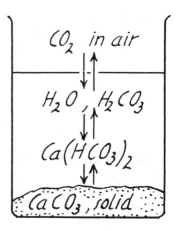

Fig. 7.11. Sketch of some of the reaction products of CO_2 and $CaCO_3$

Table 7.9. Equilibria of $CaCO_3$ and pH at Various CO_2 Pressures of Air at 25°C

CO₂ content of air		pH[a] of water plus CO₂	CaCO₃ dissolved (mg/liter H₂O) Kline (29)	pH of Ca-bicarbonate solution[b]	Environmental conditions
Volume (%)	Atmosphere P				
0.031[c]	0.00031	5.66	52	8.35	Average CO₂ content of air
0.334	0.00334	5.15	117	7.68	Average CO₂ content of soil air
1.60	0.0160	4.81	201	7.24	
11.16	0.1116	4.39	403	6.70	High CO₂ content of soil air
100	1.000	3.91	900	6.10	Water saturated with CO₂ at 1 atm pressure

[a] From Eq. (7.1).
[b] From Eq. (7.2) and column 4 after conversion to activities and using Kline's (29) ionic strengths.
[c] Or 310 ppm.

$$CO_3^{2-} \; + \; H^+O\bar{H} \; \rightleftharpoons \; HCO_3^- \; + \; O\bar{H}$$

carbonate H_2O bicarbonate
anion   anion

The hydroxyl ion imparts an alkaline reaction of about pH 10. Introduction of CO_2 from the air lowers the alkalinity and additional $CaCO_3$ dissolves to become calcium bicarbonate, $Ca(HCO_3)_2$, which dissociates into Ca^{2+} and 2 HCO_3^- ions.

A combination of the equilibrium constants k_1, k_2, k_s and the dissociation constant of water ($k_w = 1.012 \times 10^{-14}$) gives for the dependency of pH and dissolved Ca on CO_2 pressure P

$$\frac{k_s}{k_1 \, k_2 \, \alpha} = \frac{[Ca^{2+}]}{[H^+]^2} \, P$$

This equation or parts of it have been derived by many investigators. Raising the temperature has a negative effect, as shown in Figure 7.12 in which the solubility of $CaCO_3$ at temperature t is compared to that at 25°C, using for the ratio r the equation: $\log r = [830/(273° + t)] - 2.78$ from Frear and Johnston (29).

Convenient is the logarithmic form

$$pH + \frac{1}{2} \log [Ca^{2+}] = 4.88 - \frac{1}{2} \log P \tag{7.2}$$

The figure 4.88 is half the logarithm of the ratio of the constants above and the brackets denote Ca activities.

Interpretations of Solubilities. As seen in Table 7.9, elevating CO_2 pressures of air renders $CaCO_3$ more soluble in the form of calcium bicarbonate ($Ca(HCO_3)_2$). In CO_2-saturated soil microregions the solution may reach slightly acid reaction and Ca-clay in the presence of solid $CaCO_3$ might actually be converted to a partially acid clay.

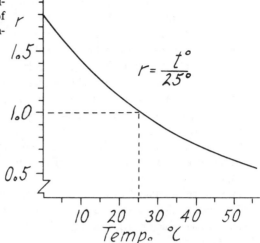

Fig. 7.12. Solubility of calcium bicarbonate rises with a lowering of temperature. After Frear and Johnston (29)

$$r = \frac{t°}{25°}$$

In the presence of Mg^{2+} and other ions of the soil solution the precipitation of $CaCO_3$ may produce crystals other than calcite (23) that may result in slightly higher solubilities. In calcareous soil samples reaction measured by the paste method may appear texture-dependent, declining from pH 7.8-8.4 in sands to pH 7.0-7.3 in clays.

Model of CaCO₃ Translocation

The solubility of $CaCO_3$ is marked in the Al horizon because of high, biological CO_2 production. When being displaced downward the salt precipitates because of low bioactivity and withdrawal of water by transpiration. Arkley (2) predicts the depth of lime horizon by combining precipitation with evapotranspiration. In the absence of surface runoff a given quantity of rainfall penetrates into a *bare, air-dry soil* until the wetted portion reaches field capacity (FC). In the presence of a *vegetative cover* the roots remove water preferentially to Ca and reduce soil moisture from field capacity to permanent wilting percentage (PWP), the difference being the available water capacity (AWC). When new rains rewet the dried soil it fills up to FC, but unless rainfall exceeds actual evapotranspiration there is no deep water penetration.

Using Thornthwaite's procedure of estimating potential evapotranspiration (Pet), which seems valid for the Great Plains area, Arkley computes the difference $P-Pet$ for every month, as explained in Chapter 2. Positive values ($P>Pet$) denote surplus water available for leaching and the annual sum is the *leaching factor Li* (Table 2.1).

How far into the soil will Li penetrate? Arkley relates the depth to the amount of available water that the horizons can hold, which depends on textures. He scales them in terms of AWC. Instead of saying that a Keith silt loam in Nebraska has a depth of 165 cm, he computes a new depth D' as 19.8 cm^3 of available water in a soil tessera of 165 cm length and 1 cm^2 cross section, or, simply, as 19.8 cm of D' depth. As surplus water flows through a soil of dryness PWP the amount of available water withdrawn is 1 cm^3 for each centimeter of D' depth traversed.

The *water penetration curve* for Keith silt loam shows the moisture depth D' in centimeters on the left-hand vertical axis of Figure 7.13 and the number of cubic centimeters of Li-water that pass in an average year through 1 cm^2 cross section at any depth D' on the horizontal axis of Figure 7.13. Whereas the mean annual precipitation at the site is 42 cm, the mean of Li is 12 cm, which also is the mean depth of penetration on the D'-axis. In a 62-year record the wettest year produces Li equal to 27 cm and in that year water descends to the corresponding D' depth. The curve is constructed from the frequency distribution of annual Li values.

Also shown in Figure 7.13 is the upper plane of the *CaCO$_3$ horizon* at D = 58 cm or D' = 9.3 cm. As read off from the penetration curve, water passes across this boundary at an average annual rate of 3.3 cm^3/cm^2. Since the measured soil pH at this upper carbonate plane is 8.0, the amount of CaCO$_3$ dissolved in water at D = 58 cm may be calculated from the aforementioned solution equilibria. For the Keith profile Arkley selects a solubility of 100 mg CaCO$_3$/liter for a temperature of 13°C (mean annual), which provides an annual flux of 0.33 mg/cm^2 of CaCO$_3$ into the zone of lime horizon.

Sampling of the Keith profile ends at 165 cm of absolute D depth which is equal to 19.8 cm of D' depth. Only 0.26 cm of water per year penetrates beyond this basal plane. Whatever soluble and dispersed materials the water acquires in the surface horizons are unloaded farther down as the moist front comes to a standstill and water volume shrinks as a result of transpiration. Very little CaCO$_3$ passes the basal plane into nonsoil below.

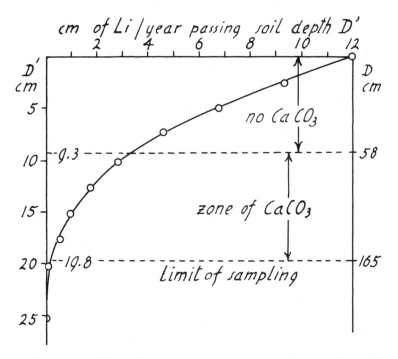

Fig. 7.13. Model of generating the carbonate horizon of the Keith profile in Nebraska. After Arkley (2)

For the Keith profile Arkley estimates the age of its carbonate horizon as follows: on the basis of 1 cm^2 area, there are 8.67 g CaCO$_3$ in the depth interval 58 to 165 cm. Assuming that this quantity was uniformly distributed at the outset, 3.24 g CaCO$_3$ were transported across 1 cm^2 at D = 58 cm. Division by the average annual influx of 0.33 mg/cm^2 gives the age of 9800 years. In this manner Arkley treats six additional profiles of late Wisconsin age and computes the plausible ages of 7200-14,600 years. Six soils from New Mexico of possibly late and mid-Pleistocene age give values of half to a million years.

Objections may be raised, such as the omission of Ca-cycling in vegetation, but, as Arkley points out, most of them tend to characterize these ages as minimum ages.

G. Stone Lines and Stone Pavements

In soil pits and along road cuts stones up to fist-size and larger may be seen lined-up in single file as more or less horizontal horizons. In three dimensions the stone lines coalesce to sheets and pavements.

In the opinion of Ruhe (e.g., 54), buried stone lines bear witness to *erosion cycles*. When pebbly soil is subjected to sheet erosion, fine earth is washed away and the coarse and heavy fragments remain, forming a pavement. In a subsequent period of deposition of soil material eroding from higher elevations the sheet of stones is buried.

A *biological origin* of fine earth resting on stone earth is advocated in the writings of Darwin who was impressed by the widespread and endless upward transport of silt and clay by earthworms. Carroll (17), intrigued by the castle-building termites in Africa (Chapter 13), credits the insects with creating subsoil stone lines all over the landscape.

The novel theory of *upward movement of stones* by Springer (63) originated in the Nevada desert where soil profiles on old mesas (table landscapes) are covered with stone pavements or "desert pavement," commonly explained as remnants of wind ablation of stony soil (Fig. 7.14). As Springer noticed, the A horizon below the desert pavement is stone-deficient, even pebble-free, compared to the rest of the solum and the parent material.

Springer reasons that a rain storm wets the sun-parched desert soil through the cracks in A that lead to the clayey, montmorillonitic B. The latter swells and pushes stones upward. When the soil dries out the large stones cannot fall or be washed into the desiccation fissures below but fine material may. The process is repeated during the next shower, months or years hence. Springer supports his idea with laboratory experiments in which 22 cycles of wetting and drying move pebbles upward, 2.0-2.2 cm closer to the surface. In a desert it takes a long time to establish a sheet of stones because a clayey B horizon must first be created through weathering and clay translocation; hence, desert pavements are features of old land surfaces.

H. Depth Functions of Organic Matter

In most well-drained soils organic carbon and nitrogen decline rapidly from surface to C horizons and to strata below.

Conventionally, the fine earth of a horizon is analyzed for organic nitrogen and carbon and the percentages found are plotted as a depth function at the midpoints of the

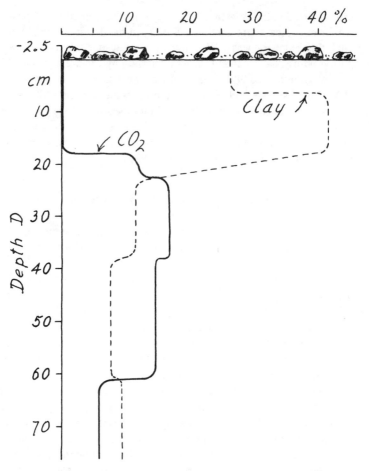

Fig. 7.14. Desert pavement of stones, and clay and carbonate (CO_2) depth functions of a high-level table landscape near Reno, Nevada. After Springer (63)

horizon intervals (Fig. 7.15, shown as open circles). When bulk densities of the soil are known the organic quantities are expressible on an area basis, e.g., kg C/m^2/cm, shown as rectangles in Figure 7.15. The area under each rectangle denotes the C content of a horizon as kg C/m^2/horizon. Summation, C_s, of rectangles generates the integral curve shown as solid line with half-filled circles.

The curves do not include forest floor and coarse roots, which may be substantial (Fig. 1.5). In soil survey work the bulk density of fine earth is measured with clods brought to the laboratory, but the conversion to bulk densities of gravelly and stony soils relies on field "estimates" of bulk densities of stone assemblies (64).

Zinke (70) plots logarithms of carbon-sum, C_s, against logarithms of depth D and often obtains near-straight lines to about 1 m depth. The equation signifies an empirical potential function, $C_s = aD^k$. Many of the k values are in the neighborhood of 0.5.

The equation may be used to couple C in the surface soil (0-20 cm), written C_{20}, with C to 1 m depth (C_{100}) by the proportionality $C_{20}/C_{100}=1/5^k$. Hence for $k=1/2$, the ratio is $1/\sqrt{5}$, or 0.447 or 44.7%, i.e., the surface soil contains almost half the carbon of a 1 m deep soil tessera, regardless of quantity of C storage. The term $1/5^k$ possesses merit for estimating subsoil organic matter from carbon in the surface soil.

Origin of C, N Depth Functions

In the Central States the prevailing native prairie grass big bluestem (*Andropogon gerardi*) used to reach 3-4 m in height and even today weighs 0.7 kg/m², air-dry, and gives rise to a mulch of 0.4 kg/m² (36). In the 1930s Weaver et al. (37) detected a linear parallelism of root material and soil organic matter with depth in Nebraskan grasslands. Figure 7.15 (inset) traces a similar correspondence of oven-dry root mass with humus content (C $\cdot$ 1.724) in a Tama prairie soil in Illinois (24). Humus in a B or C horizon reflects root growth therein.

Forest soils are covered by substantial forest floors with 01 and 02 horizons. Pedologists regard them as the source of humus that migrates into A, B, C horizons. Recent awareness of seasonal dieback and regeneration of fine roots in forest soils may necessitate a broadening of the abiotic transport scheme and include rootlet contributions.

Humus in Accruement Profiles

Since prehistoric times the river Nile has been depositing fertile silts and clays many meters in depth in the Egyptian lowlands. The mean accruement rate has been estimated as 1 mm a year, occurring during the September flood. Each increment is rich in humus that is acquired in the tropical highlands. The N depth function of the delta soils, however, is not the expected vertical straight line but an exponentially descending curve that is characteristic of many grassland soils and forested Entisols and Inceptisols (Fig. 7.16).

As a model (38) consider the imaginary, horizontal XY plane of Figure 7.16. At high flood stage the river deposits a mud layer of 1 mm thickness containing the total nitrogen content N_0 that fluctuates from 0.12 to 0.17%. During the ensuing year the lamella gains A'% of actual soil nitrogen by biological fixation and loses a certain percentage ($k_0' N_0'$) by microbiological decomposition. The total change in a year, $\Delta N/\Delta t$, is $A'-k_0' N_0'$, as outlined in Chapter 5. At the next flood stage a fresh layer is added, assumedly identical to N_0. In a year, it too becomes N_1. Meanwhile, the buried sheet N_1 continues to lose nitrogen, though at a lesser rate, and becomes N_2, as seen in Figure 7.16. As the alluvium keeps building up, a sequence of lamellas accrues with nitrogen contents $N_0 > N_1 > N_2 > N_3 > N_4 \dots$. It is easy to visualize a rising humus color profile, its newest addition having the darkest tint and the lower horizons assuming lighter and lighter shades as their ages increase.

The burials lower systematically A' and k_0' of each lamella and the two coefficients become variables. In differential notation:

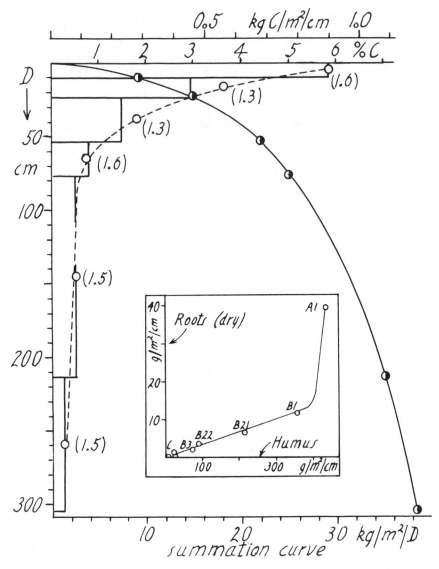

Fig. 7.15. Depth functions of organic carbon of a savanna-type Aiken loam. Dashed line: percentage C placed as open circles at midpoints of horizons (bulk densities in parentheses). Rectangles: carbon on area basis, as kg/m²/cm. Solid curve with half-filled circles: summation curve of C in horizons. Inset: proportionality of humus (C · 1.724) and root weights in Tama silt loam (24)

$$\frac{dN}{dt} = Ae^{-k_1 t} - k_0 e^{-k_2 t} N$$

change   gain   loss

Skipping derivations (38), the theoretical decline of N with depth (D) reads:

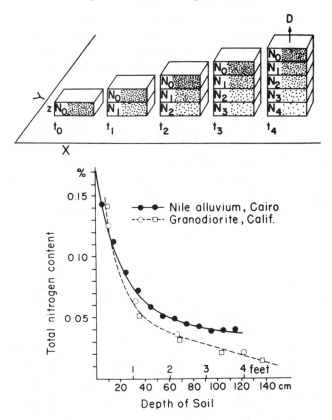

Fig. 7.16. Accruement N profile. Upper graph: model of uniform depositions of high-nitrogen Nile sediment. Lower graph: calculated N depth function with observed values (solid points). (Soil Science Soc. Am., Madison, WI)

$$\ln \frac{N}{N_0} = \frac{b}{k} (e^{-kD} - 1)$$

with b representing k_0 and $k = k_2$.

A good fit (Fig. 7.16) is obtained by setting $N_0 = 0.172\%$. At great depth N becomes constant, 0.033%. Each 30 cm of mud-overlay reduces the decomposition coefficient to one-half. The mean N fixation from air in 1500 years is only 0.0739 g N/m^2/year because annual rainfall is less than 2 cm. The losses average 1.51 g N/m^2/year from the reservoir of 3354 g/m^2 taken to a depth of 1.5 m. This yearly quantity must have been available as well to the ancient Egyptians for the growing of crops and would have supplied the protein needs of 4.3 persons/ha/year which is a substantial population intensity. It is supported by historical search of such egyptologists as Klaus Baer.

To generalize, in upland profiles derived from igneous rock the youngest horizon is at the weathering base adjacent to the rock, the oldest at the surface. The Nile profile and for that matter all young alluvial and loessial deposits are inverted, the oldest "horizon" being at the bottom, the youngest on top. Upland nitrogen profiles have their origin in positive accruement of N from the atmosphere, with gains exceeding losses, at least in the build-up stages. The Nile N-profile is caused by a preponderance of depletion of a N-rich accruement sediment, yet the two N profiles—genetic opposites—look very much alike.

On uplands, Nile-type nitrogen and carbon additions take place at foot slopes of toposequences from humus gently washed in from higher elevations. Thick organic epipedons result, as described by Riecken and Poetsch (53) for India and Iowa.

In Alpine hummocky landscapes (mima mounds) on dolomitic till the troughs are rich in organic matter that is continually being washed in from the ridges of the mounds, according to Gračanin (32). Nitrogen analyses and age determinations by the ^{14}C method again place lowest N content and highest humus age at greatest depth (Fig. 7.17). Ten C/N ratios vary from 10.5 to 12.6, regardless of depth.

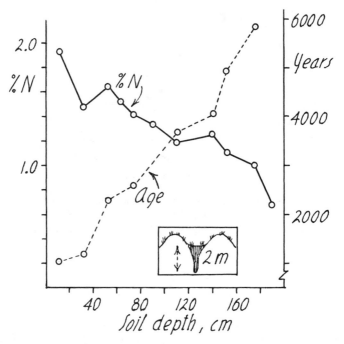

Fig. 7.17. Depth functions of nitrogen content and age of humus of an accruement profile in a trough (shown in inset) of a hummocky alpine landscape (32)

I. Review of Chapter

Seven prominent events of soil genesis are pursued with the aid of reaction mechanisms put together in previous chapters.

Textural B horizons (Bt) are explained as clay genesis in place, enhanced by *migration of colloidal clay particles* that begins with clay dispersion in A and ends with flocculation in B.

Alternating wetness and dryness initiate *gleyzation*. During the wet reduction period Fe and Mn are mobile, during the dry oxidation span they are fixed as brownish mottles surrounded by blanched domains.

In *podzolization,* insoluble iron and aluminum oxides of the surface soil are complexing with organic acids and humus molecules and Fe and Al descend, leaving behind an albic A2. Below, the chemical association is disengaged and the sesquioxides accumulate as spodic B horizon, often in the form of a Bmir hardpan.

In warm, humid climates prolonged soil formation suffers intense depletion of silica and bases, long known as *laterization*. In the residue iron and aluminum become enriched as hydrous oxides, often in brilliant reddish colors. Massive, stone-hard crusts may develop.

Throughout the arid region the *pedocal process* deposits nodules of $CaCO_3$ and $MgCO_3$ in "lime horizons," provided the parent material carries sufficient quantities of Ca and Mg.

Stone lines and stone pavements originate in erosion cycles, with faunal transport of fine material, and by upward migration of pebbles.

Organic matter profiles exhibit exponentially declining humus depth functions that provide large storages for N and C. A model of accruement of N in an alluvial plain is derived.

References

1. Alexander, L. T., and J. G. Cady. 1962. *Genesis and Hardening of Laterite in Soils.* U.S.D.A. Techn. Bull. 1282.
2. Arkley, R. J. 1963. *Soil Sci.* 96: 239-248.
3. Aubréville, A. 1947. *Agron. Trop.* 7-8: 339-357.
4. Ballagh, T. M., and E. C. A. Runge. 1970. *Proc. Soil Sci. Soc. Am. Proc.* 34: 534-536.
5. Barshad, I. 1964. In *Chemistry of the Soil,* F. E. Bear, ed., pp. 1-70. Reinhold, New York.
6. Barshad, I., and L. A. Rojas-Cruz. 1950. *Soil Sci.* 70: 221-236.
7. Beard, J. S. 1953. *Ecol. Monogr.* 23: 149-215.
8. Bennett, H. H., and R. V. Allison. 1928. *The Soils of Cuba.* Trop. Pl. Res. Found., Washington, D.C.
9. Bloomfield, C. 1957. *J. Sci. Food Agr.* 8: 389-392.
10. Bloomfield, C. 1965. In *Experimental Pedology,* E. G. Hallsworth and D. V. Crawford, eds., pp. 257-266. Butterworths, London.
11. Blume, H. P., and E. Schlichting. 1965. In *Experimental Pedology,* E. G. Hallsworth and D. V. Crawford, eds., pp. 340-353. Butterworths, London.
12. Bodman, G. B., and F. F. Harradine. 1938. *Soil Sci. Soc. Am. Proc.* 3: 44-51.
13. Bonifas, M. 1959. *Contribution à l'étude géochimique de l'altération latéritique.* Mém. Carte Géol. d'Alsace-Lorraine 17, Strasbourg.
14. Borchardt, G. A., F. D. Hole, and M. L. Jackson. 1968. *Soil Sci. Soc. Am. Proc.* 32: 399-403.
15. Brinkman, R. 1970. *Geoderma* 3: 199-206.
16. Brown, I. C., and J. Thorp. 1942. *Morphology and Composition of Some Soils of the Miami Family and Miami Catena.* U.S.D.A. Techn. Bull. 834.
17. Carroll, P. H. 1969. *Soil Surv. Horizons* 10: 3-16.
18. Coen, G. M., and R. W. Arnold. 1972. *Soil Sci. Soc. Am. Proc.* 36: 342-350.
19. Cohen, A., and J. J. van der Eijk. 1953. *Geol. Mijnbow, n.w.s.* 15: 202-214.

20. Crawford, D. V. 1965. In *Experimental Pedology,* E. G. Hallsworth and D. V. Crawford, eds., pp. 267-279. Butterworths, London.
21. D'Hoore, J. 1954. *L'accumulation des sesquioxydes libres dans les sols tropicaux.* Inst. Nat. Et. Agron. Sér. Sci. 62.
22. Dommergues, Y., and P. H. Duchaufour. 1965. *Sci. Sol* 1: 43-59.
23. Doner, H. E., and P. F. Pratt. 1969. *Soil Sci. Soc. Am. Proc.* 33: 690-693.
24. Douglas, C. L., J. B. Fehrenbacher, and B. W. Ray. 1967. *Soil Sci. Soc. Am. Proc.* 31: 795-800.
25. Ellis, B. S. 1952. *J. Soil Sci.* 3: 52-63.
26. Flach, K. W., J. G. Cady, and W. D. Nettleton. 1968. *Trans. Int. Congr. Soil Sci. 9th, Adelaide* 4: 343-351.
27. Fox C. S. 1923. *Mem. Geol. Surv. India* 49: 1-287.
28. Franzmeier, D. P., and E. P. Whiteside. 1963. *Mich. Agr. Exp. Sta. Quart. Bull.* 46: 1-57.
29. Frear, G. L., and J. Johnston. 1929. *J. Am. Chem. Soc.* 51: 2082-2093.
30. Gile, L. H., F. F. Peterson, and R. B. Grossman. 1965. *Soil Sci.* 99: 74-82.
31. Gjems, D. 1960. *Clay Min. Bull.* 4: 208-211.
32. Gracanin, Z. 1971. In *Paleopedology,* D. H. Yaalon, ed., pp. 117-127. Israel Univ. Press, Jerusalem.
33. Guillet, B., J. Rouiller, and B. Souchier. 1975. *Geoderma* 14: 223-245.
34. Hallsworth, E. G. 1963. *J. Soil Sci.* 14: 360-371.
35. Harrassowitz, H. 1926. *Laterit. Gebr. Bornträger,* Berlin.
36. Hole, F. D., and G. A. Nielsen. 1968. Soil genesis under prairie. In *Prairie and Prairie Restoration,* P. Schramm, ed., pp. 28-34. Spec. Publ. 3, Knox College, Galesburg, IL.
37. Jenny, H. 1941. *Factors of Soil Formation.* McGraw-Hill, New York.
38. Jenny, H. 1962. *Soil Sci. Soc. Am. Proc.* 26: 588-591.
39. Kha, N., and P. H. Duchaufour. 1969. *Sci. Sol* 97-110.
40. Klinge, H. 1967. *Limnologia* 3: 117-125.
41. Langmuir, D. 1968. *Geochim. Cosmochim. Acta* 32: 835-851.
42. Maignien, R. 1966. *Review of Research on Laterites.* UNESCO, Nat. Res.
43. Marbut, C. F. 1928. *Proc. 1st Int. Congr. Soil Sci.* 4: 1-31.
44. Marshall, C. E. 1977. *The Physical Chemistry and Mineralogy of Soils.* Wiley (Interscience), New York.
45. Maufe, H. B. 1928. *S. Afr. J. Sci.* 25: 156-167.
46. Mohr. E. C. J. 1972. *Tropical Soils,* 3rd ed. The Hague.
47. Morison, C. G. T., A. C. Hoyle, and J. F. Hope-Simpson. 1948. *J. Ecol.* 36: 1-84.
48. Moss, R. P. (ed.). 1968. *The Soil Resources of Tropical Africa.* Cambridge Univ. Press, Cambridge.
49. Müller, P. E. 1887. *Studien über die natürlichen Humusformen.* Springer, Berlin.
50. Parsons, R. B., and C. A. Balster. 1967. *Soil Sci. Soc. Am. Proc.* 31: 225-258.
51. Passey, H. B., and V. K. Hugie. 1963. *J. Range Mgmt.* 16: 113-118.
52. Pendleton, R. L. 1941. *Geogr. Rev.* 31: 177-202.
53. Riecken, F. F., and E. Poetsch. 1960. *Iowa Acad. Sci.* 67: 268-276.
54. Ruhe, R. V., R. B. Daniels, and J. G. Cady. 1967. *Landscape Evolution and Soil Formation in Southwestern Iowa.* U.S.D.A., S.C.S., Techn. Bull. 1349.
55. Schmidt-Lorenz, R. 1964. In *Soil Micromorphology,* A. Jungerius, ed., pp. 279-289. Elsevier, Amsterdam.
56. Sherman, G. D. 1958. *Gibbsite-Rich Soils of the Hawaiian Islands.* Hawaii Agr. Exp. Sta. Bull. 116.
57. Sherman, G. D., and H. Ikawa. 1959. *Pacif. Sci.* 13: 291-294.
58. Sherman, G. D., and Y. Kanehiro. 1954. *Soil Sci.* 77: 1-8.
59. Sherman, G. D., A. K. S. Tom, and C. K. Fujimoto. 1949. *Pacif. Sci.* 3: 120-123.
60. Sherman, G. D., Y. Kanehiro, and Y. Matsusaka. 1953. *Pacif. Sci.* 7: 438-446.
61. Sherman, G. D., J. Fujioka, and G. Fujimoto. 1955. *Pacif. Sci.* 9: 49-55.

62. Singer, M., and F. C. Ugolini. 1974. *Can. J. Soil Sci.* 54: 475-489.
63. Springer, M. E. 1958. *Soil Sci. Soc. Am. Proc.* 22: 63-66.
64. Staff, Soil Survey. 1972. *Soil Survey Laboratory Methods.* U.S.D.A., S.C.S., S. S. Invest. Rep. 1.
65. Stephens, C. G. 1971. *Geoderma* 5: 5-52.
66. Ugolini, F. C., H. Dawson, and J. Zachara. 1977. *Science* 198: 603-605.
67. Waegemans, G. 1954. Inst. Nat. Ét. Agron. Congo Belge, Sér. Sci. 60.
68. Winters, E. 1938. *Soil Sci.* 46: 33-40.
69. Wurman, E., E. P. Whiteside, and M. M. Mortland. 1959. *Soil Sci. Soc. Am. Proc.* 23: 135-143.
70. Zinke, P. J., S. Sabhasri, and P. Kunstadter. 1978. In *Farmers in the Forest*, P. Kunstadter, E. C. Chapman, and S. Sabhasri, eds., pp. 134-159. Univ. Press, Honolulu.

Part B.
Soil and Ecosystem Sequences

Knowledge of soil-forming processes has not yet advanced to the point at which the locations and behaviors of Spodosols, sodic soils, red loams or any other soil can be predicted.

A method of placing the soils and biota of a landscape into genetic view is offered by *state factor sequences* briefly mentioned in Chapter 1. The approach is phenomenological and arranges ecosystems along climatic transects, in compass directions, and along slopes. It indicates how different rocks and organisms mold different soils and how age carves its signature.

The 7 chapters of Part B show how landscape tapestries can be dissected into segments that permit their alignment along vectors (rows) of soil-forming or state factors.

8. State Factor Analysis

How ecosystems and soils evolve in time, differ from each other, and vary from one landscape to another, and why, is explored by state factor analysis. (Pedagogically speaking, the reader may wish to scan this chapter for orientation, then acquaint himself or herself with material in succeeding chapters and return to this one for critical perusal.)

A. Derivation of State Factor Formula

Visualize two metal tanks of 1 m² area and 1 m depth. The tops are open and the bottoms are closed except for drainage outlets. One tank is filled with soil of dune sand and the other with lacustrian clay soil. Both are seeded to the same mixture of grasses. The tanks, or lysimeters, are placed in the open, adjacent to a meteorological station that records the air climate cl. The event marks the initiation of genesis of two ecosystems at time zero, and the two ensembles of soil properties are labeled S_0 (from initial soil), specifically $(S_0)_{sand}$ and $(S_0)_{clay}$. Soil material S_0 includes whatever organisms are already in the tanks (e.g., decomposers).

Rain falls, contaminated with radioactive element X^* from radioactive fallout. After a while, percolating rainwater emerges at the bottoms, the sand effluent being richer in X^* than the one from clay because clay fixes X^* more strongly than sand. After 1 year of monitoring, each system contains amounts of X^* in soil and grass equal to the difference between influx (I) and efflux (E), or

$$X_1^* = X_0^* + I_1 - E_1$$

where X_0^* is the initial X^* content, if any.

After a period of t years and annual reseedings, the sums of the Is and Es are equal to their means I_m and E_m, multiplied by t. For the content of X^* in either system at time t

$$X_t^* = X_0^* + I_m t - E_m t$$

The efflux E of any element is difficult to gauge but is calculable from analyzing the ecosystem for X^* at time t. Actually, efflux is itself conditioned by the initial state and by all influxes during time t.

To generalize, for any total system properties l, for any vegetation properties v, for any animal properties a, and for any soil property s we write:

$$l, v, a, s = f(S_0, I, t) \tag{8.1}$$

Simply, the properties of the sytem depend on initial state, influxes, and duration. The dependence, symbolized by f, cannot be quantified because the amount of efflux is unknown; the values of l, v, a, and s must be secured directly by analyzing the system itself. Ecosystem, S_0, I, and t are shown in Figure 1.6.

B. Extended Equation

In any landscape key properties of the initial state S_0 are its physical, chemical, mineralogical, and organic make-up, conventionally known as *parent material* p. The surface configuration or *topography* is r, and the initial water table, insofar as it influences the moisture household, is conveniently included in it. Thus in place of S_0 we may write r, p, ..., the dots reminding us of additional properties of S_0 not necessarily embraced by r and p.

Important influxes include precipitation, light and heat radiations customarily lumped together as *climate* (cl), and macro- and microorganisms as germules, diaspores, or dissiminules. Any one organism is o_i and all of them together, including for convenience those in S_0, are named *biotic factor* ϕ (the slash through the letter o preventing its confusion with zero). Hence, influx I may be replaced by cl, ϕ, and dots ... , the latter calling attention to events such as dust influx, coastal salt spray, or fires not explicitly contained in cl and ϕ. The state of the ecosystem at time or age t is then written (Chapter 1):

$$\begin{array}{ccc} l, v, a, s & = & f(cl, \phi, r, p, t, ...) \\ \text{system} & & \text{state} \\ \text{properties} & & \text{factors} \end{array} \tag{8.2}$$

Approach and formula are known as "clorpt." In place of p the initial state may be occasionally assigned to S_0 or even to L_0, which is an entire ecosystem, as when a grassland is being plowed up or a forest clearcut. As a word equation:

Properties of		*State factors as*
total system (l), e.g., total C content, system respiration;		environment climate (cl); flora and fauna as pool of species or genes (ϕ);
vegetation (v), e.g., biomass, communities, Ca content;	f, or dependent on, related to, conditioned by,	topography and water table (r);
animals (a), e.g., work performed, reproduction, health;		initial state, as parent material (p) or as S_0, L_0; age or time (t) in years;
soil (s), e.g., pH, texture, bacteria counts, humus content;		dot factors (...);

linked in a rich and complex manner with multiple interactions and correlations

variables or groups of variables that *can* be independent, uncorrelated, orthogonal

The equation is believed to hold for compartments of ecosystems as well, but for rigor cl would have to be measured at the boundaries of the subsystem of interest.

C. Commentary to the Formula

The idea that climate, vegetation, topography, parent material, and time control soils occurs in the writings of early naturalists.

An explicit formulation was performed by Dokuchaev in 1898 in an obscure Russian journal unknown to western writers. He set down (3):

$$soil = f(cl, o, p) \, t_r$$

in which cl is regional climate, o vegetation and animals, p the "geologic substratum," and t_r relative age (youthfulness, maturity, and senility). In 1930 Shaw (9) published a "potent factor" formula which comprised erosion and deposition in addition to cl, v, p, t. Shortly thereafter Tüxen (11), a botanist, expressed vegetation (v) as a function of climate, soil (S), and man (ϕh), as $v = f(cl, S, \phi h)$.

The identity of the symbols hides serious inconsistencies. Aside from the uncertainty of whether the geologic substratum is the initial state, soil and vegetation are confounded. Dokuchaev and Shaw write soil = f(vegetation) and Tüxen writes vegetation = f(soil). The Gordian knot that vegetation acts upon soil and soil on vegetation has not been untied.

Organism properties appear on both sides of Eq. (8.2). On the right is ϕ which pertains to the genetic constitution, the genotype. Genes are acquired prior to growth, hence they can be independent of the ecosystem. On the left are v and a which are phenotypic expressions of genotypes that are strictly dependent on cl, ϕ, r, p, t, ... In botanical language ϕ represents the flora of influx and v the resulting vegetation. The ϕ may be viewed as potential vegetation which may not materialize if cl, r, p are unfavorable to growth. Crocker (2) has commented on ϕ in detail and elaborations are offered in ensuing chapters, particularly in 9 and 13.

Man is part of ϕ but deserves a special symbol ϕh, because his behavior is guided not only by genes but also by his *cultural* environment.

State factors are control variables to which the ecosystem is sensitive and that can be manipulated. They are independent of the system as it evolves and they *can* be, but may not be, independent of each other, which cannot be said of soil and vegetation. The state factors are identical with the soil-forming factors as defined in 1941 (4). In 1951 Major (6) extended the scheme to vegetation and plant communities. Equation (8.2) embraces the entire ecosystem, soil, vegetation, and animal life.

D. Ordination of Ecosystems According to State Factors

The clorpt equation is a synthesis of information about land ecosystems. In favorable landscapes the factors may be sorted out and assessed as six groups of *idealized* ordinations:

(1) $l,v,a,s = f(cl)_{\phi,r,p,t,\ldots}$ (4) $l,v,a,s = f(p)_{cl,\phi,r,t,\ldots}$

(2) $l,v,a,s = f(\phi)_{cl,r,p,t\ldots}$ (5) $l,v,a,s = f(t)_{cl,\phi,r,p,\ldots}$

(3) $l,v,a,s = f(r)_{cl,\phi,p,t,\ldots}$ (6) $l,v,a,s = f(\ldots)_{cl,\phi,r,p,t}$

The symbols in parentheses are the state factors that vary in an experiment or over a landscape, and the subscript factors are held constant or stay the same. Thus, ordination No. (5) expresses ecosystem genesis in *time* under a fixed combination of cl, ϕ, r, p, ... Ordination No. (1) studies system properties in relation to *climate,* all other factors being invariant; here "constant time" as subscript means that systems of equal ages are being compared. Subscript p refers to soils and plants developing on the same parent material (e.g., dune sand).

The single-factor ordinations are sequences of soils and vegetation. Their required factor constancies place them into the category of a "reference frame" for comparing ecosystems in time and space. Growth chambers, phytotrons, and biotrons come closest to this goal. In Figure 8.1, left, the soil property s is deterministically associated with state factor F_i, and all observational points lie on straight or curved lines. Scattering is absent in ideal ordinations.

E. Factor Variation

The best mechanical climate chambers are subject to occasional irregularities and the disturbances mar the idealized curves expected. In any landscape no two climates, rocks, or genotypes are identical. To grant nonconstancy of factors explicit recognition, a disturbance, error, or stochastic term u may be included (5) in Equation No.

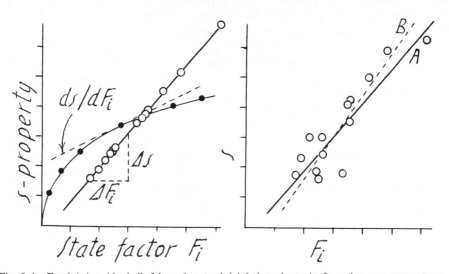

Fig. 8.1. Explaining ideal (left) and actual (right) pedogenic functions or sequences. The factor "thrusts" are given by the slopes $\Delta s/\Delta F$ and ds/dF. On the horizontal axis the row of values of a state factor (e.g., precipitation) is a vector

(5) as l, v, a, s = $f(t,u)_{cl, \phi, r, p, ...}$. It was first inserted by Olson (7). The term u changes the smooth alignment to one of point scattering, sometimes called "white noise" or background noise.

In the right-hand portions of Figure 8.1, point scattering is caused by nonconstancy of factors other than F_i, neglecting the relatively small analytical errors. The solid line A is fitted by "least-square" regression that makes the sum of the squares of the *vertical* distances from the points to the line a minimum. The correlation measure is r which has here a value of 0.898 and the square of it ($100\,r^2$) assigns 81% of the spread of s to the range of F_i and 19% to the nonconstancy of the remaining Fs. For the dashed B line, named the principal axis, curve fitting is done by minimizing the sum of squares of the *perpendicular* distances from the points to the line. It is employed in principal component analysis.

The *thrust* of factor F_i on the variation of a property in a chosen landscape is governed by the range of F_i in the area and by its effectiveness. For F_i in Figure 8.1, it is given by the slopes or gradients $\Delta s/\Delta F_i$ and ds/dF_i, which include all interactions. When the range ΔF_i is small, or when large but of low effectiveness, the factor plays a subordinate role. Moreover, constancy is relative. A factor F_i (e.g., mean annual rainfall) may slowly rise with time yet the shift may exert little impact if system properties change much faster, or not at all.

To incorporate these ideas explicitly the ordinations are written with *dominant* and *subordinate factors* as follows:

l,v,a,s = f(**cl**,ϕ,r,p,t,...)	climofunction or climosequence
l,v,a,s = f(**ϕ**,cl,r,p,t,...)	biofunction or biosequence
l,v,a,s = f(**r**,cl,ϕ,p,t,...)	topofunction or toposequence
l,v,a,s = f(**p**,cl,ϕ,r,t,...)	lithofunction or lithosequence
l,v,a,s = f(**t**,cl,ϕ,r,p,...)	chronofunction or chronosequence
l,v,a,s = f(...,cl,ϕ,r,p,t)	dotfunction or dotsequence

The first letter in parentheses denotes the dominant factor, the others are subordinate factors, their variations modulating the scatter of the l, v, a, s properties.

The state factor approach has elicited a variety of reactions: approving, disapproving, and noncommittal (1,8,10,12). Crocker (2) confines the ideal equations to "monogenetic" situations that are essentially nonexistent and contrasts them with the natural "polygenetic" ones in which factors are not constant. Once the slopes of crucial functions are known, idealized ordinations may be computed.

Many soil sequences have been established and a few of them will be recounted in ensuing chapters.

F. Review of Chapter

The historical postulate that soils are conditioned by the five factors climate (cl), organisms (ϕ), topography (r), parent material (p), and age (t), or s = $f(cl,\phi,r,p,t)$, is put into a conceptual framework that permits solving the equation when landscape configurations are favorable. Moreover, the approach is extended to entire ecosystems.

References

1. Birkeland, P. W. 1974. *Pedology, Weathering and Geomorphological Research.* Oxford Univ. Press, New York.
2. Crocker, R. L. 1952. *Quart. Rev. Biol.* 27: 139-168.
3. Dokuchaev, V. V. 1898. *Writings* (in Russian), Vol. 6, p. 381, 1951. Akad. Nauk, Moscow.
4. Jenny, H. 1941. *Factors of Soil Formation.* McGraw-Hill, New York.
5. Johnston, J. 1963. *Econometric Methods.* McGraw-Hill, New York.
6. Major, J. 1951. *Ecology* 32: 392-412.
7. Olson, J. S. 1958. *Bot. Gaz.* 119: 125-170.
8. Perring, F. 1958. *J. Ecol.* 46: 665-679.
9. Shaw, C. F. 1930. *Ecology* 11: 239-245.
10. Stevens, P. R., and T. W. Walker. 1970. *Quart. Rev. Biol.* 45: 333-350.
11. Tüxen, R. 1931/1932. *Der Biologe (München)* 1(8): 180-187.
12. Yaalon, D. H. 1975. *Geoderma* 14: 189-205.

9. The Time Factor of System Genesis

During a hundred million years a given locale on Earth may slowly sink below sea level, then rise to mountain heights, be temporarily covered by glaciers, be eroded, and finally subjected to aridity. Chronosequences deal with brief segments of geological history.

Rock material is exposed whenever a glacier retreats, a landslide descends, a volcano erupts, a sand dune is blown, or a lake is drained. Each event marks "time zero" and the initial state of the soil is identified with parent material p and its topography r. Other initial states may pertain to a plowed field, an erosion scar, or any site that we wish to choose as a starting point of ecosystem genesis.

A. Nature of Chronosequences

Human life is too short to monitor a natural ecosystem much beyond its inception. We must resort to splicing together sites that are apart in space and time. It is accomplished by selecting landscape positions that have comparable state factors, save age t, and placing them into the equations:

$$l,v,a,s, = f(t)_{cl,\phi,r,p,...} \quad \text{and} \quad l,v,a,s = f(t,cl,\phi,r,p,...)$$

$$\text{ideal chronosequence} \qquad\qquad \text{actual chronosequence}$$

Factors p and r pertain to initial states and as such remain invarient. During genesis p turns into soil and some of the r components (e.g., slopes) become soil properties that may vary with erosion and its depositions. To what extent the cl and ϕ influxes shift in time is difficult to evaluate and depends among other things on what s properties are measured.

Climate (cl) as a Subordinate Factor

Rainfall and temperature direct soil genesis in powerful ways, yet, in chronosequences they are relegated to a "subordinate" role. The word does not insinuate ineffectiveness per se; it specifies that during the span of soil formation the climatic *variations* produce no major distortions.

To measure in the laboratory the rate of a chemical process the reaction vessel is immersed in a constant temperature bath. Actually, the temperature fluctuates slightly and is characterized as, say, $20 \pm 1°C$. As long as the deviations do not upset the chemical trend they are considered negligible.

Sporadic wet years may deliver twice the annual mean of precipitation and in the long run they control the depth of leaching in arid and subhumid regions. The long-range statistical projections by Mandelbrot and Wallis (35) generate frequent drifts and clusters of wetter and drier periods that might qualify as climatic shifts. In soils they are reflected in the scatter of measurement points, the white noise mentioned in Chapter 8. To what extent cl fluctuations are acceptable rests on the demands we make on the quality of information delivered by a chronosequence.

Biota (ϕ) as a Subordinate Factor

Any organism living in an ecosystem is an effective biotic factor, but ϕ is broader, for it embraces the species that have vanished from the sytem and, theoretically at least, its future occupants. In phytotrons kinds and numbers of ϕ are controllable; outdoors, stringent growth limitations and organism competition allow wide variations in quantities of influx germules and still end up with the same vegetative cover, hence qualitative species lists commonly serve as identifiers of ϕ.

A rough estimate of the biotic factor ϕ of a chronosequence comprises the species that surround a chosen area and have access to it. What is the reach of the surrounding? It is given by the speed of migration of disseminules to the site, compared with the rate of system evolution at the site. Microflora and microfauna are considered ubiquitous, descending upon the ecosystem as a continual "microbe rain." Distribution and abundance of species growing within the ecosystem are dependent variables, being controlled by the state factors of the system.

During a long chronosequence organisms may undergo mutation and the microbes seem most susceptible to it. Geneticists rate mutation a random event and as such it preserves the genotype as an independent variable. Its perseverance in an ecosystem is again subject to factor constellations.

A late-comer, like man, may alter the ecosystem so profoundly that we may wish to start counting again with a new set of ϕ and a new t_0.

B. Buildup and Decline of Organic Matter in Chronosequences

Photosynthetic carbon fixation by plants donates oxygen to the atmosphere and subsequent storage of carbon in soil humus curtails respiratory CO_2 emission. Nitrogen fixation by bacteria and its storage in soils provides a steady, albeit slow, supply of ammonia and nitrates for the production of plant biomass that nourishes humans, animals, and most microbes. At what rates accumulation and losses of soil organic matter proceed is answered by chronosequences.

Nitrogen Acquisition of Total Ecosystems of Rhone Glacier Moraines (29)

By the year 1800 the alpine glaciers had readvanced into the narrow mountain valleys. The mighty Rhone Glacier attracted naturalists who in subsequent years recorded the glacier's retreat over a distance of 2 km. Using Mercanton's (37) historical records of the wall moraines, his elaborating personal correspondence, and the marker stone lines placed annually at the glacier fronts by glaciologists, the writer and his wife excavated 30 soil tesseras, 26 of them dated within 1-2 years, and clipped many vegetation quadrats with generous help from R. Bach, P. Dubach, N. Mehta, A. Frey-Wyssling, F. Richard, N. Winterhalter, and others (1955-1957).

State factors operating at the altitude of 1700 m are as follows:

Climate (cl): mean annual precipitation (maP) 170 cm, mean annual temperature (maT) 1.6°C. Ice retreat has been attributed to a slow 1°C rise in maT during the past century.

Biotic factor (ϕ): Low-shrub flora of alpenrose (*Rhododendron ferrugineum*), *Juniperus nana*, *Vaccinium* spp., *Calluna vulgaris*, *Erica carnea*, alder, willows, sedges, grasses (*Nardus stricta*), and herbs such as fireweed (*Epilobium* sp.) and clovers, over 50 vascular species in all.

Topography (r): The small stadial moraines are a few meters high. Selection sites are confined to SE exposures of less than 30% slope and having no water tables.

Parent material (p): boulder, gravelly and sandy till composed of granite, and metamorphic rocks. Water percolation is rapid.

Time (t): 13-353 years; Owing to the final retreat of the glacier into a rocky cliff at the end of the valley, no younger till material is available.

Within the confines of r and p specified, sampling locations are chosen at random and tesseras of 20 × 20 cm area and 50-150 cm depth (*d*) are excavated as described in Chapter 1.

Biomass Sequence

The younger moraines support forbs and grasses at densities of up to 12 specimen in a small quadrat (400 cm²). Older moraines teem with heathers and creeping, prickly junipers 20-30 cm in height. Shrub quadrats (20 × 20 cm) are clipped starting at the juniper periphery to 20 cm inward and the cut includes whatever herbs happen to grow between the branches. Oven-dry biomass in vert space increases linearly with moraine age to 1.5-3 kg/m² in 150 years. This weight persists on old, heather Podzols on weathered rock ledges above the former ice stream.

Soil Nitrogen Content

Fine earth in the 1640 A.D. moraine (Table 9.1) contains 170.5 g N/m² and a 15.8 times higher amount of C. Multiplication of C by 1.724 yields a humus content of 4.63 kg/m² which exceeds the root mass of 1.82 kg/m² more than twofold.

Three depth functions of nitrogen in fine earth, exclusive of roots, are plotted in Figure 9.1. The area of each histogram equals the absolute N content of a soil tessera

Table 9.1. Nitrogen and Carbon Assay of an Alpine Ecotessera (1640 A.D.), 315 Years Old

| Horizon | Depth (d) (cm) | Fine earth, oven dry[a] | | | Coarse organic matter[b] | | | Nitrogen in ecotessera (g/m²) | pH (H₂O) |
		Total (g/400 cm²)	N (g/m²)	C (g/m²)	Total weight (g/400 cm²)	N (g/m²)	C (g/m²)		
1	2	3	4	5	6	7	8	9	10
Vegetation (without roots)	23	—	—	—	112	25.0	1,309	25.0	
r, litter	0-2.5	70.7	25.8	553	56	17.9	476	43.7	
a, humus	2.5-9	679	63.7	1185	31	4.05	292	67.75	4.5
b, light gray	9-20	115	5.5	97	3.6	0.48	34	5.98	
c, brown	20-34	2326	33.7	519	28	3.66	265	37.36	4.4
d, dark-gray	34-60	2943	15.5	186	6.2	1.34	50	16.84	4.8
e, dark gray	60-100	5836	26.3	147	3.9	0.85	31	27.15	5.3
Totals		11,970	170.5	2687	240.7	53.28	2,457	223.78	

[a] Fine earth, < 2 mm; to obtain its % N content multiply column (4) by 4 and divide the result by column (3).
[b] Coarse organic matter (> 2 mm) for a to e pertains to roots.

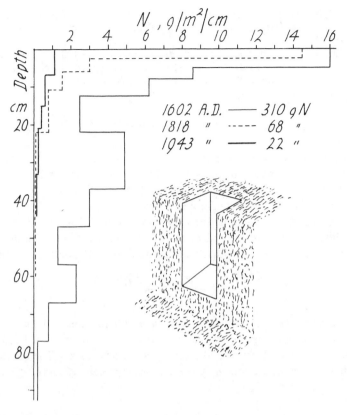

Fig. 9.1. Sketch of a soil tessera, and of three soil nitrogen depth functions, exclusive of roots, of Rhone glacier moraines

and the amounts rise from 22 to 68 to 310 g N/m^2. When roots and vegetation are added the three sums become 28, 85, and 363 g/m^2. These are "raw data," without corrections or adjustments.

The interval of 5-10 cm below the deepest visible fine roots is thought to approximate the nitrogen content N_0 of the initial state or parent material. For fine earth it is estimated as 0.00093% N. Applied to the 1640 A.D. moraine of Table 9.1, N_0 is computed as 2.8 g N/m^2 for the entire fine earth in the profile. It lowers the N content of the ecotessera to 221.0 g/m^2, which is the absolute N gain. At two sites rich in fine earth the reductions are nearly 10%.

Adjustment to Uniform Parent Material

Dividing the N content of an ecotessera by its age gives the average annual N gain. For the 1640 A.D. moraine of Table 9.1 it is 0.71 g/m^2/year. The mean of all 26 dated tesseras is 1.0 g N/m^2. Variations are substantial and are tied to age and rockiness of till because particle size of parent material is not uniform.

Fine earth (fi) is the host of humus. In Table 9.1 the concentration of fi in the entire soil tessera, $(fi)_c$, is $11970/(400 \cdot 100)$ or 0.299 g/cm^3. The mean of all tesseras is 0.530 g/cm^3.

To elicit a chronofunction for constant parent material all N gains are adjusted to a fine earth concentration 0.50 g/cm^3 using the formula $N_a = 1.25 \, t \, (0.50 - (fi)_c) + N$. Hence, the 129-year-old moraine of N content 89.3 g/m^2 and $(fi)_c$ of 0.524 g/cm^3 acquires N_a equal to 85.4 g/m^2.

N-Chronofunction

All adjusted N contents are plotted in Figure 9.2. A straight line to the origin is fitted through the 21 tessera points of the first 140 years, providing $N = 1.005 \, t$. The slope, 1.005 g/m^2, denotes the average annual nitrogen accruement at constant parent material. The source is N in rainfall and microbial nitrogen fixation.

The linear formula is unable to encompass the lowered N values of the 3-centuries-old tills. Either N fixation lessened or losses enhanced, or both. The bending, solid curve, derived in a later section [Eq. (9.4)], is fitted to all 26 points using the equation

$$N = \frac{A}{k} (1 - e^{-kt}) \qquad (9.1)$$

where N is the nitrogen content of an ecotessera of 1 m^2 area and soil depth d, A is the rate of gain of N, assumed to be the same year after year, and k is the loss coefficient or fractional loss of total N in the ecosystem at any instant. The constants evaluated on an annual basis are gain $A = 0.98$ g/m^2 and percentage loss $k' = 0.14\%$. Other

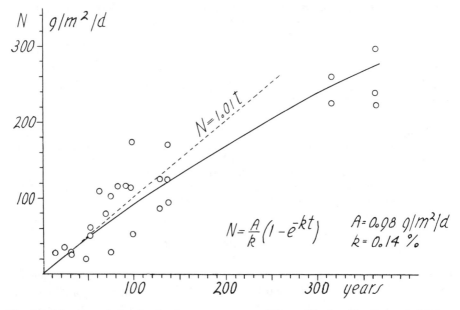

Fig. 9.2. N accumulation in shrub ecosystems on Rhone Glacier tills. On two old moraines (315, 353 years) total system C/N ratios are 20.1 and 23.0; d, depth of tessera

equations could have been used but the chosen one allows for interesting interpretations.

The curve strives toward a maximum, A/k, which is 700 g N/m². This high amount is approached in two advanced Podzols on nearby cliffs having 685 and 812 g N/m². The time required to arrive at 95% of the maximum is $3.0/k$ or about 2000 years. This age seems reasonable but it is highly speculative because of the shortness of the curve segment actually measured.

Gains in Alaskan Rock Fields

Glacier Bay in S. Alaska has a maritime climate of maP 142 cm and maT 5.1°C, recorded at Pt. Gustavus. Since 1794, when the explorer Vancouver saw the massive glaciers, the ice has been melting and retreating rapidly.

> During colonization of the exposed wasteland (39) winds sweep in microbes, spores, seeds, and insects. Birds fly across and brown bears traverse the barrens leaving dung spiked with large seeds. Only nitrogen-fixing plant species do well, among them the prostrate mat-former *Dryas* sp. and small shrubs of alder. The latter are threatened by cottonwood and Sitka spruce coming up among the clumps. The tiny deer mouse (*Peromyscus* sp.) and the wandering shrew (*Sorex vagrans*) are the first settling mammals. Locally, soil acquires a gray brown, embryonic Al horizon.
> In a century mosses, leguminous herbs, ash, spruce, and hemlocks prosper on the now acid soil. In less than 200 years the pioneers are replaced by the spruce-hemlock climax forest, the home of the stately black-tailed deer (*Dama lemionus*) and the husky black and brown bears (*Ursus americanus, U. arctos*) who debark trees and tear up the forest floor in search of root tuber delicatessen. Weathering and clay formation remain minimal.

In Crocker and Major's (10) seven profiles under alder, humus content peaks in a century to 276 g N/m² for forest floor and mineral soil together and to 126 g N/m² for 46 cm of mineral soil alone. As spruce replaces alder the curve decreases by a third, the "forest developing at the expense of N accumulated in earlier stages."

Ugolini's (51) samples accrue in 250 years 145 g N/m² in 25 cm of A, B, and C horizons (C/N=26) and 76 g N/m² in the forest floor (C/N=55), but no decline of nitrogen with age is observed at the five sites collected. He formulates the soil array as: Regosol → Podzolic soil → Brown Podzolic soil → Podzol, and considers only the first two as members of a chronosequence because of the subsequent takeover of alder by spruce and hemlock. Major (34) discounts the view.

Tundra Formation

Tundra is a treeless landscape of the arctic region. In Mt. McKinley National Park, Alaska, Viereck (54) was attracted by tundra genesis on five dated river terraces that have comparable state factors. The mixed terrace alluvium is rocky, pebbly, and gravelly and slightly calcareous, and the fine earth is very sandy and has less than 5% clay. The locations are above the timberline and have never been glaciated, and the biotic factor common to all terraces comprises over 100 plant species, many of them circumpolar and alpine. Ages, vegetation stages, climate, and soil nitrogen analyses are assembled in Table 9.2.

Table 9.2. Tundra Chronosequence in Alaska[a,b]

River terraces	Age (years)	Vegetation stages	Percentage N in fine earth, at:		
			5 cm	10 cm	20 cm
I	25-30	Pioneer	0.026	0.013	0.013
II	100	Meadow	0.129	0.079	0.033
III	150-200	Early shrub	0.161	0.119	0.038
IV	200-300	Late shrub	0.171	0.100	0.044
V	5000-9000	Tundra[c]	2.38	0.14	–

[a] After Viereck (54).
[b] maP, 52 cm; maT, $-4.6°$C.
[c] A1 (0-13) and BG (13-25 cm) horizons of Kellogg and Nygard's (32) mountain tundra No. 14.

I. Pioneer Stage (30 Years Old). This stage is characterized by much bareness and isolated, low shrubs of *Dryas* sp. (a N-fixer), bearberry (*Arctostaphylos uva ursi*), and crowberry (*Empetrum* sp.). Herbs include three genera of legumes (e.g., *Astragalus*). No alder appears. Under shrubs moss begins to grow.

II. Meadow Stage (100 Years Old). As the pioneer shrubs die out their space is taken over by scattered willows (*Salix* sp.), *Shepherdia* berry shrubs, and extensive mats of wild rye (*Elymus* sp.), *Poa* grasses, and other herbs. Above the mineral soil the organic layer of decomposing mosses and vascular plant tissue is 5 cm thick.

III, IV. Shrub Stages (150-300 Years Old). Mosses and organic detritus build up to 35 cm in height. Willows die and are replaced by shrubs of birch (*Betula* sp.) and black bilberry (*Vaccinium uliginosum*) which stay alive by sending adventitious roots into the moss layer, a feat that willows cannot accomplish.

V. Low Shrub-Sedge Tussock-Moss Tundra (Thousands of Years of Age). The wide gap in time makes succession to Viereck's climax tundra on the highest terrace conjectural, yet few botanical changes are needed. New occupants include redberry (*Vaccinium vitis-idaea*), Labrador tea (*Ledum* sp.), sphagnum mosses, and the attractive cotton grass (*Eriophorum vaginatum*) which forms the 20- to 30-cm-tall hummocks or tussocks that tax the endurance of caribou and man alike. Moss mats keep thickening and all shrubs adapt to the snow heights of 20-40 cm to avoid destruction by icy winter winds.

The youngest terrace is nearly free of small mammals, the older ones have mice (*Microtus miurus*) which are joined by lemmings (*Lemmus sibiricus*) on the tundra proper (53).

The rate of total nitrogen accumulation (Table 9.2) lags behind that of the Rhone Glacier moraines for which a value of 0.40% N at 5 cm depth is calculated for the 315-year-old soil. To characterize the millenia-old climax tundra, Viereck cites Kellogg and Nygard's (32) high value of 2.38% N (C/N 18.8) of the Al horizon (0-5 cm).

In the young gravel terraces the summertime permafrost or frozen soil stays below 75 cm depth, but in climax tundra it is at 20-25 because vegetation mats and thick humus cover act as heat insulators against melting of winter ice.

As the A horizons thicken they "drag up" permafrost, leaving buried humus layers in the ice below. These are attributed sometimes to climatic changes (50).

At Point Barrow, northern Alaska, pronounced peaks of lemming populations occurred in 1946, 1949, 1953, 1956, 1960, and 1965, but have since waned. Schultz (47) observes high Ca and P contents in grass forage at the beginning of a lemming year and low ones afterward. In 3-4 years grass quality recuperates. Grass production responds strongly to fertilizer applications, pointing to soil deficiencies.

The extensive literature does not dwell on the *principle of tundra soils* having organic accruement profiles that grow upward and are nourished by aeolian, mineral dust from the braided river beds. A heavy nutrient influx might stimulate grass production and trigger a lemming pulsation.

Shasta Chronosequence

For millennia the inactive, ice- and snow-covered volcano Mt. Shasta (4408 m) in northern California has been the source of vast, slow-moving coarse-textured masses of andesitic rock debris, known as mudflows. They overrun the valleys below, abrade luxuriant forests of pines (*P. ponderosa*), firs (*Pseudotsuga menziesii*), cedars (*Calocedrus decurrens*), and oaks (*Q. Kelloggii*), and cover the soils with meters of mineralogically uniform detritus, often without materially disturbing the existing profiles. In the early 1950s Dickson and Crocker (12) identified five flows near the town of McCloud (maP 118 cm, maT 9.7°C) at altitudes of 1000-1700 m, sketched schematically in Figure 9.3. The gently sloping deposits from young to old are labeled A, B, C, D, and E.

Years later, other parties discovered flow F at higher elevations and confirmed the crucial overlay of D on E but could not separate B and C with certainty. In 1970 the total area of 1635 ha was set aside as a Natural Research Area within the Shasta National Forest.

Flow A descended in 1924 and is the only one dated accurately. Counting tree rings, Dickson and Crocker believed flow B to be 60 years old but ^{14}C dating in 1963 of the core of a buried tree stump furnishes a reading of 450 ± 200 years (K. L. Hubbs) which leads to an age estimate of 300 years, more or less. The older flows do not exceed a few thousand years, as judged from the dark gray, humus-rich, undifferentiated profile features. Some of the huge pine specimens on old muds have ring counts of over 500 years, but no dates can be assigned.

In 1964, when the A flow was 40 years old, 42% of its area was bare, 47% was in Purshia shrubs (*Purshia tridentata*), 10% in Ponderosa pines, and 1% in miscellaneous plant species. Glauser (16) excavated 12 tesseras, and, combining them with the percentages of vegetation coverage, supplies for the entire A flow mosaic the C and N contents of Table 9.3. Because of the high proportion of bare surface area the gains in 40 years are low.

The B flow supports a nearly pure stand of ponderosa pines (563 trees/ha) of mean age of 68 years and mean height of 25.0 m. A severe tropical storm uprooted many trees which permitted Glauser to correlate organic matter contents of five soil tesseras with numerous tree parameters and construct the B flow ecotessera in Table 9.3. Mean annual N gains are 1.013 g/m2. Forest floor has pH 4.1, A1 horizon 5.3, and C horizon 6.0.

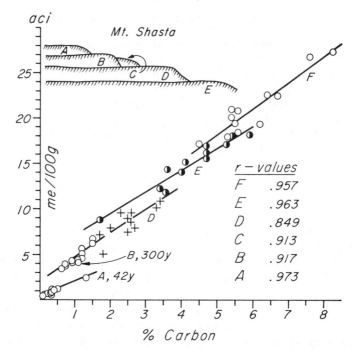

Fig. 9.3. Schematic sketch of mudflows, and correlations of exchange acidity (aci) with carbon contents of mudflows aged 42(A), 300(B), and a few thousand years (D, E, F). The older the deposit the higher percentage C and aci tend to be. Courtesy Pontifica Academia Scientiarum, Vatican City

With increasing ages of mudflows organic matter in the mineral soil keeps accumulating. In 1964 the quantities per square meter were 11,774 g C and 501 g N for D flow, 12,623 g C and 514 g N for E flow, and 22,127 g C and 1423 g N for F flow. In the younger flows Dickson and Crocker had verified the expected rises in CEC and exchange acidity with humus accession and the decline in replaceable bases (Ca, Mg, K, Na).

Table 9.3. Carbon and Nitrogen in Ecosystems of Mt. Shasta Mudflows A and B[a,b,c]

Components	A flow mosaic 40 years old		B flow, 300 years old	
	C	N	C	N
Vert space[d]	596	6.5	23,330	48.6
Roots	158	2.3	3,648	6.6
Forest floor	124	5.7	3,894	62.9
Mineral soil (0-61 cm)	632	34.7	3,567	203.3
Ecosystem	1510	49.2	34,439	321.4

[a] In grams/square meter.
[b] After Glauser (16).
[c] Initial state: C = 238; N = 17.6.
[d] Leaves, twigs, bole, all above ground; annual litterfall in B forest is 313 g/m^2 of C and 2.36 g/m^2 of N.

While C, N, and acidity are still accruing, their proportions already appear stabilized. In C, D, and E flows C/N in the first foot of mineral soil hovers between 19 and 21 (12). Organic carbon (C) and Mehlich exchange acidity (aci) in 20-cm-deep surface layers of mineral soil vary widely within and between flows, but the functions of aci = f(C) generate nearly parallel straight lines (30), plotted in Figure 9.3. The family of curves attests to rapid pedological equilibrations of humus acidity with age and environment.

Steady-State Maxima

In New Zealand's Manawatu district (maP 84 cm, maT 12.2°C) five sand dune systems are dated as 0 (fresh beach sand), 50, 500, 3000, and 10,000 years of age. Replicated tesseras of mineral soil of 1 m² area and 1 m depth on 15% south slope and under native vegetation of bracken fern (*Pteridium* sp.), manuka scrub (*Leptospermum ericoides*), and forest of *Podocarpus* species accumulate total nitrogen, as plotted in Figure 9.4 by Syers et al. (49). The build-up curve approaches a maximum, estimated by the authors as 1050-1370 g/m²/m. The steeper curve for organic carbon reaches 14.8 kg/m² at 10,000 years and is still rising.

The nitrogen-sulfur ratio widens from 1.0 in the beach sand to 3.0, 6.0, 6.8, and 5.4 in the older dunes. The authors attribute the gain in S from 75 to 194 g/m² to influxes from the outside, probably in rain.

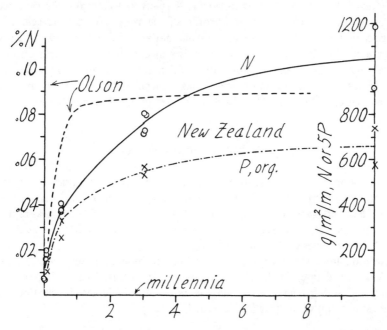

Fig. 9.4. Accumulation of total nitrogen (N) and organic phosphorus (P) over millennia of years in New Zealand soils (1 m depth), and of N in Olson's Lake Michigan dunes (0-10 cm)

A tangible N-maximum is reached in Olson's Indiana dune sequence to be described later but graphed in Figure 9.4 as % N in the surface soil 0-10 cm depth. The upper limit is 0.087%. In the Woy Woy dunes of N.S.W. Australia, organic carbon in 0-10 cm surface soil achieves a maximum of 0.55% in 150 years and keeps it for another 3.5 centuries (8).

Walker and Syers (57) explore the role of *phosphorus* in nitrogen chronosequences by fractioning total soil phosphorus (P_T) into organic (P_{org}) and inorganic forms. The ascent of P_{org} in Figure 9.4 is similar in trend to that of N and appears to peak near 150 g/m^2/m (note that for graphic considerations $5 \times P_{org}$ is plotted). The depth functions of % P in the organic fraction ($100 \cdot$ ppm P_{org}/ppm P_T) suggest that P transported from roots to tree tops accumulates in the surface soil horizons by way of litterfall.

Walker diagnoses P as a key element in the amassing of humus. Once the reserves of readily available P in the parent material (e.g., of apatite) are converted to slowly available P_{org} and to nearly insoluble P of Fe and Al-phosphates, the rate of gain of N declines and the organic matter chronosequence must pass through a maximum. In principle, any mineral nutrient that becomes growth limiting puts a halt to N and C accumulations in ecosystems.

As seen in Figure 9.4, a surface horizon maximizes sooner than a whole profile and, as to be proved later in this chapter, different soil properties reach plateaus at divergent rates. While one may recognize individual attributes as being at steady state, an entire "steady-state soil" or ecosystem is observationally elusive.

If it takes thousands of years to fill the N reservoir to steady-state capacity then some of the humus particles in the profile should possess that same age. The ^{14}C analysis cannot reveal it because old organic matter is diluted by new. In the oldest Manawatu profile the humic acid fraction is concentrated in the surface soil or immediately below it whereas the more mobile fulvic acid component extends to 100 cm depth. In the upper horizons organic C has a mean age of 670 years at 55-75 cm depth an age of 4250 years (17). Both values are minimal, mixed humus population ages.

Decline of Humus Contents

When the virgin sod of the Great Plains area was broken at the turn of the century, farsighted experiment stations established cultivation plots and maintained them for decades. Haas and Evans (19) graph the time sequences of soil N and C as reproduced in Figure 9.5. At Hays, Kansas, nitrogen declines during the first 10 years at the annual rate of 2.6% and slows to 0.7% in the fourth decade.

Cultivation lowers soil organic matter because biomass is removed and plowing and discing accelerate the biological oxidation of humus. Expectedly, practices of cropping and soil management guide the rates of soil depletion, continuous corn being most destructive, pasture and hay the least (28).

Loss of humus initiates collapse of soil structure, alterations in water and in nutrient economy, and likely enhances wind and water erosion. Crop yields decline

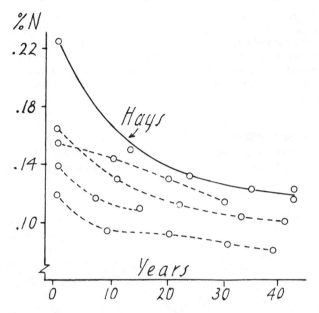

Fig. 9.5. Decline of nitrogen of virgin grassland soils upon cultivation at five Great Plains experiment stations. The curve at Hays, Kansas, is fitted to Eq. 9.5

dramatically (28) unless the land is fertilized. The modern trend of returning composted city wastes to farms helps conserving the soil resource.

Under virgin conditions the decline of soil organic matter with time is much slower and is geared to deterioration of the natural ecosystem caused by long-time soil leaching, as elaborated in the ensuing section on the Ecological Staircase. In New Zealand soils on basalt flows total nitrogen diminishes from 1547 $g/m^2/60$ cm at rock age 5000-10,000 years to 650 $g/m^2/60$ cm at rock age 300,000-600,000 years. Organic phosphorus wanes even more and insoluble types of P rise to 91% of the entire supply. Only stunted broadleaf podocarp scrubs survive. Wells (59) characterizes the declining chronosequence as decimating plant uptake of K, Mg, Ca, P, and Mo, and mildly boosting it for Al, Ti, and Fe.

Quantitative Models of Gains and Losses of Organic Matter

In the chapter on humus the idea of an equilibrium or steady-state *forest floor*, denoted here F_e, is developed quantitatively as

$$A = k' (F_e + A) \qquad (5.5)$$

where A is the mean annual litterfall and k' the annual loss or decomposition coefficient.

To model a time sequence, we remove in a deciduous forest in late summer 1 m^2 of steady-state forest floor and visualize its reconstitution. The annual litterfall A

descends in toto, it is assumed, on October 1. A year later the undecomposed remainder is the floor F_1, which is A minus the loss $k'A$, or $F_1 = A(1-k')$. Immediately after the second fall, again on October 1, the floor is $F_1 + A$ and at the end of the second year F_2 is $(F_1 + A)$ minus the loss $k'(F_1 + A)$, which is, algebraically, $A(1-k') + A(1-k')^2$. Year after year F thickens but at a slower and slower pace. For any time period t mathematical texts write the summation as

$$F_t = F_e [1 - (1-k')^t] \qquad (9.2)$$

For the oak, pine, and evergreen broadleaf trees analyzed in Chapter 5 the calculated time functions (Fig. 9.6, inset) generate at the tropical site a stable, shallow forest floor in less than 10 years. The northern pine site must wait 126 years to accrue

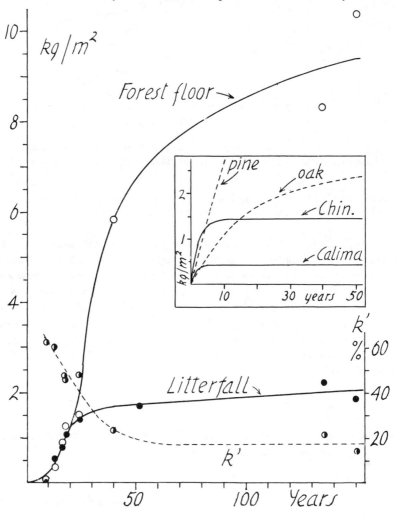

Fig. 9.6. Accumulation of forest floor and litterfall and simultaneous decline of their loss coefficients k', after Glauser (16). Inset: calculated forest floor accumulation curves based on Eq. (9.2) and Table 5.3

95% of its thick F_e. The model is unrealistic at the beginning of the sequence because young shallow floors have higher k' values than older thick ones, which retards the initial rate of accumulation.

Under pine trees growing in isolation on the Mt. Shasta mudflows the weight of forest floor rises in S-shape mode and litterfall attains a low quasiplateau. Glauser (16) computes the loss coefficient k', shown in Figure 9.6 as a dashed line. As anticipated, k' is high initially and acquires near-constancy in half a century. The sigmoid-type floor curve is confirmed by Ugolini (51) at Glacier Bay where steady-state forest floor of 8 kg/m² is attained in 150 years and is still in evidence a century later.

For assessing changes of nitrogen in the entire ecosystem or in the mineral soil the litterfall is unsuitable, for its N content is mostly soil-derived. In principle, influxes and effluxes take place continuously. In its simplest form the model assumes a constant, environmental influx gain A and an efflux that is proportional (k) to the quantity of N present in the system.

At any instant (3,28):

$$\frac{dN}{dt} = A - kN \qquad (9.3)$$

$$\begin{array}{ccc} \text{rate of change} & \text{rate of} & \text{rate of} \\ \text{in soil} & \text{gain} & \text{loss} \end{array}$$

Whether A is large or minute, it is not N in litterfall, in "clover turned under," or in "manure incorporated," but is the gain of N in the form as it exists in the soil. In the integrated equation

$$N = \frac{A}{k}(1 - ae^{-kt}), \text{ and } a = (1 - \frac{k}{A} N_0) \qquad (9.4)$$

where N_0 is the initial and A/k the steady state nitrogen content of the soil. When $(A/k - N_0)$ is positive the curve rises, when negative it descends, as shown in Figure 9.7 for $k = 0.01$, $A = 0.005$, and N_0 either 1 or 0. Both curves approach 0.5, which is the steady state. The equation is applicable to organic carbon as well.

In fitting Eq. (9.4) to the *ascending* Rhone Glacier sequence, as done in a previous section in this chapter, N_0 is zero and a becomes 1. Gain A is 0.98 g/m² per year and loss (kN) is 0.14% a year of the total N content present in the ecosystem at any instant. As the N capital accrues, the losses mount and the curve levels off.

For the Indiana curve in Figure 9.4, $N_0 = 0.003\%$ and $A/k = 0.087\%$. Olson (42) computes a loss coefficient of 0.30%. The annual gain A is small, $0.084\% \cdot 0.30\% = 0.00025\%$ N because of restriction of sampling to 10 cm of sandy surface soil. For a soil density of 1.18 g/cm³, which is a mean of 1.55 for the initial soil and 0.80 for the steady-state soil, the annual influx A amounts to 0.30 g N/m²/10 cm.

The *declining* curve of Hays, Kansas, shown in Figure 9.5, has a high initial nitrogen content ($N_0 = 0.225\%$) and losses that exceed the gains. Equation (9.4) becomes

$$N = \frac{A}{k} - (\frac{A}{k} - N_0)e^{-kt} \qquad (9.5)$$

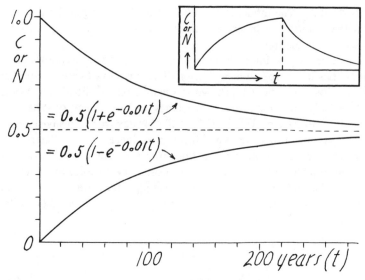

Fig. 9.7. Simple models of rising and declining curves of soil organic nitrogen or carbon, based on simultaneous gains and losses [Eq. (9.4)]. Steady state is at 0.5. Inset: sequence of build-up and decline curves of soil N or C (compare Fig. 13.9)

in which A/k is the new, lowered steady-state level. It is estimated by graphic inspection as 0.112% N. Curve fitting delivers $A = 0.0076\%$ per year and $k = 0.068$. The observed points are faithfully encompassed, but the accord is deceiving. The coefficient k implies a high annual (k') loss of 6.7% $(k' = 1 - \exp - k)$ which is partially mitigated by gain A. For the soil depth of 0-17 cm and a conventional soil density of 1.32, A translates to 17.0 g/m^2/year, which is an incredibly high input of N in absence of legumes. Evidently, a model is not "proved" by closeness of fit.

Fresh organic matter oxidizes more easily than older, more resistant humus (27). In a second model k is set as declining with time from 6-8% at the beginning to about 1% near steady state, the latter loss being balanced by a yearly accruement of about 1 g N/m^2. Gain A may also be higher initially when the sod is first broken. To complicate matters, in an analogous Canadian sequence all forms of organic matter fractions decompose at the same rate (36). A good model has yet to be described.

C. Leaching of Carbonates

Rain water passes swifly through calcareous sands and leaching of CaCO$_3$ and formation of horizons are relatively rapid. The soil alterations direct biotic successions.

Along sea and lake shores wind-blown sand dunes run parallel to the beach front. The youngest dunes are closest to it, the older ones are farther inland because the water levels had been higher in the past or the land surface had risen.

Indiana Sand Dunes

During the last 12,000 years winds on shores of Lake Michigan deposited many well-sorted calcareous sand ridges that have been dated historically and by ^{14}C technique with 10-23% uncertainty.

On gentle dune slopes 3-10 m above ground water tables in a regional climate of maP 65 cm and maT 10°C, and with no rainless month, Olson (42) recognizes invasion of bare dunes by the crisscrossing, horizontal rhizomes of marram grass. It is replaced by slow-growing little bluestem bunchgrass which slowly becomes overmature or is weakened by the shade of encroaching sand cherry and other shrubs. Jack pine and white pine follow and are reduced to minor forest components by black oak (*Quercus velutina*) and white oak with understories of blueberries, huckleberry (*Gaylussacia baccata*), and others, all growing on Plainfield fine sand, a Gray-brown Podzolic soil or Alfisol.

From the published pool of functions of vegetation and soil properties the $CaCO_3$ curve of the first decimeter of soil profile is redrawn in Figure 9.8. Most of the points scatter about Olson's negative exponential curve % $CaCO_3 = 1.48 \exp -0.0046t$, where t is time in years. Half of the initial mean carbonate content is removed in 151 years and the remainder is again halved in 151 years (half-life period). Olson concludes that about 1000 years are needed to deepen the leached zone to 2 m.

Woy Woy Sand Dunes

At Ocean Beach, N.S.W., Australia, having maP 123 cm and maT 17.2°C, and rain each month, the disappearance of the calcareous grit from the surface layer (0-10 cm) of Woy Woy dunes is very rapid (8). Half of $CaCO_3$ dissolves within 50 years and the rest in about 200 years (Fig. 9.8).

Southport Dunes, England

Salisbury (46) assesses the ages of the coastal dunes at Southport (maP 85 cm, maT 9.8°C) from old maps (1610, 1736) and descriptions. The young dunes support calciphil species of herbs and shrubs, including the abundant, tall beach grass *Ammophila arenaria*. The oldest, decarbonated dune is covered with calciphobe heather (*Calluna* sp.). The plant succession is induced by $CaCO_3$-leaching.

Mean contents of $CaCO_3$ and pH for the surface layer of 0-10 cm depth are plotted in Figure 9.9. Dune d' has been planted to pines which increases soil acidity; dunes h and i are in agricultural use and are not included in the statistical curve-fitting of the solid, exponentially declining line: % $CaCO_3 = 4.60 \exp -0.0145t$, with r of 0.98. The value of 4.60 represents the average $CaCO_3$ content of the dune parent materials. Half of the carbonate is removed in 48 years.

To comment on functions in general, whenever two or more properties are members of the same chronosequence, like $CaCO_3$ and pH presently, they also stand in functional relationship to each other, as graphed for pH = $f(CaCO_3)$ in the inset of

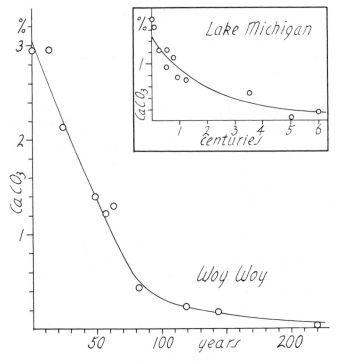

Fig. 9.8. CaCO₃ contents of sand dune chronosequences of Woy, Woy, Australia (8) and Lake Michigan (42)

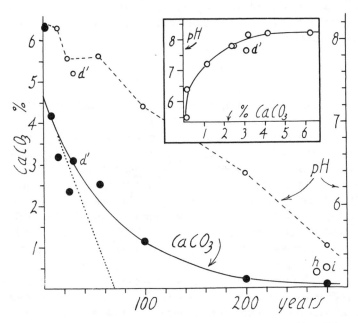

Fig. 9.9. Leaching of CaCO₃ and rise of acidity related to age of sand dunes in England (46)

Figure 9.9. In this instance the logarithmic-type curve is well defined but its rationale is hidden.

To probe the mechanism, the early leaching rate is $-0.0145 \cdot 4.60\% = -0.067\%/$ year, and for a small tessera of 1 cm^2 cross section, 10 cm length, and a bulk density of 1.5 g/cm^3 the loss is 10 mg CaCO$_3$ a year. On purely chemical grounds (Table 7.9) the Southport precipitation of 85 cm percolating through the tessera could leach out 7.5 mg CaCO$_3$, a somewhat lower figure but of the right order of magnitude.

If the initial loss of 10 mg CaCO$_3$ were to prevail, decarbonation to 10 cm soil depth would be completed in $1/0.0145$ or 69 years, indicated by the dotted line in Figure 9.9. The delay to centuries suggests that water percolation becomes channelized and moves too fast to assure solubility equilibria within horizons. Pockets of calcareous shells may persist a long time.

Microbial Successions in Dunes

In Olson's early dune sequence of marram grass and bluestem, Wohlrab and Tuveson (63) note a parallel succession of soil fungal flora. Webley et al. (58) count soil microbes in Scottish dunes and arrange their chronosequence as shown in Table 9.4. Bare sand is definitely not sterile. The acid humus derived from heather plants strongly depresses bacteria counts but fungi keep rising. Microbe populations in the root vicinity (rhizosphere) are much higher than in the bulk of sand.

Saitô (45) corroborates the sequence for the Japanese coast. In young, bare, and grassy dunes of pH 7.2-8.2 bacteria and actinomycetes outstrip fungi whereas in old, pine-covered sand hills of pH 5.5 the reverse condition holds. Pine pioneers sustain root mycorrhiza. Among the nitrogen-fixing bacteria *Azotobacter* lives in young dunes only. *Clostridium* shows no preference, and neither do the abundant denitrifying organisms. Nitrifying bacteria are present in the early but not in the late dunes that support pine forest.

D. Clay Formation and Profile Features

Humus accumulation and leaching of carbonates rapidly endow soils with distinguishable horizons. The deep-seated alterations that set soils apart from their parent materials are the redox reactions and clay syntheses that help mobilize silicon and sesquioxide ions into characteristic depth signatures. These processes are time consuming.

Table 9.4. Microorganisms in a Scottish Dune Chronosequence[a]

Nature of dune	pH	Microorganism per 1 g of oven-dry soil	
		Bacteria	Fungi
1. Open sand	6.80	18,000	270
2. Yellow dune	6.68	1,630,000	1,700
3. Early fixed dune	5.14	1,700,000	68,470
4. Dune pasture	4.80	2,230,000	209,780
5. Dune heath	4.27	127,000	148,000

[a]Webley et al. (58).

Rates of Rock Decay

In Great Britain dated tombstones weather 2-3 cm in 250-500 years. Hirschwald (28) gauges rates of decay of century-old buildings that can be traced to a common stone quarry. Constructed of Brochterbeck sandstone, the 100-year-old church of Risenbeck is still in good condition, in Osnabrück the 550-year-old St. Catherine's Church shows slight alteration, the 770-year-old St. Mary's Church is strongly marred in parts, and most of the ruins of 900-year-old Tecklenburg Castle are severely decayed. Buildings made of Rothenburg sandstone weather much faster than those of Nahetal porphyry.

During 17 years of outdoor weathering at Göttingen University fine particles accumulated rapidly from sandstone and schist and slowly from limestone (28). At Versailles, granite pieces of 2-4 mm size weathered (23) in 30 years to coarse sand (31%), fine sand (62%), silt (4.6%), and clay (2.4%).

Rates of Clay Formation

In the European Alleröd period, 9000-10,000 years B.C., volcanic activity in West Germany deposited extensive areas of coarse trachytic pumice, which is a gray, hardened glass froth.

In a beech-oak-pine forest about 1.5 km southwest of the west end of the town of Nickenich a granular to pebbly profile on a 1% slope has a gray-brown surface soil (10 YR 5/2, dry) that turns to light brown and reddish yellow (7.5 YR 6.6, 6.4, dry) in the subsoil. Percentage of organic carbon in the fine earth diminishes from 4.71% at 0-3 cm to 0.8-0.9% at 30-40 cm and to 0.1-0.2% at 100-120 cm.

For a soil tessera of dimensions 20 × 20 × 120 cm, oven-dry quantities of coarse (> 2 mm) and fine (< 2 mm) material without roots are plotted in Figure 9.10 in grams per square centimeter area per centimeter depth. Below 30 cm the coarse frac-

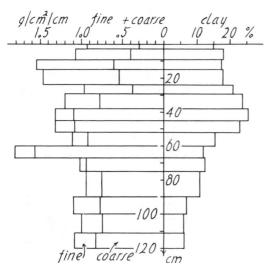

Fig. 9.10. Texture tessera of pumice soil at Maria Laach, West Germany. Coarse (> 2 mm) and fine (< 2 mm) fractions and percentage of clay in fines

tion is the chief component of the soil mass. Clay ($< 2 \mu m$) content in percentage of fine earth begins at 6% below 1 m depth and enlarges to 25.6% at 38-45 cm. Concentrations of 15-18% in the A horizons are corroborated by two partial tesseras located 0.5 km southwest, one of them having 15.6-15.9% clay in 21 cm of surface soil, the other 16.4-16.7% to 35 cm depth.

Weights per square centimeter area and to 120 cm depth range from 141.91 g of total soil to 45.55 g of fine earth to 7.21 g of clay. On the assumption that the original pumice strata are clay-free, the rate of clay accumulation in 11,000 years is, yearly, 0.66 mg/cm^2/120 cm. It represents the balance of clay formation and clay losses by dissolution and erosion.

Based on biotite alteration during the Holocene in Central Europe, Meyer and Kalk (38) estimate for young loess a clay genesis of 15 g/cm^2 to a depth of 180 cm. For the postglacial time period of 10,000 years the gain is 1.5 mg of clay/cm^2/180 cm/year.

In a semi-evergreen, seasonal forest on the volcanic island of St. Vincent, B.W.I. (maP 230 cm, maT 27°C), a yellowish-brown soil 1.2-2.4 m in depth contains 4-5% clay ($< 2 \mu m$) and rests on pyroxene andesite ash having 0.5-1% clay. Hay (22) finds that during 3600-4100 years part of the fine vitric ash transformed to halloysite and allophane clays and traces of iron oxide. When a bulk density of 1.5 g/cm^3 is assumed, the acquired clay content (3.75%) of a tessera of 1 cm2 area and 1 m depth is 5.6 g or 1.4 mg/cm^2/m/year.

To be elaborated later (Fig. 9.13), in 10,000 years clay genesis in sands averages 0.17 mg/cm^2/m/year in Michigan and 0.73 mg/cm^2/m/year in New Zealand.

Red Soils of Sonoran Desert

In the arid, southwestern United States and in the adjacent Sonoran Desert of Mexico fanglomerates from granite mountains turn into reddish-colored soils. Walker (56) distinguishes the following geological and pedological age sequence:

Geologic formations	*Soil colors*
Recent deposits	Light gray
Young Pleistocene deposits	Reddish yellow (6.5 YR 6/6)
Older Pleistocene deposits	Light reddish brown (2.5 YR 6/4)
Pliocene (Tertiary) deposits	Intense red (10 R 4/6)

The deepening reddish tints are matched by seams of Ca-cementation and by C$_{ca}$ horizons (caliche). Reaction remains above pH 8 in all horizons of all ages. Clay formation is sparse and does not go beyond montmorillonite and illite-montmorillonite intergrades. The red ferric oxide pigments of the Pleistocene soils are amorphous to X-rays, but in Pliocene deposits well-crystallized hematite appears.

Present-day climate has maP 8 cm and maT > 20°C and the few rains fall as cloudbursts that cause flash floods. Fossil evidence and evaporates (halides, gypsum) indicate that desert climate prevailed throughout the Pleistocene and Pliocene periods. The iron stains originate from *in situ* alteration of biotites and hornblende minerals and in early stages form reddish halos around the grains. Walker rules out tropical source materials and lengthy humid environments though some vegetational fluctuations appear to have taken place (52).

Nevada Glacial Outwash Terraces

Turbulent Truckee River runs from scenic Lake Tahoe (1900 m) in the Sierra Nevada eastward through Reno (maP 18 cm, maT 9.5°C) into Pyramid Lake in the desert. During the Pleistocene mountain glaciations named Tahoe, Donner, and Pre-Donner, the river deposited granitic and basaltic outwashes as sand, gravel, and boulder terraces that were stream-incised during interglacial periods, as depicted by Birkeland (4) in Figure 9.11 for the vicinity of the hamlet of Verdi (maP 60 cm). The open forest of tall Jeffrey pines (*Pinus jeffreyi*), largely undisturbed except by occasional ground fire, has understories of native bunch grasses and fragrant sagebrush (*Artemesia tridentata*).

Soil features in Table 9.5 indicate progressive weathering of pebbles, clay formation, and profile differentiation, with halloysite being the key clay mineral; near drier Reno montmorillonite becomes more abundant (5). The soils on Tahoe and Donner outwashes are Typic Argixerolls and are low in extractable iron, 0.20 to 0.53% in oxalate and 1.1 to 1.6% in citrate-dithionate solutions (1).

In an overall view, Birkeland and Janda (6) interpret the profiles as a chronosequence that evolved under a relatively constant climate throughout the entire middle and late Quaternary weathering episodes.

San Joaquin Hardpan

In the 1920s C. F. Shaw recognized in semiarid San Joaquin Valley alluvial fans and terraces that are alike in all state factors except age. Arkley (2) identifies the chronosequence of soil series:

Tujunga   →   Hanford   →   Greenfield   →   Snelling   →   San Joaquin
less than   1000-10,000 years   10,000-140,000 years
1000 years

All soils develop on granitic alluvium of medium sandy loam texture and light brownish-gray color. Slopes are 0-2%. The climate is hot and dry in summer and cool

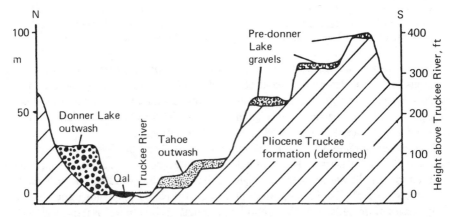

Fig. 9.11. Positions at Verdi, Nevada, of outwash gravel terraces belonging to the Sierra Nevada glaciations Tahoe, Donner and Pre-Donner, from Birkeland (4)

and moist in winter, with maP 28-33 cm, except Snelling (45 cm), and maT 15-16°C. It supports grasses, forbs, and oak-grass savanna.

The brown B horizons of the younger series convert to reddish brown (2.5 YR, 4.4 dry) in the older soils. Reactions in A and C horizons are pH 6.0-7.3, in B horizons 5.4-6.2. Total nitrogen is low, 0.01-0.04%. In pot tests available phosphorus is ample in young and deficient in old soils, in spite of total P remaining constant throughout the chronosequence.

Clay depth functions are plotted in Figure 9.12. The San Joaquin soil series acquires an impressive, indurated iron-silica hardpan with spots of purplish-black manganese oxides. It is impermeable to roots and water, but softens with depth.

Below it, at depths of 2.9-6.2 m, bones of camel (*Camelops hesternus*), horse (*Equus* sp.), and mammoth (*Mammuthus* sp.) have been discovered. In the fossils Hansen and Begg (20) measured uranium content and actinium series nuclide activities and concluded that soil formation time zero occurred 103,000 + 6000 years ago. (Older ages are offered by D. E. Marchand and A. Allwardt (1977) in USGS Open-File Report No. 77-748.)

Much of the surface of the old, nearly level fan is modulated by Mima mound or hogwallow microtopography. Winter rains fill the depressions and create myriads of ponds, 5-10 m in diameter. During spring they become "vernal pools," their drying-out creating concentric rings of colorful flowers, discussed in Chapters 11 and 13 on topography and biotic factor.

Valders Chronosequence

In northern Michigan (maP 70 cm, maT 6.2°C) deglaciation left three beaches at altitudes of 179-195 m and adjacent Valders drift at 266 m. The deposits have carbondated ages of 2250, 3000, 8000, and 10,000 years B.P., and all parent materials are sands with combined silt and clay contents of less than 3%. The water tables are at least 2.4 m below the soil surfaces.

Table 9.5. Semiarid Soils on Glacial Outwash Terraces at Verdi, Nevada[a]

Soil properties	Outwash deposits (boulders, gravels, sands)		
	Tahoe	Donner Lake	Pre-Donner Lake
Approx. age (years)	10,000	40,000	> 200,000
Pebble condition			
Granite	Fresh	Grus	Decomposed
Basalt	Fresh	Rinds, 1-2 mm	Thick rinds
Percentage clay in A	16	23	22
Percentage clay in B	18-21	25-46	25 (thicker)
Percentage clay in C	10-20	21-25	25
Color of B	Dark brown	Dark yellow brown	Redder
pH of B	6.5-6.0	6.5-5.5	6.0-5.5
Clay minerals[b]	H, I	H, K, I	H, K, I, M

[a]After Birkeland (4,5).
[b]H, halloysite; K, kaolinite; I, illite; M, montmorillonite.

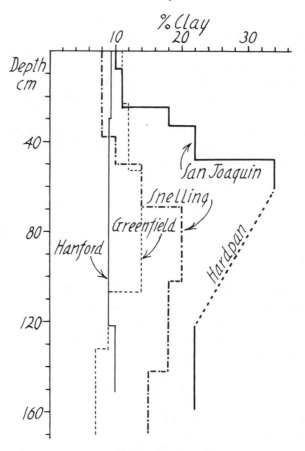

Fig. 9.12. Clay depth functions of four soil series of the San Joaquin chronosequence (2) on alluvial fans. Note cemented hardpan in San Joaquin profile on the oldest geologic Riverbank terrace

Plant succession is believed to have advanced from dune pioneers to heath to jack pine (*Pinus banksiana*) associations, then to the present-day coniferous-broadleaf forests and to the maple-beech climax forest on Valders moraine dominated by *Acer saccharum, Fagus grandifolia,* basswood (*Tilia americana*) birch, elm (*Ulnus* sp.), and others. All soils are Podzols but only the oldest (Valders) qualifies as a Spodosol.

Franzmeier and coworkers (e.g., 14) use quartz as an index mineral and calculate horizon changes in tesseras of 1 cm² surface area and about 1 m depth for numerous soil constituents. The gains in grams of clay and in millimoles of extractable sesquioxides (Fe_2O_3, Al_2O_3), plotted in Figure 9.13, are derived from weathering of primary minerals and from transformations of allophanes, illites, and chlorites to vermiculite in Bh and to dominant montmorillonite in both Bh and A2 horizons.

In entire tesseras the CECs follow aging as 0.7, 1.2, 2.8, and 4.0 meq/100 g, and the degrees of base saturation diminish from 74 to 47 to 48 and to 38%.

Included in Figure 9.13 are the much higher clay contents of the sand dunes of New Zealand's warmer Manawatu district, mentioned in a preceding section. Differentials in temperature, 12.2 versus 6.2°C, and possibly in dune mineralogy may be steering the spread in clay genesis.

Means and Variability within a Chronosequence

Soil series are *relatively* uniform soil bodies that have areal extents of a few to hundreds of hectares. Variability of profiles within a series area is made visible in the sketches of Figure 7.4. As a rule, soil surveyors select but one "typical" profile for laboratory analysis, as illustrated in the previously mentioned Figure 9.12 of the San Joaquin chronosequence in which no account is taken of soil variability.

Heterogeneity may be expressed by the statistical device of "standard error" (SE) which is attached to the mean value (M) of a property as M ± SE. The word error is a misnomer, for soil variability is caused mainly by variations of state factors within the series area.

Harradine's (21) chronosequence of Yolo → Zamora → Myers → Hillgate series explicitly recognizes variability. The soils occupy alluvial terraces that have their source materials in upland strata of shales and sandstones. Recent Yolo is the youngest series and Hillgate the oldest. It has a clay-pan B horizon. The random collections of *16 tesseras* within each series provide the means and their errors, arranged here according to soil age. For water-soluble phosphorus (ppm P) in A horizons of 27 cm thickness the values are 0.76 ± 0.142 → 0.52 ± 0.069 → 0.16 ± 0.035 → 0.17 ± 0.024; for bulk densities (g/cm^3) of B1 horizons; 1.58 ± 0.02 → 1.70 ± 0.02 → 1.79 ± 0.02 → 1.88 ± 0.02. Phosphorus decreases and density increases with age.

As judged by criteria offered by statisticians the means are significant in that resampling would produce similar values. Further, *the means* of the end members of the sequence, and of some of the intermediates, *differ from each other* in a statistically highly significant way. No doubt, Yolo and Hillgate are unlike bodies. But no statistical inference can be drawn regarding the cause of the differences, which Harradine identifies with soil age.

Mineral Influxes

In five Victorian (Australia), quartz-free basalt flows of ages 6200 to 4 million years and maP 70-90 cm the extractable amorphous material in *A horizons* declines from 51% in the youngest to 36% in the oldest soil. Simultaneously, its SiO_2/Al_2O_3 ratio narrows from 3.8 to 1.8-2.2. In clays freed of amorphous substances kaolinite plus halloysite increase from 21 to 35% (40).

Unexpectedly, the gravel and sand fractions of A horizons augment with age, contributing up to 81% of the soil mass. Examination identifies the particles as pedogenic,

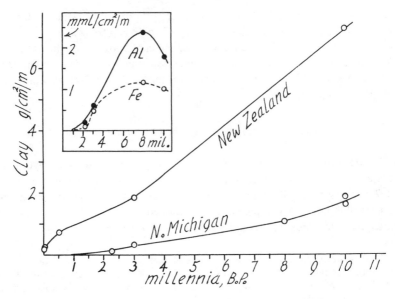

Fig. 9.13. Clay gains in New Zealand (48) and in North Michigan sand dunes, and readily extractable Al and Fe in the latter (14)

ferruginous, concretionary nodules, and fine, foreign quartz grains, the latter amounting to 20% of weight of the oldest soil. Jackson et al. (25) believe that silica is being blown in from adjacent coastal sand dunes and that the accumulations depend on duration of influx, that is, soil age, rather than distance from the source, as in midwestern loess deposits.

Hawaiian basic igneous rocks also are quartz-free and their old soils too are infused with fine-grained quartz, yet the dune source is missing. Marshalling oxygen isotope ratios, Jackson et al. (24) postulate aeolian origin of Hawaiian soil quartz.

Silt-size mica in Hawaiian soils, once attributed to *in situ* genesis, is likewise suspect of continental, aeolian origin, the more so as the age of the mica platelets exceeds that of the Islands by many millions of years, as determined by K-A dating by Rex et al. (44). Tropospheric dust sedimentation in rainfall over the North Pacific Ocean is reported to average 1 m in a million years (26) or 1 μm/year.

E. Pygmy Forest Ecological Staircase

An extraordinary chronosequence of ecosystems that evokes either admiration or disbelief is known as pygmy forest ecological staircase.

Along the Pacific Coast of Mendocino County, California, a series of elevated marine terraces are clad in a mosaic of expanses of grasslands, pygmy trees, pine groves, and tall redwood-Douglas fir forests. Many sites and particularly their soils are still in condition prior to white man's arrival in the 1850s. For over a century naturalists have debated the origin of the stunning vegetational scene and have attached prominence to fires, climatic shifts, and edaphic factors. Here, the entire staircase is viewed as a long-time chronosequence of ecosystems (15,31). Climatically, the area has rainy, snow-free winters and dry, cool and foggy summers, with maP 97 cm and maT 11.6°C.

Landscape Features

The west (left) to east transect in Figure 9.14 outlines schematically the well-preserved sequence of *terraces* that face the incised *canyon* of Jug Handle Creek, one of several that flow east to west into the sea. In Pleistocene times, when the continental ice sheets formed and melted, the world-wide ocean levels sank and rose as much as 115 m. The rising sea cut terraces into the local graywacke sandstone bedrock and during retreat covered the platforms with *beach sands, gravels,* and *clays* several meters thick. Slowly and steadily tectonic forces uplifted the landmass some 200 m without tilting the terraces, thus keeping them intact.

Gardner (15) assigns major terrace altitudes of 30, 53, 91, 130, and 198 m, corresponding to first, second, third, fourth, and fifth terraces. The heights cited scale the hidden, buried nickpoints, not the terrace surfaces. If the cutting of the first platform was completed at the onset of the Wisconsin glaciation, its age would be about 100,000 years. Higher terraces may span a million years. The higher a marine plateau the older it is likely to be.

Besides the platforms covered with beach material, large *sand dunes* mold the topography. Wind is presently blowing graywacke-derived sand from the beaches onto the first terrace, thereby placing fresh dunes upon an older, weathered and plant-covered beach deposit. What is happening today also happened in the past, for extensive sand dunes rest on all terraces. Dune size, shape, and orientation of slip faces are largely preserved and are the same as those of modern counterparts. As parent materials, the beach sand and dune sand on all terraces had initially the same mineral make-up, as verified by analysis of deep C horizons as shown in Figure 9.18 later in this chapter.

Flora and Vegetation

A brief list of conspicuous plant species includes the following:

Trees

Coast redwood (*Sequoia sempervirens*)
Douglas fir (*Pseudotsuga menziesii*)
Grand fir (*Abies grandis*)
Western hemlock (*Tsuga heterophylla*)
— large commercial specimen on fertile soils of moderately weathered dunes and beach materials, and on canyon slopes.

Beach pine (*Pinus contorta*) — on recent dunes near the coast.

Bishop pine (*Pinus muricata*) — closed cone, throughout the staircase but dwarfed on old terraces.

Bolander pine (*Pinus contorta* var. *Bolanderi*) — closed cone endemic of pygmy forest on Blacklock soil on higher terraces.

Mendocino cypress (*Cupressus pygmaea*) — strongly dwarfed endemic in pygmy forest.

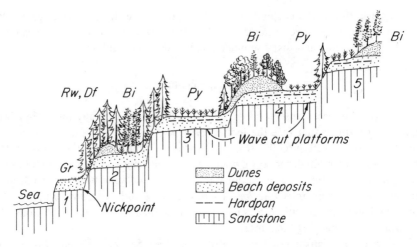

Fig. 9.14. West to east sequence of elevated marine terraces covered with beach material and sand dunes along the Pacific Coast of N. California. Gr, grassland; Rw, redwood; Df, Douglas fir; Bi, bishop pine; Py, pygmy forest

Shrubs

Rose-bay (*Rhododendron macro-
  phyllum*)
Labrador tea (*Ledum glandulosum*)   understory species, wide-spread;
Huckleberry (*Vaccinium ovatum*)   dwarfed in pygmy forest.
Salal (*Gaultheria shallon*)

Ft. Bragg manzanita (*Arctostaphylos   an endemic on older terraces.
  nummularia*)

Two Sequences

Based on the terrace (T) transect sketch shown in Figure 9.14, two ecosystem chronosequences are recognized, one on dunes and the other on beach materials:

Dunes:

near coast	→	on T_1	→	on T_2, T_3	→	on T_4, T_5
bare sand, beach pine (salt vector)		redwoods, firs		redwoods, firs bishop pines		bishop pines on deep Podzolized soils (Ultisols)

Beach materials:

T_1	→	T_2	→	T_3, T_4, T_5
grassland, Mollisol (salt vector)		redwoods bishop pines		pygmy forest on hardpan Podzols (Spodosols)

1. Ecosystem Evolution on Dunes. Near Ten Mile River where the absence of steep cliffs promotes advance of *recent dunes* onto the *first terrace,* the sands become stabilized by the deep-rooted, nodulated yellow Bush Lupine and the beach pine. Accelerated weathering, fixation of nitrogen, and nutrient cycling slowly enhance soil ferility. About 1-2 km inland, *dunes on the second terrace* (Lv site) are tree-covered and several thousand years old, but no ^{14}C dates are available. *Dunes on the front of the third terrace* might have been blown at the end of the last interglacial period or sometime thereafter. At site Nm well drillers encountered Sequoia wood at 7 m depth.

In these relatively young dunes (Lv, Nm) oxidation converts the original gray color into a rich brown. Carbonates are leached out and primary minerals transform to clay contents of 10-20%. Exchangeable Ca + Mg + K are plentiful, especially in the surface horizons (Fig. 9.15). Humus is abundant and at pH 5 is mild. The soils are highly fertile and support magnificent forests of redwood, Douglas and grand firs, and hemlock, with a sprinkling of tall bishop pines.

The very old dune sites on the fourth and fifth terraces are labeled Wi at 125 m altitude and Dr at 171 m. The sands are strongly weathered to great depth and exhibit red-white reticulate mottling which is a sign of profound chemical alteration. The soils, known as Noyo, are Podzolized, and in the Dr dune area have a bleached, whitish A2 horizon underlain by a yellow-brown Bt horizon speckled with hard, pea-sized iron concretions. Exchangeable base content is low (Wi, Dr in Fig. 9.15), acidity high, and of the clay-derived Al-type. To a depth of 10 cm organic matter is enriched; below it the humus reservoir is half-depleted (Fig. 9.16) owing to reduced biomass production.

The fertility-demanding commercial tree species are replaced by bishop pines, the more so the older and more leached the dune. On the highly weathered sands of the Ultisol order the pines and shrubs are undersized. Yet, 50 m away redwoods prosper on the steep-walled Jug Handle Canyon where slow, natural erosion rejuvenates the graywacke slopes to Hugo soils of the Inceptisol order.

2. Ecosystem Evolution of the Terraces Proper. Since sandy beach deposits are only a few meters thick the platform rock below generates a fluctuating water table during the rainy season (Fig. 9.17). It guides soil genesis toward iron hardpan and pygmy forest.

The *first terrace* (100 ft or 30 m) is a grassy plain, the coastal prairie, covered with perennial bunch grasses mixed with colorfully flowered species of the lily, pea, buttercup, and sunflower families. The grasslands awaited the settlers who used them for grazing. Prior, Indians left small kitchen middens, which admits burning as a possible cause of prairie origin.

Above the estuary of Jug Handle Creek *salt injury* from wind-blown sea spray keeps a Sitka spruce bent to the ground. The needles on the seaside are burned brown, and when immersed in water (1 g/30 ml) release 34.1 me/liter chloride (Cl). Green, leeside leaves deliver only 8.2 me Cl/liter. Domestic grasses and surface soils (0-20 cm) also are Cl-contaminated, but less so as the distance from the shore is greater. A prairie transect of soil profiles, analyzed by R. Glauser for exchangeable Na, K, Ca, and Mg, produces for surface soils the plant-favoring trend of Ca versus Na of Figure 9.16. In its inset, depth functions of exchangeable Na$^+$ echo the same course. For comparison, a degree of saturation (DS) of Na above 15% is considered detrimental to domestic crops. The observed salinities are balances of salt influx by wind and soil leaching by rainfall.

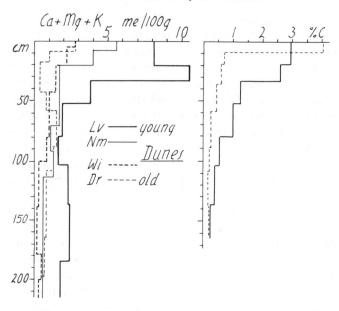

Fig. 9.15. Depth functions of exchangeable bases (Ca + Mg + K) and organic carbon
(C) of young and old sand dunes in Figure 9.14

To uncover the impact of salt on trees, 2-year-old seedlings of redwood, Douglas
fir, beach pines, and bishop pines were planted in 1969 in five replications (and
watered during two summers) at the following distances (meters) inland from the
sea cliff: 6, 27, 58, 73, 144, 220, 337. Within the first 60 m all trees soon died in
spite of plot replications and replantings. Beyond, redwood and fir lingered on but
succumbed 5 years later; only pines, mostly *P. contorta,* are surviving. Their seaside
needles are bronzed and their statures are a diminutive 50 cm at 73 m and a modest
200 cm at 327 m. Because ocean sprays are conditioned by cliff configurations and
storm patterns and since leaching adjusts to rainfall, the salt-modulated swath of
coastal grassland may expand and contract.

The first terrace has dark-colored Prairie soils or Mollisols that require thousands of
years of genesis. The common, very dark brown (10 YR 2/2, moist) Baywood loamy
sand (Pachic Ultic Haploxerolls) is of the accruement type, receiving minute annual
additions of wind-blown particles from the narrow sand beaches below the cliff. Soil
surface and vegetation are slowly rising together.

On the much older *second terrace* the heterogeneous beach material is well
weathered. Dark gray loams are underlain by pale brown sandy clay loams. The soils
belong to the many-faceted, acid Noyo series of the Alfisol and Ultisol orders. Tree
vegetation is diverse. At some sites a few redwoods mingle with tall bishop pines, at
others hemlocks and slender redwoods occupy fine-textured profiles with A2, iron
concretions, and water tables, or 11- to 18-m-tall cypresses and bishop pines with an
occasional redwood tree may grow on bleached soil.

The *third* and *higher terraces* carry predominantly pygmy forest on Blacklock soil
which is a seasonal ground water *Podzol,* a *Spodosol* of suborder Typic Sideraquod.

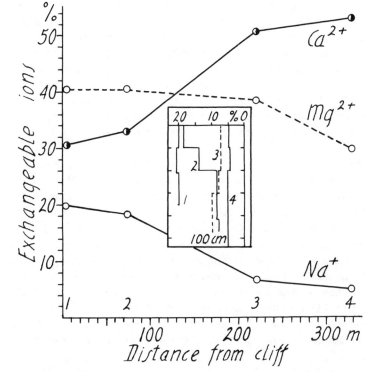

Fig. 9.16. Exchangeable Na$^+$, K$^+$, Mg^{2+}, and Ca^{2+} along a west (left side) to east transect on the first terrace grassland of Figure 9.14. Inset: depth function of exchangeable Na$^+$ at sites 1, 2, 3, 4

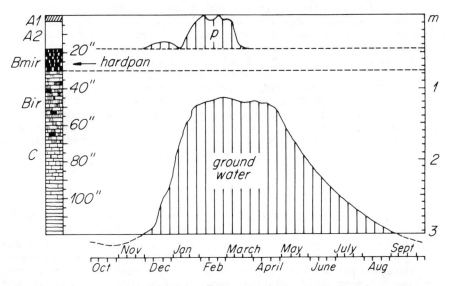

Fig. 9.17. Seasonal water table fluctuations in Blacklock Spodosol (Podzol). Note perched water table on hardpan during the rainy season

The dark-gray, 10-cm-thick A1 horizon rests on 36 cm of bleached, white A2 horizon which is thixotropic and strongly acidic having pH as low as 2.8. Below A2, the concrete-like iron hardpan Bmir spans the depth interval of 46-76 cm. Under the pan are rusty, mottled, and weakly cemented sands and sandy loams. At depths of 1.5-3 m unaltered beach sand is reached and at 3-6 m the sea-cut platform of hard sandstone. Over horizontal distances of a few meters the profiles may or may not have a humus tongue in A2, a humic Bh horizon on top of the hardpan, or a thin ($<$ 5 cm) clay layer on or within, but never below, the stone-hard subsoil crust (Fig. 7.4). Possibly, the clay lenses are remnants of pedogenic mineral weathering in A and B.

Water Regimen (Fig. 9.17). When a 3-m-deep test hole is dug at the end of the dry summer no free water is encountered. Following the first 10-20 cm of winter rains water begins to pile up on the basal rock plane and a water table ascends. Long before it reaches the upper strata, rain water accumulates above the hardpan and creates a perched water table that floods the surface soil. The standing water is colored coffee brown from dispersed humus. In late spring water in the surface soil disappears by evapotranspiration and lateral seepage and the A horizons become extremely xeric. In October the descending ground water table sinks below the 3 m mark. In swales at the foot of dunes wetness persists, giving rise to sphagnum bogs. Podzolization moves Fe from A2 to B in both Noyo dunes and Blacklock soils. In the latter the rise and fall of the water table during winter time, followed by severe desiccation during the rainless summer, seems to encourage "fusion" of iron hydrous oxides and quartz grains to a hardpan.

Pygmy Forest Vegetation. Extreme pygmy forest is species poor and space unsaturated, with as much as one-quarter of the ground area bare or covered with colonies of lichen (e.g., *Cladonia* sp., *Usnea* sp.). Trees too are lichen-encrusted and they are infected by gall-forming fungi that stunt leaders and turn bishop and Bolander pines into gnarled and twisted cripples with high death rates. Though decades and centuries old most trees are only 1.5-3 m tall. Associated with stunted growth is poverty of soil nutrients. The sum of exchangeable Ca, Mg, and K is less than 1 me/100 g, a pitiful quantity. Its function is worsened by high exchange acidity (Al^{3+}, H^+) and by shallow root depth. Available N and P also are low, shown by leaf analysis and by pot tests in which N places first among limiting elements.

On fertilizer plots response of pine and cypress growth to a mixture of dolomite, K, S, P, and urea nitrogen is minimal though statistically significant. When urea is replaced by nitrates biomass production rises dramatically. As elucidated by A. D. McLaren and A. Pukite, ammonification of urea occurs readily in these soils, inferentially by urease enzyme sorbed on soil colloids, but nitrification on treated and untreated stands is negligible because bacteria are sparse.

Fire

The endemic Bolander pine is serotinous, its cones remaining closed for about 3 years. Closed-cone pines are perceived by ecologists as fire adaptations and since the subspecies congregates on Blacklock soils the entire ecosystem is suspected of being pyrogenic, but lightning is rare and fires are not known to make A2 horizons, iron hardpans, and reticulated C horizons. How many millennia ago the cones began closing is unknown. Maybe selection pressures other than fire, e.g., pedogenic stress, played a role, though no handy mechanism comes to mind.

Numerical Chronosequence

In sand fractions of A2 Gardner (15) counted the slowly weathering potassium feldspar crystals (F) and the highly resistant quartz grains (Q). The ratios F/Q are plotted in Figure 9.18 against terrace heights which reflect age differences. The two large half circles on the vertical axis are F/Q of recent dune and beach materials. The white dots characterize the C horizons of the Blacklocks. Their mean is 17.5 feldspars to 100 quartz grains. The dashed horizontal line expresses the mineralogical conformity of the parent materials on the various terraces.

The black dots record F/Q of the Blacklock A2 horizons. The first terrace soil pertains to a localized site of a Blacklock precursor north of Ft. Bragg. The profiles of the older surfaces have low ratios, less than 0.0003 for the fifth terrace. They signify fargone weathering, soil antiquity, and, by inference, stability of land forms. The almost horizontal base branch of the curve defines a near-steady-state condition in which minute rates of losses by solution and surface erosion, e.g., triggered by invertebrate turbations, balance gains of weathering and aerial influxes. Actually, the hardpan may be descending vertically at nanorates by dissolution at its upper face and reformation below.

Past climatic shifts are not disputed. Their severity is unknown. A higher rainfall accelerates, a lower one retards the leaching process without changing its character. Soil and plant cover come close to a terminal, steady-state ecosystem and are expected to persist unless the state factors cl and r change drastically, the latter by tectonic tilting.

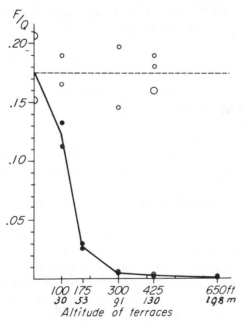

Fig. 9.18. Ratios of K-feldspar (F) to quartz (Q) in the A2 horizons of terrace soils (black dots). White circles pertain to ratios in fresh sands and in C horizons of older deposits

Summation

Westman and Whittaker (61) measure soil (s) and vegetation (v) properties and comment on the enormous spread of biomass in the staircase that rivals any forest comparison anywhere. The Blacklock Spodosol supports a scant 21 vascular plant species whereas the other notorious problem soil, the serpentine derivative (Table 10.8), shelters a signal 113. Westman (60) singles out pH of A horizons as an "environmental axis" or pH "gradient" on which he ordinates v properties with considerable success though no causality is implied. On the axis the role of hardpan becomes confounded with low pH.

To focus on pedogenic trends (Chapter 7), the chronosequence does not end in laterite because the siliceous parent material is too low in Fe and Al. Frugal pines displace redwoods, firs, and hemlocks because longtime leaching depletes the nutrient supply and acidifies the soil. Spodosols develop because the surviving plant and microbe species produce humus that favors Fe translocation from A to B horizons.

Note during printing: An extreme Quaternary chronosequence on dune sands in Australia generated a pygmy forest on a gigantic Podzol that possesses a 12 m thick A2 horizon resting on a cemented iron-humus B horizon of 5 m thickness (C. H. Thompson and G. D. Hubble. 1977. *Proc. Clamatrops* 179-188).

F. Chronosequences and Plant Succession Theories

H. C. Cowles formulated in 1899, in what was to become a paradigm, the idea of a temporal plant succession that culminated in Clements' (9) visionary construction of the vegetation climax in a mature landscape. The maturity expression is borrowed from geomorphologist Davis's (11) tenets of erosion cycles in which mountains are reduced to undulating, "mature" terrains. Later, Marbut, physiographer and one-time chief of American Soil Survey, and also a follower of Davis, coined the "normal" or "mature" soil profile in a mature landscape (28).

Ever since, and even before (7), the notion of plant succession and soil genesis being synchronized and *progressing toward the plant climax on the mature soil* has held sway in many quarters (13,62). This envisioned nexus of soil and biota invites scrutiny, the more so as the "mature soil" has failed to pass the test of general acceptance.

Some Conventional Ecological Beliefs

In basic outline, plant successions proceed orderly from species-poor associations of low stature (prostrate plants, grasses, sedges) to communities of tall plants having many species and complex structures. The driving force of a succession is plant interaction through shading and chemical repression and site modification in such a way that a species can no longer grow as successfully as others. Advancement is community-initiated.

Ecological energetics (41) brings into focus energy flow, which is the caloric value of organic mass transport from leaf producers to consumer animals to microbes, and it underscores gross photosynthesis (P) and community respiration (R). A plant succession begins with P/R > 1 and ends in the climax ecosystem in which P/R is equal to 1.

Climax is said to be the endpoint of succession, a steady-state system in permanent equilibrium with environment, if undisturbed, and having maximum total biomass, highest species diversity, and highest stability. Further, an advancing succession "tightens" mineral cycling and loss by entrapping and holding nutrients within the biota. Implicit in these definitions is constancy of gross environment during succession and while the climax persists.

It should not be construed that all ecologists accept all of these postulates. Loucks (33) and Drury and Nisbet (13) among others dwell on the reasons for discordance. Since half of the plant lives underground and impresses its seal on the soil, ecopedology amasses its own observations and thoughts.

Role of Initial States

"Primary succession" starts on land not previously occupied by vegetation and responds to parent material (p) and ensuing soil genesis. From a pool of available germules p selects those destined to be pioneers, for a given species may be successful on acid material but not on a calcareous one. On a purely mineral substrate the pioneer has to be a nitrogen accumulator, alone or in symbiosis.

"Secondary succession" begins in a clearcut forest area, in a burned watershed, or on abandoned agricultural land (old-field succession). The initial state is S_0 and a reservoir of soil nitrogen and available nutrients had been built up previously, enabling succession to be fast. It need not recapitulate the primary succession.

Affecting the Balance of Minerals

With advancing succession the leaching of nutrients becomes accelerated by biologically induced higher CO_2 pressures in the soil pores, by chelate mobilization of microelements, by nitric acid solubilization of replaceable cations, and by microbial mineralization of plant and animal residues. At the same time essential elements are being preserved by uptake into biomass, by reduced water percolation stimulated by transpiration, and by lower hydrolysis rates of exchangeable bases in the presence of exchange acidity. The balance of gains and losses is upgraded by weathering and aeolian influxes.

Steady States Reexamined

Some of the imagined Clementsian and Marbutian sequences last millions of years, but latter-day plant succession studies are confined to decades, centuries, and a few millennia. Although steady-state and chemical equilibrium differ in entropy production (Chapter 1) they share the independence of time, written for system property l:

$$\Delta l/\Delta t = 0, \quad \text{or} \quad dl/dt = 0$$

average rate   instantaneous rate

which, supposedly, lasts a long time, or, until a state factor changes.

For a given r, p, and ϕ the idea of maximum biomass being at steady state has appeal in hilly and mountainous terrain. On the steeper slopes of Jug Handle Canyon minute, natural erosion at the soil surface is seemingly balanced by rock weathering below and a luxuriant climax plant cover coincides with an Inceptisol that may have a steady-state profile.

Contrastingly, on level and gently sloping lands of the humid-temperate regions a long-time pedogenic plant succession may obey the functions drawn in Figure 9.19. Minor climatic shifts may have altered species frequences but barely biomass production and standing stock. Random perturbations (e.g., fires, windthrow) are not drawn. As the ecosystem unfolds soil transformation induces displacement of species quite separately from community control and maximum biomass emerges as a temporary, early episode during which soil properties are still changing. Thus, if humus is included in biomass (55), its steady-state condition will have to be ascertained separately from that of plants. The P/R of a climax is not equal to 1 unless soil respiration too is at steady state.

The plant-soil dissonance may be partly resolved by discounting slow reactions. If vegetational (v) changes outpace soil (s) changes, that is, if $\Delta v/\Delta t \gg \Delta s/\Delta t$, then $\Delta s/\Delta t$ is unimportant and the soil may be taken as invarient. Annual cycling of nutrients may be much faster than the rate of leaching. The size of the error committed depends on what v and s properties are being studied.

Terminal Steady States

Beyond the biomass maximum, P/R becomes < 1 because the humus capital is being oxidized in excess of its renewal. Eventually, P/R = 1 again, but for lower magnitudes of P and R. Individual soil properties advance at their own beat and reach steady states in thousands to 1 million years, often as substrates that are detrimental to plant growth and animal niches. The pauperized vegetations are "edaphic climaxes." As end members of longtime chronosequences they may be very stable ecosystems.

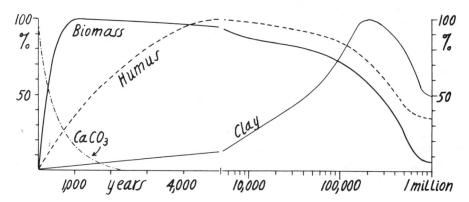

Fig. 9.19. Hypothetical chronosequence of soils and vegetation during a million years on level topography. Vegetation climatic climax arrives first, humus maximum later, and clay peaks last, followed by an edaphic plant climax. Absolute scale to 5000 years, logarithmic scale afterward. Humid climate

Edaphic Climax

It was A. F. W. Schimper who in 1898 introduced the Greek word edaphon (soil) into the botanical literature. Today, an edaphic climax is a stabilized plant community not fitting into temporal succession because of an unusual substrate. Thousands of soils are unusual, for various reasons. The botanically featured serpentine barrens are an imprint of parent material, whereas the biorestricting Blacklock soil is a time-Methuselah. In the light of chronosequences climatic and edaphic plant climaxes may belong to the same time arrow, the former converging to the latter in a pedogenic plant succession. Within the same macroclimate, as in the Staircase condition, both climatic and edaphic climaxes are substrate-conditioned, namely, fertile versus infertile soils.

Clementsian succession permits a reversal of the time arrow, the edaphic climax eventually turning into a climatic one, a view that still commands partial support. In a chronosequence such a turn-about presupposes a major alteration of a state factor, e.g., tectonic tilting that erodes the A and B horizons and starts a new sequence.

Speciation

End members of long-time chronosequences carry vegetations impoverished in species but relatively enriched in endemics, subspecies, and ecotypes that suggest evolutionary trends. The dominant cause of evolution by selection is conventionally attributed to climatic shifts (43) but since soil genesis may transform a fertile substrate into one of nutrient stress, even in a stable climate, the impact of soil in fostering speciation through selection and DNA disturbance (18) may be greater than hitherto imagined.

G. Review of Chapter

In chronosequences pieces of landscapes that differ mainly in age are strung together, as documented for sand dunes, moraines, terraces, and the Shasta mudflows.

Beginning on wasteland the soil processes fabricate humus and clays that promote diversity and productivity of soils and plant life. The few data on hand reveal yearly gains of nitrogen and clay amounting each to a few grams/square meter.

In arid regions soil genesis bolsters the base status (e.g., $CaCO_3$) and perpetrates near-neutral reactions or enhances alkalinity. In humid regions bases are leached out and soil acidity accrues.

Soil properties experience gains and losses simultaneously and appear to be verging toward steady states. These are approached slowly compared to the rapid progress of vegetation toward climaxes. In a long-time chronosequence climatic plant climaxes may merge into edaphic climaxes.

Conversion of a natural ecosystem to agricultural practice initiates humus decline and horizon disorders. Fortunately man's endeavors and large inputs of energy can maintain biomass productivity, even escalate it.

References

1. Alexander, E. B. 1974. *Soil Sci. Soc. Am. Proc.* 38: 121-124.
2. Arkley, R. J. 1964. *Soil Survey of the Eastern Stanislaus area, California.* U.S.D.A., S.C.S. Soil Surv. 1957, No. 20.
3. Bartholomew, W. V., and D. Kirkham. 1960. *Trans. 7th Int. Congr. Soil Sci.* 2: 471-477.
4. Birkeland, P. W. 1968. In *Means of Correlation of Quaternary Successions,* R. B. Morrison and H. E. Wright, Jr., eds., pp. 469-500. Congr. Int. Ass. Quat. Res. 8.
5. Birkeland, P. W. 1974. *Pedology, Weathering and Geomorphological Research.* Oxford Univ. Press, New York.
6. Birkeland, P. W., and R. J. Janda. 1971. *Geol. Soc. Am. Bull.* 82: 2495-2514.
7. Braun-Blanquet, J., and H. Jenny. 1926. *Denkschr. Schweiz. Nat. Ges.* 63: 183-349.
8. Burges, A., and D. P. Dover. 1953. *Aust. J. Bot.* 1: 83-94.
9. Clements, F. E. 1928. *Plant Succession and Indicators.* Wilson, New York.
10. Crocker, R. L., and J. Major. 1955. *J. Ecol.* 43: 427-448.
11. Davis, W. M. 1909. *Geographical Essays.* Ginn, Boston.
12. Dickson, B. A., and R. L. Crocker. 1953/1954. *J. Soil Sci.* 4: 123-154; 5: 173-191.
13. Drury, W. H., and I. C. T. Nisbet. 1973. *J. Arnold Arb.* 54: 331-368.
14. Franzmeier, D. P., and E. P. Whiteside. 1963. *A Chronosequence of Podzols in Northern Michigan.* Mich. Agr. Exp. Sta. Quart. Bull. 46.
15. Gardner, R. A. 1967. Sequence of podsolic soils along the coast of northern California. Ph.D. Thesis, Univ. of California, Berkeley. Univ. Microfilms, Ann Arbor, MI.
16. Glauser, R. 1967. The ecosystem approach to the study of the Mt. Shasta mud flows. Ph.D. Thesis, Univ. of California, Berkeley.
17. Goh, K. M., T. A. Rafter, J. D. Stout, and T. W. Walker. 1976. *J. Soil Sci.* 27: 89-100.
18. Grant, V. 1975. *Genetics of Flowering Plants.* Columbia Univ. Press, New York.
19. Haas, H. J., and C. E. Evans. 1957. *Nitrogen and Carbon Changes in Great Plains Soils as Influenced by Cropping and Soil Treatments.* U.S.D.A. Techn. Bull. 1164.
20. Hansen, R. O., and E. L. Begg. 1970. *Earth Planet. Sci. Lett.* 8: 411-419.
21. Harradine, F. F. 1950. *Soil Sci. Soc. Am. Proc.* 14: 302-311.
22. Hay, R. L. 1960. *Am. J. Sci.* 258: 354-368.
23. Hénin, S., and G. Pedro. 1965. In *Experimental Pedology,* E. G. Hallsworth and D. V. Crawford, eds., pp. 29-39. Butterworths, London.
24. Jackson, M. L., T. W. M. Levelt, J. K. Syers, R. W. Rey, R. N. Clayton, G. D. Sherman, and G. Uehara. 1971. *Soil Sci. Soc. Am. Proc.* 35: 515-525.
25. Jackson, M. L., F. R. Gibbons, J. K. Syers, and D. L. Mokma. 1972. *Geoderma* 8: 147-163.
26. Jackson, M. L., D. A. Gillette, E. F. Danielsen, I. H. Blifford, R. A. Bryson, and J. K. Syers. 1973. *Soil Sci.* 116: 135-145.
27. Jansson, S. L. 1958. *K. Lantb. Högsh. Ann.* 24: 101-361.
28. Jenny, H. 1941. *Factors of Soil Formation.* McGraw-Hill, New York.
29. Jenny, H. 1965. *Z. Pfl. Düng. Bod.* 109: 97-112.
30. Jenny, H., A. E. Salem, and J. R. Wallis. 1968. In *Soil Organic Matter and Soil Fertility,* P. Salviucci, ed., pp. 5-37. Pont. Acad. Sci. Scripta var. 32, Vatican.
31. Jenny, H., R. J. Arkley, and A. M. Schultz. 1969. *Madroño* 20: 60-74.
32. Kellogg, C. E., and I. J. Nygard. 1951. *The Principal Soil Groups of Alaska.* U.S.D.A., Agr. Monogr. 7.
33. Loucks, O. L. 1970. *Am. Zool.* 10: 17-25.
34. Major, J. 1974. In *Vegetation Dynamics,* R. Knapp, ed., pp. 207-213. W. Junk, The Hague.
35. Mandelbrot, B. B., and J. R. Wallis. 1968. *Water Resourc. Res.* 4: 909-918.

36. Martel, Y. A., and E. A. Paul. 1974. *Can. J. Soil Sci.* 54: 419-426.
37. Mercanton, P. L. 1940s-1950s. Les Alpes, Berne, Switzerland.
38. Meyer, B., and E. Kalk. 1964. In *Soil Micromorphology,* A. Jongerius, ed., pp. 109-129. Elsevier, Amsterdam.
39. Mirsky, A. (ed.). 1966. *Soil Development and Ecological Succession in a Deglaciated Area of Muir Inlet, Southeast Alaska.* Inst. Polar. Stud., Res. Found. Rept. 20, Columbus, OH.
40. Mokma, D. L., M. L. Jackson, and J. K. Syers. 1973. *J. Soil Sci.* 24: 199-214.
41. Odum, E. P. 1971. *Fundamentals of Ecology,* 3d ed. Saunders, Philadelphia.
42. Olson, J. S. 1958. *The Bot. Gaz.* 119: 125-170.
43. Raven, P. H. 1964. *Evolution* 18: 336-338.
44. Rex, R. W., J. K. Syers, M. L. Jackson, and R. N. Clayton. 1969. *Science* 163: 277-279.
45. Saitô, T. 1955. *Tohoku Univ., 4th ser. Biol.* 21: 145-157.
46. Salisbury, E. J. 1952. *Downs and Dunes: Their Plant Life and Its Environment.* G. Bell, London.
47. Schultz, A. M. 1969. In *The Ecosystem Concept in Natural Resource Management,* G. M. Van Dyne, ed., pp. 77-93. Academic Press, New York.
48. Syers, J. K., and T. W. Walker. 1969. *J. Soil Sci.* 20: 57-64.
49. Syers, J. K., J. A. Adams, and T. W. Walker. 1970. *J. Soil Sci.* 21: 146-153.
50. Tedrow, J. C. F. 1977. *Soils of the Polar Landscapes.* Rutgers Univ. Press, New Brunswick, NJ.
51. Ugolini, F. C. 1968. In *Biology of Alder,* J. M. Trappe, et al., eds., pp. 115-140. Pacific N. W. For. Range Exp. Sta., Portland, OR.
52. Van Devender, T. R., and W. G. Spaulding. 1979. *Science* 204: 701-710.
53. Viereck, E. 1959. Small mammal populations in Mt. McKinley National Park, Alaska. Ph.D. Thesis, Univ. Colorado, Boulder, CO.
54. Viereck, L. A. 1966. *Ecol. Monogr.* 36: 181-199.
55. Vitousek, P. M., and W. A. Reiners. 1975. *BioScience* 25: 376-381.
56. Walker, T. R. 1967. *Geol. Soc. Am. Bull.* 78: 353-368.
57. Walker, T. W., and J. K. Syers. 1976. *Geoderma* 15: 1-19.
58. Webley, D. M., D. J. Eastwood, and C. H. Gimingham. 1952. *J. Ecol.* 40: 168-178.
59. Wells, N. 1959. *N. Zeal. Inst. Agr. Sci. Proc.* 40-44.
60. Westman, W. E. 1975. *Ecol. Monogr.* 45: 109-135.
61. Westman, W. E., and R. H. Whittaker. 1975. *J. Ecol.* 63: 493-520.
62. Whittaker, R. H. 1970. *Communities and Ecosystems.* Macmillan, New York.
63. Wohlrab, G., and R. W. Tuveson. 1965. *Am. J. Bot.* 52: 1050-1058.

10. State Factor Parent Material

Fresh sand dunes, moraines, lava flows, alluvium, and mine tailings possess the composition p and the shape configuration r. Their description as the initial state of landscape evolution is a benchmark like the birth record of a child. The infant develops, but the registry remains and is basic to growth studies.

In soil genesis p is the parent material of soils, or mother rock in French (*roche mère*) and German (*Muttergestein*). As soon as the parent material body is exposed to the environment of the site the pedogenic processes of weathering, translocation, and humification begin operating, and the more refined the instrumentation available the sooner alteration from p to soil is detectable. Moisture and temperature may be the first to change.

A. Conceptual Topics

To early pedologists, p was the "geologic substratum" of soils. When pedogenic functions were formulated, p needed redefinition and precision.

Parent Material and C and R Horizons

Soil science began in northern Europe where many soils are of postglacial age, hence relatively shallow and young. Their A and B horizons are distinctly different from the underlying, fresh, unweathered C horizon which is viewed as the initial material of the upper horizons.

Where, as in the United States, soil survey is geared to agricultural production and soils are old and deep, soil depth is confined to root depth, more or less. Parent material and C horizon advert to weathered rock or saprolite. The undecomposed rock at greater depth is labeled R or D, the "grandparent material" in Milne's (32) facetious remark. Either weathered or unweathered rock may be chosen as representing p though the two may lead to discordant conclusions. For state factor sequences the choice is preferably D or R because C may already be a function of cl, ϕ, r, and t and thus the ps could not be compared or "kept constant." When a former soil or an entire ecosystem is chosen as initial state, descriptions of entire depth functions not merely of a lower horizon are required.

The present-day state of an ecosystem possesses the property groups l, v, a, s and the four letters include the specialized climates of the subsystems. The initial state is written l_0, v_0, a_0, s_0 where 0 denotes the condition of each property at start. For precision and consistency the initial soil system is referred to a chosen standard state, say, 25°C for temperature, air- or oven-dryness for moisture, and 1 atmosphere for pressure. Standardization puts all parent materials into the same "climate" and the need for it is most obvious for swelling shales and clay deposits.

Rock versus Soil

By Webster's definition a rock is any solid mineral matter occurring naturally in large quantities, hence geologists consider soil a soft, unconsolidated rock. Soils are unique rocks because their properties are functions of the atmospheric environment. Actually, exposed, hard rocks too are governed by season and climate, at least as far as their moisture, temperature, and related properties (e.g., refractive indices) are concerned. In that sense they are soils. They are low in fine earth and will not sustain commercial crops, but neither will strong saline soils or indurated laterite surfaces. The deduction that hard rocks on the Earth's surface are soils is at odds with tradition but eases conceptual dilemmas.

Mineral Influx

Ecosystems may receive aeolian dust and aquatic sediments. The sudden massive influx of 30 m of ash from the Krakatoa eruption in 1883 created a new parent material and set a new time zero. The soil it buried became a fossil soil or paleosol. A swollen river may unload gravel and sand and later top it with a thick layer of silt. Machette et al. (30) describe a soil profile on multiple parent materials resulting from several deposition episodes: aeolian sand, colluvium, loess, and alluvium on shale, identified with Roman numerals. Its buried, calcareous B horizon is identified as III Bca.

When thin additions are separated by decades or centuries, each layer undergoes successive transformation and a complex soil profile arises in the manner of the Nile profile of Figure 7.15. To generalize, if the rate of soil formation is negligible compared to the rate of mineral influx, parent material builds up; if the rate of transformation is rapid, soil accrues. In such accruement profiles the distinction between soil and parent material is no longer expedient because both may accumulate simultaneously. The process becomes ecosystem genesis under a mineral influx regime.

Nature of Lithosequences

The quality of a soil stands in rapport with the original rock material below it. State factor analysis explores the dependency inductively by comparing soils derived from a variety of parent materials under comparable conditions of the other state factors, written formally as

$$l,v,a,s \ = \ f(p)_{cl,\phi,r,t,...} \qquad \text{or} \qquad l,v,a,s \ = \ f(p,cl,\phi,r,t,...)$$

$$\text{ideal case} \qquad\qquad\qquad\qquad \text{actual case}$$

The simplest lithosequence comprises adjacent rocks and their soils and vegetation. In the locality shown in Figure 10.1, about 5.5 km north of summer-dry Geyserville, California (maP 100 cm), oak trees (*Q. agrifolia*) occupy the rocky outcrops of schists; native bunch grasses (*Stipa* sp.) and wild oats (*Avena* sp.) populate the extensive slopes (Raynor series), and a digger pine forest (*P. sabiniana*) is confined to adjacent serpentine soil (Montara series). Across the sharp boundary of 1 m width grass biomass drops precipitously from 40-110 to 5-15 cm in height. The sites experience identical macroclimates, the seeds reach both soils, grass fires find no obstacles among pines, and the landscape setting is very old. This qualitative lithosequence and grass-forest boundary is considered pedological and is not attributed to climatic change or to fire history.

Many lithosequences are practically discontinuous in space because the parent material changes abruptly. When wind and water segregate particles according to size, their settling across the landscape brings forth influx-sequences, typified by continuous texture gradations along alluvial fans and diminishing loess thicknesses at lengthening distances from the river flats.

Fig. 10.1. Two-member lithosequence of native ecosystems with sharp contact boundary. Oak-savanna on schist-derived soils and Digger pine forest on serpentine soils

B. A Primer of Rock Compositions

Most rocks are mixtures of minerals and from Streckeisen's (1967) compilation it is clear that rock classification is beset with conceptual and practical difficulties. No general consent is at hand.

Igneous Rocks

An erupting volcano ejects hot ashes and disgorges liquid lava streams that may scorch vast areas of land. As the mass cools it hardens, quickly at the surface, slowly at great depth. Slow crystallization favors coarse-grained varieties of rocks that glitter with large crystals of quartz, feldspar, mica, hornblende and olivine.

Chemical analyses of rocks are reported as percentages of weights of oxides of elements, the OH ions appearing as H_2O. Because numbers of chemical species are more revealing than weights the percentages are divided by the molecular weights which are the sums of atomic weights. For a granite rock the calculation of molecular values, MV, as moles (%/MW) or as millimoles, is illustrated in Table 10.1. The

Table 10.1. Calculation of Molecular Values in Millimoles of a Granite Rock, Oven-Dry[a]

Oxides	Weight (%)	Molecular weights (MW)[b] (g)	Molecular values (MV)[c] (mmoles)
SiO_2	74.51	60.1	1239.8
TiO_2	0.12	79.9	1.5
Al_2O_3	14.45	101.9	141.8
Fe_2O_3	0.89	159.7	5.6
FeO	0.22	71.8	3.1
MgO	0.07	40.3	1.7
CaO	0.58	56.1	10.3
Na_2O	1.76	62.0	28.4
K_2O	6.57	94.2	69.8
H_2O	0.92	18.0	51.1
F	0.04	19.0	2.1
P_2O_5	0.22	142.0	1.6
	100.35		1556.8

[a] Clarke (9).
[b] Additional MW that may be needed are CO_2 (44.0), MnO (70.9), SO_3 (80.1).
[c] (%:MW) · 1000.

MV of total silica includes the mineral quartz, SiO_2, which is the dehydrated, crystalline form of silicic acid. Silica combined with Al, Fe, and bases may be estimated from chemical mineral models. The calculation assigns 37.3% quartz (9) to the granite rock in question.

The amount of quartz in a rock is pedologically important because the weathering-resistant mineral accumulates in the soil as sand and silt particles which are determinants of soil texture.

Analyses of a few igneous rock individuals are arranged in Table 10.2 according to their quartz and total silica contents. A quartz-rich rock is an *acid igneous rock* (Ai), an unfortunate term because silica is an exceedingly weak acid. A rock low in quartz or entirely free of it is a *basic igneous rock* (Bi). In the extreme it is an *ultrabasic igneous rock,* low in Al, Ca, K, and Na but surpassing the other rocks in Mg.

Chemical variation echoes mineral suites. Granites are light colored because of dominance of colorless quartz, whitish K- and Na-feldspars, and lustrous muscovite. In basic igneous rocks the profusion of black biotite, blackish hornblende and augite, and green olivine lends darkness. Their high base and phosphorus contents favor the genesis of productive soils and the abundance of iron may bring forth reddish soil colors.

The most widespread igneous rocks are the granites and basalts.

Sedimentary Rocks

When soil material erodes from hillsides and is carried away in creeks and rivers the turbulent waters segregate and sort the mass according to particle sizes and weights: ions, molecules, clays, silts, sands, and gravels. Clays and coarser particles are redeposited as *alluvium* in valleys, in basins, and at seashores. *Colluvium* may pause on

Table 10.2. Molecular Values of Major Constituents of Selected Rocks[a] (Oven-Dry Basis)

Rock type[b]	Quartz (wt%)	Molecular values (MV) (mmoles)						
		Total SiO_2	Al_2O_3	Fe_2O_3	FeO	MgO	CaO	Na_2O plus K_2O
Acid igneous rocks								
Granite (c.gr.)	37.3	1240	142	6	3	2	10	98
Rhyolite (f.gr.)	35.1	1267	121	5	13	4	2	112
Granodiorite or quartz diorite (c.gr.)	27.1	1171	152	6	16	22	57	97
Dacite	25.0	1093	153	13	29	61	65	80
Tonalite[c]	20.0	1060	150	10	59	79	101	69
Basic igneous rocks								
Andesite (f.gr.)	2.2	949	181	23	33	58	76	119
Basalt (f.gr.)	0	815	156	38	64	149	146	87
Gabbro (c.gr.)	0	786	142	10	192	130	145	63
Ultrabasic rocks								
Peridotite[d]	0	715	16	9	93	1070	25	0.5
Serpentinite	0	693	8	26	29	956	7	6
Sedimentary rocks, averages								
Sandstones (114 CO_2)	—	1313	47	7	4	29	99	21
Shales (105 CO_2)	—	942	139	26	25	67	109	59
Limestones (945 CO_2)	—	87	8	3	—	197	761	4

[a] From the collection of Clarke (9) and others.
[b] c.gr., Coarse grained; f.gr., fine grained.
[c] From Larsen (27).
[d] From Hotz (16).

lower slopes or valley heads. During long geologic periods silica and iron solutions may cement the soft rocks to hard ones. Gravel beds become *conglomerates,* sandbeds *sandstones,* and clay deposits turn into *shales.*

The dissolved cations Ca^{2+}, Mg^{2+}, K^+, and Na^+ accompanied by bicarbonate anions (HCO_3^-) reach the oceans. There, many of the divalent ions are precipitated chemically or biologically as *limestone* ($CaCO_3$) and *dolomite* (Ca, Mg, CO_3).

Advancing continental glaciers plow up ancient soils, excavate bedrock, and push the debris ahead of them and within them. In alpine valleys glaciers collect and carry upon them weathered rock fragments rolling down the bare mountain slopes. Melting and evaporating ice leaves in its wake a jumble of drift composed of *till, moraines,* and *fluvioglacial* outwash containing boulders, stones, pebbles, sands, and clays.

On bare land surfaces, strong winds move sand grains that pile up as *dunes* and blow silt particles hundreds of kilometers until they settle down as aeolian *loess.*

Chemically, sedimentary rocks vary widely. Means of sandstones, shales, and limestones are listed in Table 10.2. The presence of CO_2 is indicative of carbonates.

Metamorphic Rocks

In mountain making the igneous and sedimentary rocks are compressed, heated, and folded. Recrystallization of minerals and banding and flattening to platy structures take place. Granites may become *gneisses* and shales and clays may turn into *schists, phyllites,* and the finer *slates.* Sandstone transforms to *quartzite,* pure limestone to *marble,* and peridotite to *serpentine.* In reverse sequence, some of the granites are suspected of being metamorphized sedimentary rocks.

Phosphate Resources

Except for nitrogen, the bulk of nutrient elements in a soil is inherited from the parent rock. A small fraction is acquired by wind, rain, and snow, but even that, barring ocean sprays, is dust from soils and rocks. Organisms consume calcium (Ca), potassium (K), and phosphorus (P) in relatively large quantities; among the three, P is ecologically crucial because of its erratic occurrence in rocks and its fixation in soils.

Expressed as the oxide P_2O_5, Clarke's (9) average phosphate content of all *igneous rocks* is 0.37% or 162 ppm P, and the element is present mostly as the mineral apatite. Acid igneous rocks (Ai) rank lower than the basic (Bi) ones which reach 0.89% (390 ppm P) in Clarke's tables. Correspondingly, in Scotland basaltic soils have nearly twice as much P, both organic and inorganic, as granitic soils (57). The availability of P to plants is reversed in highly weathered soils of low latitudes because the copious iron oxides inherited from Bi rocks immobilize P.

In sedimentary rocks P is associated with Al and Fe oxides which lower the solubility. Shales average 0.20% (87 ppm P) and limestones 0.04% (17 ppm P). As $CaCO_3$ weathers away P concentrates in the residue, benefitting plants. Moreover, individual limestone and dolomite facies may be P-rich, like the 2.76% P_2O_5 (1205 ppm P) of the parent rock that underlays the fertile Maury soils of Tennessee (1). Phosphate rocks of Florida, North Africa, and China assay to 35% P_2O_5 and they are the coveted staples of the phosphate fertilizer industry.

Uniformity of Parent Material

To assess manifestations of horizons in profiles the parent material must be known and it must have been uniform throughout the soil depth. As *index minerals* for all horizons, Marshall (31) advocated in 1940 determination of heavy, weathering-resistant zircon, tourmaline, anatase, and rutile. Earlier, Rode (31) had used quartz, and Barshad (3) added the light albite and microcline to the list. The mineral grains must be sufficiently large not to be transported by percolating water. Marshall's criteria of uniformity demand that proportions of *two* index minerals, or the particle size distribution of *one,* remain invariant throughout the profile.

Sheridan sandy clay loam, derived from granodiorite, was sieved by Barshad (3) into particle size fractions and in each he isolated readily weathering mica and resistant quartz and albite. For the size class 2-0.1 mm his figures are recalculated in Table 10.3 such that in each horizon the weights of mica and of quartz and albite are set equal to 100% and the four size subfractions are expressed as percentages thereof.

Table 10.3. Comparison of Nonresistant Mica and Resistant Quartz (Q) and Albite (A) in the Sheridan Soil Profile[a,b]

depth, cm	Very coarse sand 2-1 mm Mica (%)	Q+A (%)	Coarse sand 1-0.5 mm Mica (%)	Q+A (%)	Medium sand 0.5-0.25 mm Mica (%)	Q+A (%)	Fine sand 0.25-0.10 mm Mica (%)	Q+A (%)
0-15	0	23.1	0	33.1	0	21.3	0	22.6
30-46	0	20.3	5.1	32.6	20.4	22.4	74.5	24.8
46-61	0.4	23.7	14.8	34.0	19.6	21.8	65.2	20.5
61-76	0.4	23.6	13.7	33.8	28.7	19.6	57.2	23.0

[a] After Barshad (3).
[b] For either mica or (Q + A) their weights in the 2 to 0.1-mm fraction are set equal to 100%.

Mica vanishes from all fractions of the surface horizon. Coarse flakes disappear in the subsurface and some of the fragments accumulate in the finer fraction. The sums of quartz and albite remain stable throughout the profile, save for random variations. Barshad computes the ratio of albite to quartz and finds it remarkably constant. Both criteria attest to the uniformity of the initial state of the Sheridan profile.

Strict uniformity of parent materials with depth is the exception rather than the rule, but one may treat the irregularities as being randomized and choose the mean index of all horizons as the criterion of parent material as suggested in Table 10.4 of Siderius (41). Moreover, variations may not materially affect the profile properties of interest, as is the case of Singer and Ugolini's (42) Podzols on volcanic ash layers.

In Hawaii, andesite and basalt have similar chemical compositions, except for the rare elements among which *zirconium* (Zr) exceeds *nickel* (Ni) in andesite and lags in basalt. The derived soils are indistinguishable members of Lahaina silty clay, a red Oxisol. The profiles are deep and bedrock is not readily accessible. Since both of the elements resist leaching and accumulate in the soil, Kimura and Swindale (22) are able to reconstruct the kind of parent rock by soil analysis of Zr and Ni.

Table 10.4. Minerals in the Fine and Medium Sand Fraction (50-200 μm) in an African Aeolian Sand Profile (Typic Ustipsamment)[a]

Depth (cm)	Light minerals Quartz (%)	Feldspar (%)	Heavy minerals[b] Zircon (%)	Tourmaline (%)
0-23	94	6	52	21
23-52	94	6	43	26
52-80	94	6	41	27
80-120	94	6	44	28
120-165	94	6	49	20
165-220	92	8	43	28
Means	93.7	6.3	45.3	25.0

[a] After Siderius (41).
[b] These and the remaining eight heavy minerals add up to 100%.

C. Soils Derived from Igneous Rocks

Many igneous rocks are hard. Rock surfaces polished by glaciers millennia ago may appear still fresh. Deep soils on granites and basalts may date back to early Pleistocene (1-2 million years B.P.).

Rock Weathering

Many granitic rocks are permeated by joints or crack-planes spaced up to several meters apart and extending to depths of 100 m and more. The blocks decay at corners, edges, and faces and leave behind fresh, ellipsoidal "woolsacks" or core stones, up to meters in diameter and buried in coarse-particle grus.

> Russell (39) discusses granitic grus formation as breakdown along cleavages and deduces an asymmetrical, S-shaped distribution curve for particle-size that looks like a poised cobra. Grain sizes of 1-10 mm predominate. Extremes include coarse silt and fist-sized chunks. Wahrhaftig (49) stresses the conversion of mica flakes to swelling clays as a cause of subterranean rock shattering. He holds that weathering proceeds much faster underground than aboveground.

In arid and semiarid landscapes the sparse vegetation encourages erosion of grus material and canyon backcutting and the core stones become exposed on summits, ridges, and flanks. A half-buried boulder may have a wind-polished protruding face and a crumbling ventral base. Hot days and cold nights may split the surfaced boulders into two or more pieces. On steep canyon slopes the woolsacks may slowly converge downhill and become large rock streams, as on Mt. Vitosha, Sofia.

Soils of Ai and Bi Rocks in a Moisture Transect

In a belt stretching from the arid to the perhumid region of the foothills of the Sierra Nevada, California, to be described in Chapter 12, 52 randomly selected surface soils (0-20 cm) derive from acid igneous rocks (Ai), mainly granites and granodiorites, and 45 from basic igneous rocks (Bi), mainly basalts and andesites. The collection permits direct soil comparisons because the geographic distributions of rocks and rainfall are independent of each other ($r = -0.08$). The soils had not been touched by glaciers and they evolved in Pleistocene and Holocene times.

At high rainfall deep, reddish soil mantles rest on basalts and shallower ones on the more slowly weathering granites. Owing to abundance of quartz the lighter colored Ai soils are sandy loams; all Bi soils are dark clay loams. On the average, Bi soils contain twice as much clay and exchangeable bases as Ai soils (Table 10.5). Total acidity, CEC, and humus, that is, C and N, also are higher in Bi soils, and C and N correlate (r values) as 0.90 for Ai and 0.95 for Bi soils. Mean C/N ratio is 23 and rock-indifferent.

Over the entire transect the surface soils are abodes for illite, vermiculite, montmorillonite, kaolinite, halloysite, gibbsite, quartz, and hydrous iron oxides. Their proportions are rock- and climate-dependent. In each rock province statistical analysis singles out the four minerals listed in Table 10.5 as closely linked and orders them

Table 10.5. Mean Values of Properties of Soils Derived from Acid Igneous (Ai) and Basic Igneous (Bi) Rocks[a,b]

Soil properties	Ai soils	Bi soils
Clay (%)	11.6	21.2
Silt (%)	21.2	33.0
Sand (%)	58.0	34.5
C (%)	1.74	2.88
N (%)	0.074	0.121
Bases (sba) (me/100 g)	5.33	10.86
1PC of clays[c]	Vermiculite	Montmorillonite
	Illite	Illite
	Quartz	Gibbsite
	Montmorillonite	Halloysite

[a] Jenny et al. (20).
[b] All Ai – Bi differences are significant at the 1% level or less.
[c] 1PC, First principal component.

into the first principal components (1PC). Vermiculite and quartz are characteristic of the Ai mineral cluster and gibbsite and halloysite of Bi. Montmorillonite ranks first in Bi and last in Ai soils.

Many of the properties impose limits on *soil fertility* that become explicit in plant growth. The role of nitrogen emerges from Klemmedson's (26) 2000 randomized pots filled with 97 soils and planted to barley.

> One set is fertilized with ammonium nitrate (N), monocalcium phosphate (P), and sodium sulfate (S), designated as full treatment, NPS; another set is fertilized only with P and S, leaving it up to the soils to deliver soluble nitrogen by way of ammonification and nitrification. Potassium is not limiting.
>
> For each soil sample a relative mean yield is calculated as % RY = 100 · (yield of PS)/(yield of NPS) and is plotted in Figure 10.2. It measures a soil's supplying power of nitrogen to barley plants under the conditions of the experiment.

Since N and C in soils are in high correspondence, r being 0.90 for Ai and 0.95 for Bi soils, the higher humus resources located in the Bi soils provide higher levels of nitrogen useful to plants. The scatter of points is due in small part to experimental error in pot test technique, in large part to interplay of fertility determinants other than N. Thus, the above-average soil No. 13 is less acid and has much more Ca than the depressed soil No. 12. The point scattering may result from insufficient control of the parent rocks in field sampling because no appraisal of the quantitative mineral design was made or of the chemical constitution, including the biologically important microelements.

Basalt and Rhyolite Ecosystems in Semiarid Arizona

On the Colorado Plateau, Arizona, Welch and Klemmedson (53) established N and C inventories for an ecosystem lithosequence of augite basalt, andesite, rhyolitic alluvium (sandy loams), and limestone. The extensive region supports a ponderosa pine

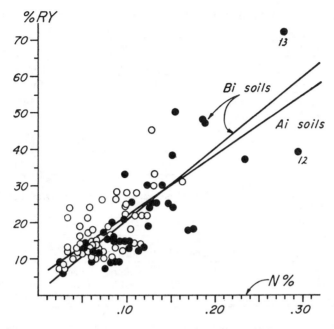

Fig. 10.2. Relative yield RY plotted against total nitrogen content of soil. RY reflects the rate of mineralization of organic N. Outdoor pot experiment conducted at Berkeley during rainless summer by J. O. Klemmedson (26). Courtesy Pontifica Academia Scientiarum, Vatican City

forest (*P. ponderosa*) composed of mature trees, dense sapling-pole stands, and grass openings, the latter serving as the major source of forage for game and livestock.

N and C means of five replications of pole stands of pines (8 cm tree diameter) and of grass enclaves dominated by fescue (*Festuca arizonica*) and muhly (*Muhlenbergia montana*) are compared in Table 10.6. To a depth of 60 cm the mineral soil on basalt contains 1.5-2 times more nitrogen and carbon than the rhyolite soil. On the other hand, standing total biomass, which is the sum of needles, leaves, branches, boles, and roots, does not differ significantly because of high variability of tree stands.

Lithology and Forest Site Index

In the humid, cool mountains of Oregon, Parsons and Herriman (34) excavated pits in forest soils on (a) granitoid rock, ranging from granite to diorite; (b) schist and gneiss, with some quartzite; (c) pyroclastic rocks, composed of massive andesite tuffs and basic volcanic breccia. All soils support Douglas fir (*Pseudotsuga menziesii*) and grand fir (*Abies grandis*) as the dominant tree species, but productivities are keyed to soil-rock lithologies (Table 10.7).

The high ranking soils on pyroclastics have argillic horizons, strong ones on 35% slopes and thick ones (120 cm) on 7% slopes. Colors are dark reddish brown and contrast with the gray browns and browns of granite- and schist-derived soils. All profiles have a complicated genesis because of downslope movement of soil materials, but not severe enough to greatly retard soil development.

Table 10.6. Effect of Basaltic and Rhyolitic Parent Materials on Ecosystem Properties[a,b]

Parent material		Basalt		Rhyolite	
Soil		Argic Cryoboroll		Lithic Haplustoll	
		(g/m^2)	(%)	(g/m^2)	(%)
Soil nitrogen (0-60 cm) { Forest	354	0.102	241	0.059
(C/N ratios 19-20) { Grass	584	0.113	324	0.058
		(kg/m^2)	(C/N)	(kg/m^2)	(C/N)
Plant biomass[c] { Forest	24.6	281	34.8	291
{ Grass	1.38	57	1.70	105

[a] After Welch and Klemmedson (53).
[b] cl, maP 52 cm, maT 7.4°C; ϕ, pine forest and grassland; r, east exposure, 1-10% slopes, well drained; t, thousands of years.
[c] Leaves, wood, roots.

Table 10.7. Forest-Site Lithosequence in Mountains of Oregon[a,b]

Lithology and names of soils	Soil texture	Argillic horizon	Exchange capacity of horizons CEC (me/100 g)	Forest Productivity (site index)
Granite (Typic Xerochrept)	Coarse loamy	Absent	15-28	Poor (5)
Schist (Ultic Argixeroll)	Fine loamy	More or less	20-27	Fair (4)
Pyroclastic (Ultic Argixeroll)	Fine, mont- morillonitic	Strong	33-49	High (3)

[a] Parsons and Herriman (34).
[b] cl, maP 110 cm; ϕ, Douglas fir and grand fir forest; r, 5-7% on ridges and saddles; also steep north slopes; t, equivalent ages.

D. The Serpentine Syndrome

Soils derived from the green-colored serpentine rocks, or serpentinites, and from their ancestors, the peridotites, have long intrigued agriculturists who often tried in vain to cultivate them.

Rock and Soils

In Clarke's serpentine rock of Table 10.2 magnesia and silica make up the bulk of mass. The sesquioxides, R_2O_3, calculated as $Al_2O_3 + Fe_2O_3 + 0.45$ FeO, are trailing and Ca, K, and Na are very subordinate. Millimoles of MgO, SiO_2, and R_2O_3 are plotted in the triangle of Figure 10.3 as a star having the coordinates 40.9% for SiO_2, 56.3% for MgO, and 2.8% for R_2O_3. These percentages are obtained by dividing the individual molecular values by their sum which is 693 + 956 + 47, or 1696. Had the rock been pure serpentine, which is the magnesium silicate $Si_2O_5 Mg_3(OH)_4$, also written $2SiO_2 \cdot 3MgO \cdot 2H_2O$, its locus would be on the SiO_2-MgO line 3/5 from the base.

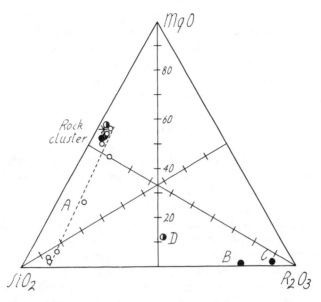

Fig. 10.3. Serpentine triangle, each apex representing 100% of a component. Cluster of eight rocks (star from Table 10.2) and spread of seven derived soils. A, Atlantic Coast (4·2 open circles); B, Borneo; C, Cuba; D, Eight Dollar Mt., Oregon

Serpentinites are slightly alkaline and are unstable in acid solutions (Fig. 4.10). At $25°C$, from low to high CO_2 pressures the crushed rock of Wildman et al. (55) increases its solubility from 0.41 to 1.65 mmoles/liter for Mg and from 0.25 to 0.71 mmoles/liter for Si.

In the composition triangle of Figure 10.3 the small cluster of worldwide serpentinite rocks bursts into a wide scatter of soil points, each representing a surface soil 20 to 30 cm in depth. The gray and light-brown horizons of the moderately deep (50-100 cm) Connowingo series (36) of the humid-temperate Atlantic Coast cling to the fitted, dashed trajectory which points to a severe removal of Mg. Near the right-hand basal apex of Figure 10.3 the deep, red tropical Nipes of Cuba (16,36) and the Borneo profile (40) achieve narrow soil SiO_2/R_2O_3 ratios of 0.09 to 0.38. Between these extremes lies the 3-m-deep, reddish "laterite" (16) of old, late-Tertiary land surfaces (Eight Dollar Mountain) on the humid-cool Pacific Coast at about $43°$ of latitude north. Its SiO_2/R_2O_3 ratio of surface soil is 0.91, which exceeds the tropical values but is still low compared to the Connowingo range of 6.0 to 10.6. To what extent climate and soil age determine the tableau of loci in the triangle is not yet understood.

In serpentinites the biologically potent ion quotients Mg/Ca are unfavorably high, 8 to >4000. Soil genesis scales down every one of them, but not below 4. In California the reddish serpentine soil series Hennecke and Dubakella of maP 110 cm have in their A horizons Mg/Ca ratios of exchangeable ions of 2.2-3.7, and Ca/CEC, which is the degree of Ca saturation, spans from 15 to 19%, according to Wildman et al. (56). In B22t horizons the <0.2 μm clay fraction shows 70-85% iron-rich montmorillonite and 4-8% vermiculite. In the tropical Borneo soil, however, the clays are mainly goethite, gibbsite, and, surprisingly, kaolinite (Fig. 4.4). In the surface of Eight Dollar Mountain X-rays respond to colloidal talc, chlorite, and goethite (16).

Serpentinites generate silicate clays insofar as they carry aluminum and aluminosilicate impurities. The petrologic aluminum deficiency and the concomitant retardation in clay formation explain the erosiveness, shallowness, and mineral freshness of many of the soils and their conspicuous impact on vegetation.

Serpentine Vegetation

As if to compensate for their notoriously low productivity the serpentine soils possess unusual native floras characterized by many endemics, that is, species restricted to locales with specific habitats. Within expanses of luxuriant forests and grasslands the serpentine enclaves present a brush and bare-land physiognomy.

In the humid Siskiyou Mountains of southern Oregon Whittaker (54) compares floras on shallow, stony soils derived from quartz diorite and from serpentine (Table 10.8). Serpentine soils have only half of the tree species and these are inferior pines widely spaced. Herbs other than grasses proliferate in speciation.

In any soil a multitude of variables are interlocked and the task of sorting out those responsible for a given biotic product involves formidable experimental and conceptual hurdles.

The barrenness and the attenuated stature of vegetation have been blamed on the deficiency of the nutrients N, P, K, and possibly Mo (51); on the presence of toxic elements, notably Cr, Ni, and Co (35,36), and on MgO/CaO ratios exceeding 4/5 in total soil analyses, announced by Loew in 1901 (36).

In the greenhouse commercial Romaine lettuce (*Lactuca sativa* L., var. Romaine) planted on serpentine soil suffers a "rosette" disease of curled and stunted inner leaves that is correlated with low degrees (<20%) of saturation (DS) of exchangeable Ca. Rosette can be induced in normal soils, or prevented in serpentine soils, by changing DS of Ca with ion exchangers (47). Though Mg is the natural complementary ion of Ca, it is not decisive. The DS of Ca is crucial, not its absolute amount (21). Walker (51) and Kruckeberg (24) find native serpentine endemics able to secure Ca at very low degrees of Ca-saturation.

The herb *Phacelia californica* inhabits both serpentine and "normal" soils. Seeds of Phacelia plants growing on serpentine soil and seeds from plants on "normal" soil are sown in pots containing either serpentine soil alone or amended with ferti-

Table 10.8. Distribution of Floras on Diorite and Serpentine in Siskiyou Mountains at Sameness of Other State Factors[a]

	Number of species		Number of endemics	
	On diorite	On serpentine	On diorite	On serpentine
Grasses	8	17	0	0
Other herbs	58	71	2	25
Shrubs	19	17	0	5
Trees	16	8	0	0
Total	101	113	2	30

[a] From Whittaker (54).

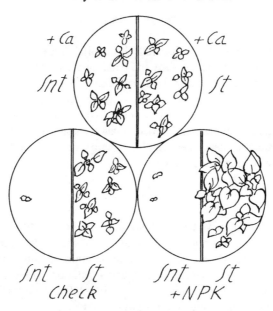

Fig. 10.4. Bisected pot experiment (24) with serpentine soil and with serpentine-tolerant (St) and serpentine nontolerant (Snt) races of *Phacelia californica*. Lower left: on untreated soil Snt does not grow. Lower right: on heavily fertilized soil (N, P, K) Snt still fails to grow. Upper pot: upon addition of Ca, both races grow equally well (27-day-old plants)

lizer (N,P,K) or with gypsum ($CaSO_4 \cdot 2H_2O$). From the all-or-none performances in Figure 10.4 one may conclude that the strain from "normal" soil will not grow on natural serpentine soil, even if heavily fertilized, unless Ca is added and its DS rises. Vice versa, the serpentine-tolerant strain also does well on "normal" soil, when seeded to itself, but succumbs in the presence of competing "normal" soil variant. The species is an aggregate population of serpentine-tolerant and intolerant biotypes that are morphologically indistinguishable.

E. Diversity of Limestone-Derived Soils

Calcite and the less common aragonite are crystals of pure $CaCO_3$ containing 56% CaO and 44% CO_2. Limestone rocks are impure $CaCO_3$. Dolomite rock is rich in magnesium carbonate according to the mineral equation

$$CaCO_3 \quad + \quad MgCO_3 \quad \rightleftharpoons \quad CaMg(CO_3)_2$$

calcite   magnesite   dolomite

In pure form it is composed of 30.4% CaO, 21.9% MgO, and 47.7% CO_2.

Field assay of carbonates employs cold, dilute hydrochloric acid (HCl). Calcite effervesces (fizz), magnesite and dolomite do not. Solubilities of $CaCO_3$ are listed in Table 7.9.

Caves and Karsts

When tectonic movements tilt the naturally horizontal limestone and dolomite strata (Fig. 10.5) water and roots follow joints and bedding planes, and differential solubility of strata gives rise to rock outcroppings and bizarre landscape features. Dissolutions may lead to underground caves and caverns of grotesque shapes and huge dimensions. Karst landscapes are rolling plains that are pitted with solution depressions similar to rude cisterns, hornlike funnels, and broad, shallow basins up to 100 m in diameter. In Europe they are best developed on the eastern coast of the Adriatic Sea; in the United States they abound in Kentucky.

Talus Lithosequence

In Bach's (2) drawing (Fig. 10.6) of a talus slope at the foot of a cliff of hard, pure limestone rock the largest stones roll farthest down the incline. The limestone fragments are graded in proportion to distance from the source and they initiate a continuous lithosequence of parent materials, soils, and vegetation. The site is typical of the Jura mountains in Central Europe having maP of 100-200 cm and maT of 3.5-6°C. Ecosystems on north-facing slope are described as follows:

Blocky, Raw Carbonate Soil at Foot of Slope. Spruce forest (*Picea abies*) with *Hylocomium* mosses. No fine earth exists; a 30-cm-thick surface layer of acid mor humus rests on the boulder mass. Some of the organic matter is washed into the large cavities below where it nourishes roots. Spruce does not ascend the slope because it cannot endure the steady gravel influx and soil motion higher up.

Humus Carbonate Soil (Rendzina) on Upper Portion of Slope. The 60-75% slope supports a forest of mountain maple (*Acer* sp.) with shrubs of ash (*Sorbus* spp.)

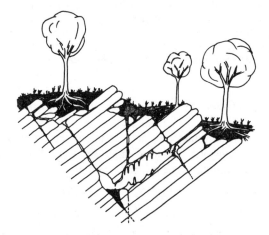

Fig. 10.5. Idealized cross section of landscape on limestone in Appalachian Mountain region, from Fanning (11). Soil depth is variable with certain beds of rock outcropping along the strike. Plant roots, percolating waters, and caves are channeled along joints and bedding planes. Courtesy Soil Science Society America, Madison, WI

and hazelnut (*Corylus* sp.) as well as ferns and crucifer herbs. The productive soil is a deep, gravel-rich, sandy loam with granular structure and with up to 20% of co-progenous mull humus to a depth of 60-100 cm; earthworm casts are numerous; pH varies from 6.5 to 7.8, depending on the position in profile and reactions of $CaCO_3$ with metabolic CO_2 production.

Terra Rossa

The famed Terra rossa of the Mediterranean Region of Europe, North Africa, and the Middle East is a brick-red earth sharply set off from the white, hard limestone or dolomite below. The color contrast appeals to scientists, poets, and painters alike.

The soils are neutral clay loams low in carbonates (6,25). North of Varna on the Black Sea Coast Terra rossa is buried under a Chernozem soil in loess, the double profile bearing witness to a long history of the red soil species in general. Oak forest (*Q. ilex*) is now climax in southern France (maP 77 cm, maT 14.4°C). The red color is turning brownish, leaching is minimal, and, expressed in kilograms/square meter, the aerial biomass of 150-year-old tree stands is 26.9, the root biomass 4.6, the soil organic matter 16.0, and the forest floor 1.4, its coefficient of decomposition k' being 26%, according to Lossaint (29).

The often vivid reds of Terra rossa are attributed by some to the low humus contents caused by the dry, hot summers and by others to a previous humid, tropical climate. Neither explains the dissimilarity with the geographically related black and gray Rendzina soils which also are limestone-derived, though sometimes from an impure kind.

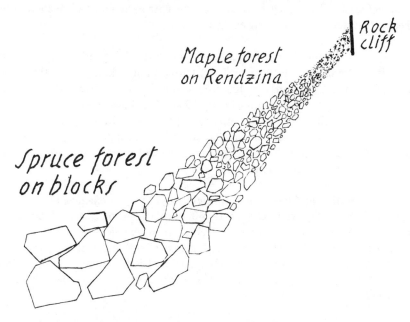

Fig. 10.6. Bach's (2) lithosequence of a limestone talus slope. Soils and vegetation are controlled by particle sizes, the blocks having diameters of 20-50 cm

A prominent school views Terra rossa as the solution residuum of the underlying carbonate rock. Clay mineralogy lends weight to this theory (8,58). X-Rays (4) of soil clays and remains of fresh bedrock dissolved in HCl furnish for both the diffraction peaks that belong to kaolinite, mica, quartz, and the rare attapulgite, the latter functioning as trusted tracer from rock to soil.

> The Tiger Creek soil series is an American kin to Terra rossa. It it a Typic Rhodoxeralf (pH 7-7.5) and the B2 horizon may be as red as 10R. It forms on steep slopes in the Mediterranean climate of California (maP 84 cm, maT 15°C), is underlain by 95% pure dolomite, and is covered with mixed conifer-oak forest. Montmorillonite is a prominent clay mineral in both rock and soil.

To seek the source of redness, Barshad et al. (4) divide the HCl-soluble Fe_2O_3 of the rock by its insoluble Fe_2O_3. Terra rossa tends to have higher ratios than Rendzina, but Yaalon (58) considers the form of iron oxide more important than its quantity.

Limestone Province of the Central and Eastern United States

Limestones reach from Pennsylvania south into Alabama and west to Kentucky, Tennessee, and the Ozark Plateau (Fig. 12.19). In Missouri, at maP 114 cm and maT 13°C, a dolomite-derived Hagerstown silt loam has acidic (pH 4.8), red (2.5 YR 3/6) B horizons. The upper Bs are fine subangular blocky, the lower ones and C are plastic and waxy and average over 85% clay, three-fourths of it colloidal. Bedrock, reached after 2 m depth, holds in its HCl-insoluble residue 42% clay, half of it fine (< 0.2 μm). Brydon and Marshall (7) conclude that soil clay and its mineralogy are inherited from the parent rock. The strong resistance of the illite-glauconites to decomposition is linked to their tenacity of holding exchangeable Ca.

Clay Depth Functions. In mixed hardwood-pine forests on gentle slopes Hagerstown soil in Pennsylvania (maP 100 cm, maT 9.2°C) and Frederick in Virginia (maP 104 cm, maT 12.0°C) derive from massive limestones, and Fullerton gravelly loam in Alabama (maP 125 cm, maT 16.0°C) from a cherty facies. The three soils have silty A horizons and reddish and yellow-brown Bt horizons of pH 4.4-4.8. In Figure 10.7, percentages of < 2 μm clay rise sharply and steadily as bedrock is approached (1). All B and C subdivisions qualify as argillic horizons, but the familiar textural B horizon that surpasses the clay content of C is missing.

SiO_2/Al_2O_3 Ratios of Clays. Nine profiles analyzed by Alexander et al. (1) have a spread of SiO_2/Al_2O_3=sa of clay of 1.95-3.24 that is properly accredited to heterogeneity of rock composition because the values are independent of climate.

The depth functions of sa of clays in Figure 10.7 are nearly straight verticals that point to clay stability. However, for relative sa, that is, sa of any horizon divided by sa of the C horizon, a subdued pattern emerges (inset of Fig. 10.7) that differs radically from either podzolization or laterization.

Silt Influx. From bedrock to soil surface the Harpeth pedon on Tennessee limestone experiences a widening of silt/sand ratios but a narrowing of clay/sand ratios (Fig.

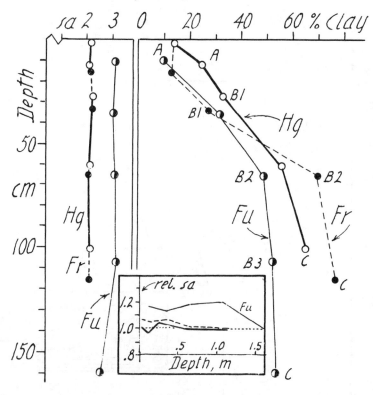

Fig. 10.7. Depth functions of percentage clay, SiO_2/Al_2O_3 (sa) of clay, and relative sa of soils derived from limestone rock, from data of Alexander et al. (1). Hg, Hagerstown soil, PA; Fr, Frederick soil, VA; Fu, Fullerton soil, Al

10.8). Further, the proportion of feldspar to quartz in the silt fraction is higher in the surface than at lower depths. The silt-sand configuration is interpreted by Edwards et al. (10) as resulting from gradual influx of wind-blown silt or loess. It divides the profile into two stories, the upper one ending at about 65 cm. Why the ratio of coarse silt (0.05-0.02 mm) to fine silt varies little throughout the profile in Figure 10.8 is still puzzling.

Soil Fertility. In the prefertilizer era soils on limestone were valued for their cornucopian productivity and they are still mainstays of agriculture. The Maury soils in particular enjoy a high reputation for bluegrass (*Poa pratensis*) pastures because of high calcium and phosphorus supplies that reach 5.34% P_2O_5 in the B2 horizon, its lineage descending to the bedrock endowed with 2.76% P_2O_5. As awarders of prosperity the limestone soils played a key role in the early colonization of the United States (e.g., by the Pennsylvania Dutch), as vividly told by historian Hulbert (17).

By no means are all of the carbonate-derived soils lastingly fertile, a lesson the early immigrants to the Ozarks learned. In a few decades after settlement the very old and highly leached Clarksville soils on cherty limestone gave out. Today for-

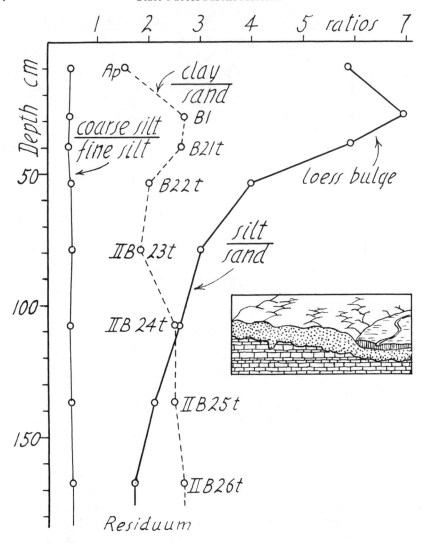

Fig. 10.8. Depth functions of ratios for Harpeth soil (Humic Hapludult) on limestone in Tennessee (10). The surface bulge of the percentage silt/percentage sand curve is attributed to wind-blown influx of silt (loess). Inset: soil mantle on limestone, after Buntley (10)

gotten graveyards in brushy scrub oak woodlands are the sole reminders of defeated pioneering toil.

Biotic Calcium Discriminations

The peculiarities of vegetation on dolomite mountains and chalk downs led early botanists to coin physiologic plant attributes such as calciphile (Ca-liking), calcicole (Ca-dwelling), calciphobe (Ca-hating), and calcifuge (Ca-avoiding). Agricultural chemists like Hilgard soon learned that some calciphiles do not confine themselves to carbonate substrates as long as the soil yields substantial amounts of HCl-extractable Ca, that is, exchangeable Ca. Leaching depauperizes soil in calcium and the calciphile species are eliminated, as recounted in the chapter on Time.

Gigon's (15) two-member lithosequence of range lands on fairly steep south slopes in the Alps scrutinizes the Ca-phile-phobe query. The two soil bodies, one on Ai moraines and the other on dolomitic limestone, have near-identical state factors, including time since deglaciation. The till soil is acid (pH 4.8-5.1) and carries exchangeable Al, the calcareous type is slightly alkaline and its step-like solifluction surface interfers with plant growth and creates bare spots (Fig. 10.9). Mineralization of soil organic nitrogen in incubation tests delivers mainly NH_4 in the acid and NO_3 in the alkaline fine earth, without eliciting differential plant growth in pot tests.

Together, the two sites house over 100 species of vascular plants, the joint biotic factor. The plants are segregated into the two communities *Nardetum* on the siliceous soil, the grass *Nardus stricta* being dominant, and *Seslerietum* on the calcareous soil where the grass *Sesleria coerulea* reigns. Discrimination comes to light in *Seslerietum* which harbors seven species, including *Sesleria* sp., that never appear on the siliceous soil. They are the calciphile plants. Similarly, 23 of the *Nardetum* species, inclusive of *Nardus* sp., are absent on the calcareous substrate. They are the calciphobe members.

To separate root milieu and plant competition as causes of segregation Gigon plants *Sesleria* sp. and *Nardus* sp. in pots filled with either silicate or carbonate soil. It turns out that calciphile *Sesleria* sp. alone in pots grows equally well on both substrates, hence is not suffering disadvantages. Calciphobe *Nardus* sp. alone in pots is less prolific on dolomite than on silicate soil, presumably because of nutritional handicaps. When the two species are planted together in the same pot, *Sesleria* sp. overcomes the inherently vigorous *Nardus* sp. on the alkaline soil and *Nardus* sp. displaces *Sesleria* sp. on the siliceous soil.

White clover (*Trifolium repens*) prospers on a gamut of soils that spans very acid to calcareous reactions, indicating wide physiological tolerance. The species qualifies as a soil facultative. In the hands of Snaydon (44), however, specimens transplanted from acid soils to calcareous soil develop severe iron chlorosis that can be cured by supplying Fe-chelate. Evidently, the White Clover species is a mixture of physiological types. In Texas, Little Bluestem (*Andropogon scoparius*) behaves identically, as reported by Nixon and McMillan (33).

F. Glacial Till Variations

In the Middle West the continental ice sheets advanced south to central Missouri and into southern Illinois and Indiana. The massive till, up to 5-m-thick, and outwash deposits that the glaciers left behind constitute the parent materials of present-day soils.

Tills in Illinois

In northeastern Illinois (maP 90 cm, maT 10.5°C) the glacial debris of Wisconsinan age is predominantly dolomitic calcareous (52). In some localities the gravels and cobbles congregate to over 80% by weight, in others they are absent. In turn, abun-

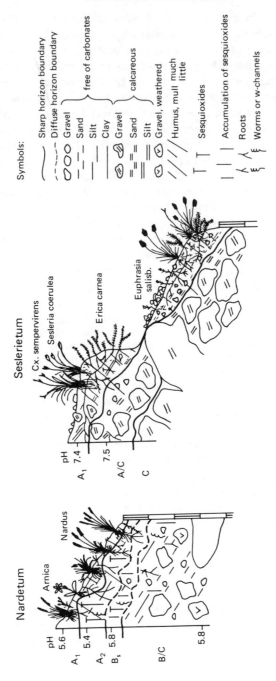

Fig. 10.9. Gigon's (15) alpine ecosystems on acid, glacial till (schist and gneiss materials) and on dolomite limestone at equal cl, φ, r, t

dance of the illitic clay in fine earth ranges from 1 to 70%. Lowest bulk density is a high 1.52 g/cm³, highest is 1.88 g/cm³, possibly from compaction by glacial ice that may have stood 2000 m high.

Soil Textures and Water Economy. Waterholding capacity of fine earth is below 4% by weight in the well-oxidized, brown, *loamy gravel till* and ascends to 35% in olive-gray *clay till*. Percentages of plant-available water (AWC), computed as the difference in water held at 1/3 and 15 atmospheres of suction, average as follows: loamy gravel 3.5, sandy loam 6.7, loam and silt loam 8.7, silty clay loam 9.2, silty clay 10.4, and clay 14.7. To 1 m depth the AWC expressed in centimeters of water provides storage capacities of 28-61 cm. In low AWC soils shallow-rooted plants often experience water stress in spite of frequent summer rains. In laboratory tests of till cores the rates of water movement extend from only 0.25 cm/hr in silty clay to 13 and 36 cm/hr in gravelly sands.

Leaching and Gleying. The hydraulic conductivities govern leaching. In the Cary glacial substage area (13,000-14,000 B.P.) sandy loam soils on noneroded slopes of 1-4% have the *carbonates removed* to an average depth of 91 cm. In clayey tills leaching progresses to only 61 cm. In gravelly tills partially rotten limestone pebbles still linger on the surface and the soils are droughty.

The expectation that the texture of B horizons is homologous to the amount of clay in the parent till is examined in Figure 10.10, which is based on the 33 profile

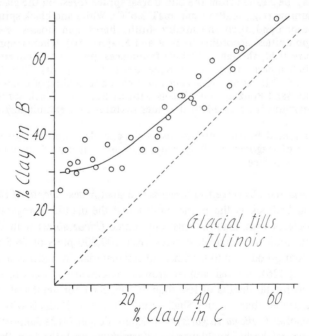

Fig. 10.10. Percentage of clay in B horizons related to percentage of clay in C horizons of N.E. Illinois tills (52). Each point represents a profile, its highest percentage of clay in B and its mean percentage of clay in C. The dashed line denotes equality of percentage of clay in B and C

analyses of Wascher et al. (52). Maximum clay content ($< 2 \mu m$) in B is plotted against clay percentage in C. Every B has more clay than its C, and regression of clay in B to clay in C is positive. Further, clay gains of B lessen as C gets finer.

In *Mollisols* the black surface soils (epipedons) on *loamy tills* are underlain by brown, well-oxidized subsoils; on *clayey tills* by grayish and gleying zones, not because of low degrees of slope or shallow depth to a permanent water table, but due to sluggish water flow and cognate poor aeration. Similarly, coloration of the B horizons of the lighter tinted *Alfisols* switches from reddish-brown and yellow on the loamy till to grayish-brown on the clayey till. Illite is the dominant clay species.

Tills and Vegetation

In the early nineteenth century northeastern Illinois was still covered by tallgrass prairie (*Andropogon* spp.) on Mollisols (Prairie soils) and by deciduous oak-hickory forest on Alfisols (Gray-Brown Podzolic soils). Pertinently, each of the broad floras covered the entire scope of till textures, but detailed knowledge of how the plant subcommunities had accommodated to them is scattered among old manuscripts (e.g., in Hilgard's archives) and records (28) or is lost. The forest soils are less fertile than the prairie soils. Whether nutrient status is cause or effect of forest vegetation is answered by Runge (38) in favor of soil being the cause.

How virgin forest vegetation adjusts to the *spectrum of textures* of the parent material may be gleaned from the Sub-Boreal spruce forest on the glaciated plain of British Columbia (e.g., maP 66 cm, maT 2.6°C). White and black spruce, fir, lodgepole pine, aspen and birch and sundry shrubs, herbs, and mosses assemble into six upland communities, according to Wali and Krajina (50). Whereas spruces span the entire texture field, though at various frequences, pines with understories of bearberry and lichen concentrate on extreme sands of 96% purity. The climatic climax community of white spruce and *Vaccinium* with a sprinkling of pines and aspen occupies loamy sand Podzols and Alfisols with Bt horizons. Black spruce with birch, fir, and sphagnum is confined to the finer textures that exhibit gleying at the foot of slopes.

As documented by the authors, rising silt and clay contents not only augment the reservoir of stagnant capillary water but also delivery of nutrients in solution and on particle surfaces.

In *Scandinavia* the tills verge to rockiness and shallowness. *Depth of glacial material* resting on solid bedrock is the prime modifier of the mantle of vegetation. Pine (*P. sylvestris*) associates with shallow soils and spruce (*Picea abies*) with deep ones. In forests of coniferous and broad leaf trees, over 100,000 plots of 78.5 m² area each serve to tie timber production to thickness of soil material. In Agder counties, Norway, according to Låg (26), normal annual growth increments of trees in cubic meters/ hectare are 2.26 for till depths of less than 20 cm, 3.61 for depths of 20-70 cm, and 5.03 for drifts deeper than 70 cm. Shallow moraines have Podzols with ground cover abundant in heather (*Calluna vulgaris*) and lichen; deeper drifts support Brownearths with understories of herbs, huckleberries (*Vaccinium* spp.), the fern *Dryopteris,* and mosses.

G. Loess Influx Sequences

In the Americas and in Eurasia a yellow-colored deposit of silty texture spreads over millions of square kilometers. Farmers in Europe and in China have been cultivating the fertile, mellow silt loams for countless centuries. This "loess" feels soft and forms upright bluffs tens of meters high. Shells of land snails are buried in it and, occasionally, bones of prehistoric, large animals (*Elephas, Mastodon, Equus*) are unearthed.

Origin

Earlier geologists identified loess as a water-laid deposit and their "aquatic theory" is still receiving support. In 1877 the geographer von Richthofen advanced the "aeolian hypothesis" which views the immense loess sheets of China as accumulations of dust carried by the winds from the deserts to the humid zones. The source of the extensive Mississippi Valley loess was placed into the Great Plains area and the deserts of the Great Basin.

In 1897, Chamberlin, at the University of Chicago, singled out the ancient glacier-fed river flood plains within the loess belt itself as the source of aeolian silts. Over a thousand deep borings enabled soil surveyors Smith (R. S.) and Norton (19) to construct a map of the thickness, D, of the loess mantles in Illinois. Extent and depths are clearly related to present-day major stream channels and their widths, which supports Chamberlin's thesis.

Traverses

In the late 1930s Guy D. Smith (43) staked out on level uplands of $< 1.5\%$ slope two traverses leeward of the major wind directions. With long extension augers he ascertained the depths of loess blankets down to the pebbly, glacial till. A younger loess (Peorian) rests on an older, more weathered and redder silt deposit (Sangamon paleosol). Smith plots his D reading of Traverse I in relation to distance d from the river bluffs (Fig. 10.11) and notes that thinning progresses as the logarithm of the distance, written in the metric system (centimeters for D, kilometers for d) as $D = 1214\text{-}511 \log d$. The equation fails near the bluffs where depths reach 2800 cm and gradients are much steeper. The broad patterns of depths are confirmed in Iowa (18,46) and along the shores of the lower Danube in Europe (14). The descending curve has been ascribed to scarcity of ancient boreal vegetation away from the river that permitted dust loads to be picked up again; thus, D is seen by Smith as the resultant of loess deposition and removal.

Aeolian Models

In Waggoner and Bingham's (48) quantitative visualization of loess formation silts picked up by wind puffs from dry river flats disperse into clouds and are diluted in the blowing to the lee, simulating a pattern of diffusion from a constant source (Fig. 10.11). Low-cloud particles lodge in vegetation along a dust-concentration gradient, which explains the thickness-distance curve. D becomes proportional to

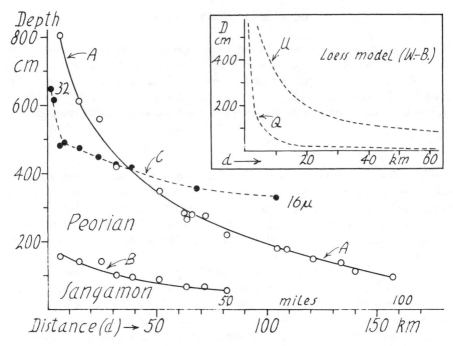

Fig. 10.11. Smith's (43) loess traverse I from river bluff toward the lee. Curve A: thickness of loess sheet; B: thickness of buried Sangamon loess; C: mean particle size of silt and clay decreasing from 32.3 to 16.4 μm. Inset: loess model of Waggoner and Bingham (48), based on air turbulence theory

$d^{-m/2}$ in which m characterizes air turbulence. Additional parameters account for channel width of source and settling of dust. In the inset of Figure 10.11, theoretical curve U starts from a 10-km-wide river bottom and in curve Q the dust cloud is being depleted by rapid sedimentation.

In deeply buried, unweathered loess strata Smith detected a decrease in mean particle size with distance from source, fast near it and slow farther away (Fig. 10.11). It accords with Stokes' law of settling velocities of airborne particles which operates when eddy movements are randomized. Clay fallout is least conspicuous, possibly because in the calcareous river bottoms the particles are flocculated and are not significantly segregated by wind. Combining the roles of dilution, particle size, and wind source, Frazee et al. (13) furnish empirical equations with several constants that describe the depth-distance dependency $D = f(d)$ with great accuracy.

Layering of Loess

Deposition of aeolian silts on top of Sangamon paleosol began in Illinois about 70,000 years ago. Peoria loess proper commenced accumulating 25,000 B.P. Siltation did not progress uniformly, according to Fehrenbacher (12) and Kleiss (23), and dusts varied in their proportions of montmorillonite, illite, and kaolinite depending on the drift materials which the ancient shifting creeks and streams drained.

During periods of wind relaxation vegetation and microbes generate humus bands and weathering incipient soils and paleosols, like the Farmdale soil of 25,000 B.P. Willman and Frye (12) recognize four buried soils of the Wisconsinan stage, and a bewildering 10 are classified in Europe where loess was wedged in between the Scandinavian and the Alpine glacier fronts and where the many buried ice wedges and solifluctions chlieren point to drastic periglacial disruptions.

Effective Soil Age

Smith learned that the carbonate contents of the lower portions of the loess mantle diminish with distance from the source, from over 35 to below 10% of $CaCO_3$ equivalent. He contends that the thin loess sheets laid down far from the river were being leached of carbonates as fast as the layers accumulated.

The concept of effective age or time, t_{ef}, of soil genesis (18) is diagrammed in Figure 10.12 (inset) for average dust influxes along a traverse. Time t_{ef} is the sum of the time-period of loess accumulation, t_a, and time since cessation, t_e. From the measurements of Kleiss (23) along Traverse 1, the latest Peorian loess increment was laid down at the rate of 20 cm per century near the bluff and at 3.8 cm/100 years at a distance of 50 km. Hence, 1 m of tessera depth required 500 years of accumulation time close to the source and 2500 years at the distant site. Time t_c since the terminus of accruement is about 12,000 years, and t_{ef} is, therefore, 12,500 and 14,500 years, respectively. One may hesitate, however, adding the two ages t_a and t_c that differ so profoundly in their operational mechanisms. Ruhe (37) questions the entire idea of effective age.

Soil Sequences

In the natural grasslands of Illinois, Missouri, and Iowa four to five soil series are mapped in each traverse area and the end members of each sequence are Argiudolls and Albaqualfs or equivalent taxa:

Deep loess (> 2m)	Intermediate	Shallow loess (< 1 m)
Muscatine (IL)	⟶	Cisne (IL)
Marshall (MO)	⟶	Putnam (MO)
Minden (IA)	⟶	Edina (IA)
Udolls ⎫		⎧ *Albaqualfs*
Prairie soils ⎬	⟶	⎨ Claypan soils
Brunizems ⎭		⎩ Planosols

The sequences of individual soil properties were early diagnosed by Bray (38) and subsequently by many others (5,18,46).

As distance from source lengthens the ratios of quartz to feldspars, the percentage of clay in B horizons, and the exchange acidity enlarge, and exchangeable bases, pH, humus, and Ca/Zr ratios diminish. In profile sketches of Udolls and Albaqualfs (43) the *deep* loess has granular, black and dark brown A and B1 horizons; the *shallow* kind

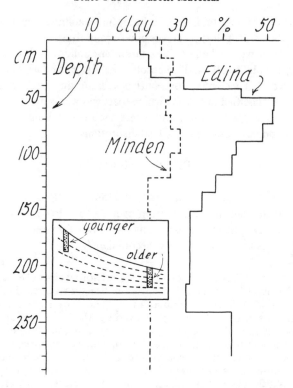

Fig. 10.12. Clay depth functions of Minden silt loam near Missouri River bluffs in Iowa and Edina silt loam some 300 km inland in S.E. direction (46). Inset: Loess blankets along a traverse produce younger tesseras of 1 m depth nearer to the dust source than away from it

has an ashy gray subsurface underlain by a heavy, plastic zone of clay accumulation. The latter's signature is etched in the clay depth functions (46) of Figure 10.12. Up to 63% of the $< 2 \mu m$ clay is ultrafine, having sizes below 0.06 μm. To a depth of 125 cm the Minden tessera has 40.6 g clay/cm^2, the Edina tessera 63.5 g/cm^2.

Loess Fertility

Owing to their high humus and N, P, K nutrient contents and desirable, granular structures, the Argiudolls, which are the classic Prairie Soils, are famous for the production of corn (*Zea mays*). In the prefertilizer days of the 1920s and 1930s, they yielded over 3800 kg/ha (60 bu/A), then hailed as magnificent. The intractable claypan soils brought forth but 1267 kg/ha (20 bu/A). Their corn crops suffered moisture stress during summer dry spells, but when it was discovered that heavy fertilization encourages roots to penetrate the dense clay horizon and tap the moisture reserve below, yields began to climb. From 1956 to 1965 fertilization of hybrid corn coupled with intensive management practices doubled the yields on Muscatine silt loams and more than trebled them on Cisne silt loams (13).

Outlook

Build-up of loess blankets and their thinning with distance from river source are influx sequences but the trend of the profile features from evenness to multiform is not readily understood. Smith (43) chooses effective soil age, t_{ef}, as the leading cause, the Brunizems slowly turning into claypan soils, or Arguidolls into Albaqualfs, and the arrows in the aforementioned list denoting a type of chronosequence.

If the aforementioned discovery of postloess time t_c greatly exceeding build-up time t_a is confirmed elsewhere, an alternative explanation has to be sought. Choosing 12,000 B.P. as time zero, the profiles along a traverse follow an initial-state sequence of variable slopes and parent materials, but the roles of r and p as topo- and lithosequences have not yet been sorted out quantitatively. Bray's old theme that in shallow loess the underlaying paleosol influences soil development above it is detailed by Runge (38) as creating a fluctuating water table that enhances leaching of $CaCO_3$ and confines clay accumulation to a narrow band.

H. Review of Chapter

On level land the ordinations of soils and organisms with parent material (at comparable cl, ϕ, r, t) tend to be *discontinuous,* with rather sharp boundaries at the geological contact zones. Rock maps and soil maps reflect the earth's mineral mosaic. *Continuous* lithosequences originate on talus slopes, in colluvium, in thinning loess blankets, and in other sedimentary deposits.

Save nitrogen, the initial stockpile of nutrients in the soil is inherited from the parent rock. Leaching of ions eventually lessens the diversity, which fosters similarity among soils and their plant covers. Still, within a given climate and on chosen topographic sites soil fertility and production of biomass point toward the initial state.

The highest position in the classification hierarchy used to be vested in the *zonal soils* that are said to be dominated by climate and plant life and that evolve on "normal," chemically balanced parent materials such as granites, basalts, loess, and medium-textured alluvium and glacial tills. The *intrazonal* soils of equal rank develop instead on "abnormal" substrates, like the Ca-deficient serpentinites and the carbonaceous limestone and dolomite rocks. Actually, the geographic distribution of normal and abnormal rocks is a product of geologic earth history far removed from pedology, and the new Soil Taxonomy discards zonality.

References

1. Alexander, L. T., H. G. Byers, and G. Edgington. 1939. *A Chemical Study of Some Soils Derived from Limestone.* U.S.D.A. Techn. Bull. 678.
2. Bach, R. 1950. *Schweiz. Bot. Ges. Ber.* 60: 51-152.
3. Barshad, I. 1964. In *Chemistry of the Soil,* F. E. Bear, ed., 2nd ed., pp. 1-70. Reinhold, New York.
4. Barshad, I., E. Halevy, H. A. Gold, and J. Hagin. 1956. *Soil Sci.* 81: 423-437.

5. Beavers, A. H., J. B. Fehrenbacher, P. R. Johnson, and R. L. Jones. 1963. *Soil Sci. Soc. Am. Proc.* 27: 408-412.
6. Blanck, E. 1930. In *Handbuch der Bodenlehre,* E. Blanck, ed., Vol. 3, pp. 194-257. Springer, Berlin.
7. Brydon, J. E., and C. E. Marshall. 1958. *Mineralogy and Chemistry of the Hagerstown Soil in Missouri.* Mo. Agr. Exp. Sta. Res. Bull. 655, Columbia, MO.
8. Cecconi, S. 1955. *Ann. Speriment. Agr. n.s.* 9: 77-86.
9. Clarke, F. W. 1924. *The Data of Geochemistry,* 5th ed. U.S. Geol. Surv. Bull. 770, Washington, D.C.
10. Edwards, M. J., J. A. Elder, and M. E. Springer. 1974. *The Soils of the Nashville Basin.* Tenn. Agr. Exp. Sta. Bull. 499.
11. Fanning, D. S. 1970. *Soil Sci. Soc. Am. Proc.* 34: 98-104.
12. Fehrenbacher, J. B. 1973. *Soil Sci.* 115: 176-182.
13. Fehrenbacher, J. B., G. O. Walker, and H. L. Wascher. 1967. *Soils of Illinois.* Ill. Agr. Exp. Sta. Bull. 725.
14. Fotakiewa, Em., and M. Minkow. 1966. *Eiszeit. Gegenw.* 17: 87-96.
15. Gigon, A. 1971. Geobot. I. Rübel, Veröff. 48, Zurich.
16. Hotz, P. E. 1964. *Econ. Geol.* 59: 355-396.
17. Hulbert, A. B. 1969. *Soil: Its Influence on the History of the United States.* Russell and Russell, New York.
18. Hutton, C. E. 1948. *Soil Sci. Soc. Am. Proc.* 12: 424-431.
19. Jenny, H. 1941. *Factors of Soil Formation.* McGraw-Hill, New York.
20. Jenny, H., A. E. Salem, and J. R. Wallis. 1968. In *Soil Organic Matter and Soil Fertility,* P. Salviucci, ed., pp. 5-37. Pont. Acad. Sci. Scripta var. 32, Vatican.
21. Jones, M. B., W. A. Williams, and J. E. Ruckman. 1977. *Soil Sci. Soc. Am. J.* 41: 87-89.
22. Kimura, H. S., and L. D. Swindale. 1967. *Soil Sci.* 104: 69-76.
23. Kleiss, H. J. 1973. *Soil Sci.* 115: 194-198.
24. Kruckeberg, A. R. 1954. *Ecology* 35: 267-274.
25. Kubiena, W. L. 1953. *The Soils of Europe.* Th. Murby, London.
26. Låg, J. 1974. *Acta Agr. Scand.* 24: 13-16.
27. Larsen, E. S. 1951. *Crystalline Rocks of Southwestern California.* Calif. Div. Mines Bull. 159.
28. Lindsay, H. A. 1961. *Ecology* 42: 432-436.
29. Lossaint, P. 1973. In *Mediterranean type Ecosystems,* F. diCastro and H. A. Mooney, eds., pp. 199-210. Springer-Verlag, New York.
30. Machette, M. N., P. W. Birkeland, G. Markos, and M. J. Guccione. 1976. In *Studies in Colorado Field Geology,* R. C. Epis and R. J. Weimer, eds., pp. 339-357. Colorado School Mines.
31. Marshall, C. E. 1977. *The Physical Chemistry and Mineralogy of Soils.* Wiley (Interscience), New York.
32. Milne, G. 1940. *A Journey.* E. Afr. Agr. Res. Sta. Amani, Tanganyika Terr.
33. Nixon, E., and C. S. McMillan. 1964. *Am. Midl. Nat.* 71: 114-140.
34. Parsons, R. B., and R. C. Herriman. 1975. *Soil Sci. Soc. Am. Proc.* 39: 943-948.
35. Proctor, J., and S. R. J. Woodell. 1975. *Adv. Ecol. Res.* 9: 255-366.
36. Robinson, W. O., G. Edgington, and H. G. Byers. 1935. *Chemical Studies of Infertile Soils Derived from Rocks High in Magnesium and Generally High in Chromium and Nickel.* U.S.D.A. Tech. Bull. 471.
37. Ruhe, R. V. 1969. In *Pedology and Quaternary Research,* S. Pawluk, ed., pp. 1-23. Edmonton, Canada.
38. Runge, E. C. A. 1973. *Soil Sci.* 115: 183-193.
39. Russell, D. A. 1976. *Soil Sci. Soc. Am. J.* 40: 409-413.
40. Schellmann, W. 1964. *Geol. Jahrb.* 81: 645-678.
41. Siderius, W. 1973. Soil transitions in central East Botswana (Africa). Thesis (print), Univ. Utrecht.

42. Singer, M., and F. C. Ugolini. 1974. *Can. J. Soil Sci.* 54: 475-489.
43. Smith, G. D. 1942. *Illinois loess.* Illinois Agr. Exp. Sta. Bull. 490.
44. Snaydon, R. W. 1962. *J. Ecol.* 50: 439-447.
45. Streckeisen, A. L. 1967. *N. Jahrb. Min. Abh.* 107: 144-240.
46. Ulrich, R. 1950. *Soil Sci. Soc. Am. Proc.* 14: 287-295.
47. Vlamis, J. 1949. *Soil Sci.* 67: 453-466.
48. Waggoner, P. E., and C. Bingham. 1961. *Soil Sci.* 92: 396-401.
49. Wahrhaftig, C. 1965. *Geol. Soc. Am. Bull.* 76: 1165-1190.
50. Wali, M. K., and V. J. Krajina. 1973. *Vegatatio* 26: 237-381.
51. Walker, R. B. 1954. *Ecology* 35: 259-266.
52. Wascher, H. L., J. D. Alexander, B. W. Ray, A. H. Beavers, and R. T. Odell. 1960. *Characteristics of Soils Associated with Glacial Tills in Northeastern Illinois.* Illinois Agr. Exp. Sta. Bull. 665.
53. Welch, T. G., and J. O. Klemmedson. 1973. In *Forest Soils and Forest Land Management,* B. Bernier and C. H. Winget, eds., pp. 159-178. Laval Univ. Press, Quebec.
54. Whittaker, R. H. 1954. *Ecology* 35: 275-288.
55. Wildman, W. E., M. L. Jackson, and L. D. Whittig. 1968a. *Am. Mineral.* 53: 1252-1263.
56. Wildman, W. E., M. L. Jackson, and L. D. Whittig. 1968b. *Soil Sci. Soc. Am. Proc.* 32: 787-794.
57. Williams, E. G. 1959. *Agrochimica* 3: 279-309.
58. Yaalon, D. H. 1955. *Res. Council Israel, Bull.* 5B: 168-173.

11. State Factor Topography

The immense tectonic plates that make up the Earth's outer shell may slide beneath one another and initiate volcanos or collide with each other and push up and fold mountain ranges. Precipitation decimates the peaks, erodes the hill soils, and covers the valleys with new parent materials.

Topography, r, pertains to the configurations of the landscape and includes the positions of water tables insofar as they are not themselves the products of pedogenesis (e.g., perched water table).

Toposequences are written

$$l,v,a,s = f(r)_{cl,\phi,p,t,\ldots} \quad \text{or} \quad l,v,a,s = f(r,cl,\phi,p,t,\ldots)$$

$$\text{ideal case} \qquad\qquad\qquad \text{actual case}$$

Quantitatively, r may refer to inclination in percent degrees, length of slope in meters, concavity or convexity, and exposure (aspect) in points of the compass. Strictly speaking, r is r_0 the initial configuration at the chosen time zero.

Any happening on a slope that affects the soil, including erosion and deposition, is a pedogenic process. A geomorphologist may designate the same event a geologic process. No conflict need arise unless an attempt is made to distinguish the two in the field, for example, considering soil creep a geologic phenomenon and downslope solution and suspension flow inside the soil as pedogenic.

A. Points of Compass Orientation

Soil and vegetation react to aspect, whether facing north, south, east, or west.

Local Climates

Isolated Albany Hill, a block of sandstone 90 m high, faces San Francisco Bay and the long axis (1 km) of the elliptical shape is oriented north-south. Viewed in Figure 11.1 from the east, the north-facing hill slope carries a live oak forest (*Quercus agrifolia*) rooting in deep dark soils; the south slope is an open grassland on shallow, gray-brown soil comparatively low in organic matter. The bifurcation coincides with aspect-

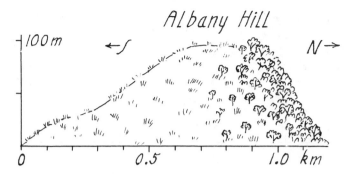

Fig. 11.1. Situation sketch of Albany Hill of the Pacific Coast Range showing distribution of natural grassland and forest with aspect

controlled evapotranspiration in a mediterranean-type climate having maP 58 cm and maT 13.3°C. The face turned to the reader is the seat of a vegetation transition or ecotone, the oak forest thinning to a miniature oak-grass savanna, its trees being smaller and younger and their spread being halted, assumedly, by moisture stress and occasional grass fires. (An uninformed spectator may not immediately recognize the pattern because of superimposed eucalyptus trees planted in the 1900s and view-homes erected recently.)

Many miles north, a twin hill at higher rainfall (maP 90 cm) is entirely clothed in deciduous-coniferous forest while in the southward arid portion of the same Coast Ranges (maP 18 cm) all slopes are grass covered.

Ellipsoidal, 61-m-high gravelly "Snake Hill" in south-central Idaho (maP 22 cm, maT 11°C) maintains sagebrush (*Artemesia tridentata*) and grasses, both native (*Stipa* spp.) and introduced. In Klemmedson's (29) north-south transect of Figure 11.2 soil nitrogen and organic carbon to a depth of 10 cm (grams/square meter/10 cm) are higher on north than south-facing slopes while biomass (grams/square meter) of the 1962 herbage behaves in an opposite manner, except for root mass which is 37% lower on the southern exposures. There, the root scarcity is considered responsible for the natural erosion that subdues the N contents.

Contour Function

Half-way up the hill a contour transect across slopes of about 37% delivers in Figure 11.2 a sinusoidal soil carbon function and the sites (No. 4, 11) common to the two transects occupy the maximum and minimum. The smoothed herbage curve (grams/square meter) appears as a dashed line.

Aspect Profile Sequence

In the State of Washington (33) windblown, crescentic loess hills of Wisconsin age support forbs and grasses on gentle, south and west-facing slopes, and shrubs on the steeper (30-50%) north-east exposures. Climate, considered normal (maP 53 cm) on south-facing slopes, is drier at the crest and colder and wetter on north slopes because

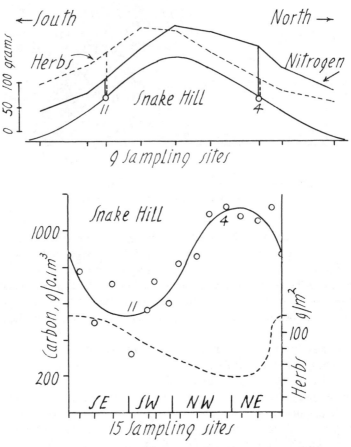

Fig. 11.2. Ecosystem properties related to aspect on Snake Hill, Idaho, after Klemmedson (29). Above: north-south transect with herbage (grams/square meter) and total soil nitrogen (grams/square meter/10 cm). Below: contour transect of herbage (dashed line) and soil carbon (solid line and open circles)

of low-angle incidence of sun rays and accumulation of drifting snow from ridges and opposite flanks.

In the new Taxonomy all soils are Mollisols of the Xeroll suborder (Fig. 11.3) having pH values above 6. The Athena soil series on the crests has a calcareous B horizon and was classified a Chernozem; Palouse soil on the sunny slopes has an argillic B horizon and qualified as a Prairie soil. Shady Thatuna acquired a clay pan with a light A2 horizon above it. This soil is richest in organic matter. Its parent material may be an older loess.

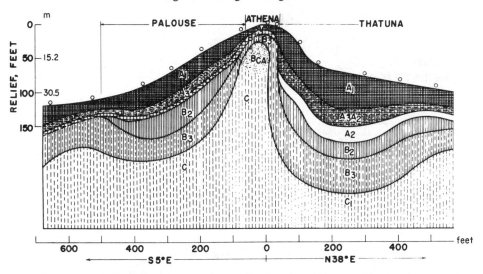

Fig. 11.3. Horizon sequences through a loess hill in the state of Washington, from Lotspeich and Smith (33). Circles on soil surface indicate profile sites. Courtesy The Williams and Wilkins Co., Baltimore, MD

B. Ridge and Trough Configurations

Relief, as elevation differences between high and low points in landscapes, is measured in decimeters in mound and hillock country, in meters on undulating plains, and in kilometers in the Alps, Himalayas, and Andes.

Water Regime

In Ellis's (25) slightly altered Figure 11.4, rolling loess or moraine topography has a macroclimate of 50 cm average rainfall. Surface runoff and internal downslope flow of water in saturated and unsaturated state maintain relatively dry knolls and humid troughs. In wet climates the depressions become bogs, marshes, and swamps and in very dry ones salty soils and alkali flats. When a ground water table lies at depths exceeding 2.5 m in sands and 3.5 m in clays, capillary rise will not reach A and B horizons, but deep roots penetrate the moisture fringe and multiply biomass and humus contents.

Catena Model

Mentally place the ridge-trough configuration in the arid region and let it have a salty water table a foot or so beneath the depression. Sprinkle annually and randomly seeds of gray saltbush (*Atriplex cordulata*), medium-tolerant Bermuda grass (*Cynodon dactylon*), and salt-intolerant brome grass (*Bromus rigidus*). In the depression capillary rise and evaporation create a salt playa that kills all germinating seeds. The ridge tops will carry brome grass on a well-drained soil. Along the slopes salt concentrations vary in proportion to capillary rise and dilution by occasional rains.

KNOB AND BASIN TOPOGRAPHY

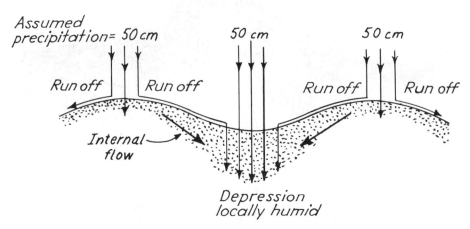

Fig. 11.4. Water regimen in knob and basin topography. After Ellis (25)

Salt-tolerant *Atriplex* thrives on the lower slopes, Bermuda grass on the upper portions. The symmetrical configuration is a catena, a chain hanging between two points of support, but the author of the expression, G. Milne, did not confine it to uniform parent material, as is done here. Each half of the catenary is a toposequence.

The biotic factor, or potential flora, is the same for all parts of the slope and is the seed mixture supplied; even the playa has the same ϕ. The species growing and the soils beneath are positioned in a continuous ecosystem toposequence with zero biomass in the trough.

Elgabaly's (18) calcareous sand dune of Arish fits the model closely. The crest is 13 m high and chlorides and sulfates of sodium ascend from the brine and descend from wind influx, as shown in Figure 11.5 by the salt percentages to a soil depth of 25 cm. The low concentration near the crest is due to dissolved Ca, Mg-bicarbonates at pH 7.7. The dominant desert plant species are lined up according to their salt tolerances and competitive behavior.

Sukatchev's (49) botanical "type complexes" on deep sand dunes in eastern Europe illustrate toposequences in the absence of a water table. Podzol on the ridge nurtures short, crooked pines and is decorated with nests of lichen. On the slopes the trees do better and thick pads of mosses soften the ground. The deep, moist trough adds grasses and linden trees (*Tilia*) to the flora and holds more humus. Coaldrake (12) sketched an analogous podsolized ecosystem sequence in an Australian evergreen, multistoried rainforest of tall, vined trees. Mattson's moraine toposequence in Sweden was reproduced previously (25).

Mima Mound Topography

West of the Mississippi River old, alluvial plains often are "pimpled" by hummocks or Mima mounds, the adjective referring to the Mima prairie in the State of Washington. The circular and oval domes are up to 1 m high and have diameters of 4-30 m, and as

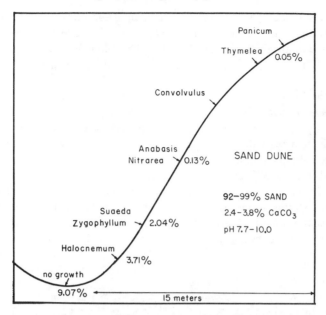

Fig. 11.5. Elgabaly's (18) toposequence of plant species and salinity on calcareous sand in North Africa. Percentage figures along slope denote salt content

many as 4000 crowd into 1 km² area (5). In summer-dry climes mounds and troughs are underlain by pedogenic, water-impermeable iron-silica hardpan, as in the San Joaquin series of Chapter 12, or by clayey B horizons set forth in Figure 11.6 for the Antioch series.

On the assumption that clay pan antedates mound construction, pedological thought conjectures the Antioch Bt initiated by the presence of alkali which encourages translocation of Na clay from A to B, Na⁺ then being displaced by H⁺ through vertical and lateral leaching, meting pH 6 and less to the epipedon. The ion conversion is known as solodization. The 5-6 month summer drought shrinks the clay mass into columns 30 cm long and 5-10 cm wide, and the rainy season induces swelling of the clay to a water-tight seal. The landscape becomes dotted with innumerable ponds, and forbs and grasses sprout on mounds and under the water in the sinks. As the "vernal pools" evaporate in springtime garlands of flowers unfold, as if the plains were offering the spectator a myriad of golden bouquets (23).

In the western Great Plains native short grasses dominate the south faces and tall grasses the north mound exposures, the two simulating climates that are hundreds of miles apart (51). In Colorado, herbage production is 92 g/m² on the knolls and 29 g/m² in the depressions and soil humus follows suit (4.8 versus 2.9%). According to McGinnies (35) mounds store more available water than shallow troughs.

Mound building has been linked with wind transportation and deposition of sand around shrub clumps in a previously arid climate, with channeled erosion, with frost heaving during past glacial periods, with hydrostatic pressures of ground waters (39) with swelling of Vertisolic B horizons (10), and with gopher activities (mentioned in Chapter 13).

The mounds, hillocks, or hummocks of the humid *Coastal Plains of Louisiana* belong to the Alfisol order and exhibit high acidity (pH 4.6-5.5) and low base saturation (5-33%). Al is the dominant exchangeable cation. Three-fourths of the vegetation cover is made up of 30-m-tall loblolly pines (*Pinus taeda*) and the remainder of sundry broad-leaf species.

Pines in the fine-silty intermound depressions suffer massive invasions of bark beetle (*Dendroctonus frontalis*) but trees on the coarse-silty mounds are mostly bypassed. Infestation is geared to tree physiology and is resisted by exudation flow of sticky oleoresin which raises with turgor pressure of living cells. Turgor, in turn, falls with severity of soil and plant moisture stress.

Lorio and Hodges (32) educe microrelief control of root development. The wet flats are deficient in fine and mycorrhizal roots because of attacks by water molds and root-let pathogens. During severe drought spells, when crown demand for water is high, pines in depressions experience more moisture stress than trees on mounds, and therefore succumb to beetle invasion.

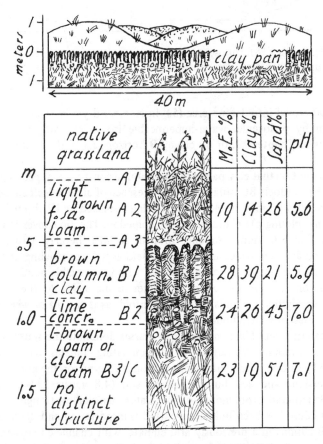

Fig. 11.6. Mima mound or hogwallow microtopography (height exaggerated compared to length) and profile of Antioch clay pan soil. Adapted from U.S.D.A. Soil Survey of Suisun Area, Ser. 1930, No. 18

The Closed Basin

When the last midcontinental ice sheet melted in Iowa 13,000 years ago, it left the extensive Cary drift pocketed with topographic depressions up to 10 m in depth, hundreds of meters in diameter, and without outlets (Fig. 11.7). The bog in the center stores and chronicles minutely all materials that erode from the surrounding watershed. Combining soil profile studies with pollen analysis and ^{14}C dating, Walker (56) discovered five episodes of landscape history in the Colo bog basin (Table 11.1).

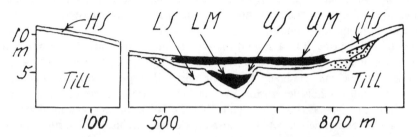

Fig. 11.7. Profile of Colo bog basin in Iowa (56). HS, Hillslope surficials; LS, lower silt; LM, lower muck; US, upper silt; UM, upper muck

Table 11.1. Profile in Center of Colo Bog Depression in Iowa[a]

Depth (m)	Colo profile	Years B.P.	Clay (%)	Organic Matter (%)	Pollen	Accumulation	Erosion
	UM (muck)		43	47			
			36			2.87	0.18
			35	54	Grasses		
1 –		3,100		25	and		
			45	26			
	US				herbs		
2 –	(silt)		44	9		4.80	1.04
			36	2			
				21			
3 –				48	Ferns		
			31				
		8,300		60			
	LM		22	64	Hardwoods	2.69	0.02
4 –	(muck)						
		13,800	29	32			
			30	38			
5 –			29	18	Spruce		
	LS						1.37
	(silt)			5	No		
			21				
6 –			35	4	pollen		
			14	2			
	Till			1			
			15	1			
7 –							

[a]After Walker (56).

At the base the calcareous gravelly loam of *Cary till* is covered with 130 cm of gray *Lower silt* (LS), a sticky, silty clay loam, mostly calcareous, washed in from the surrounding hill slopes. Because of its low content of organic matter and scarcity of pollen in its lower portion, Walker concludes that the watershed was covered by meager vegetation that encouraged the relatively rapid erosion rates of 1.37 cm/100 years, as estimated from the volume of LS sediment.

Above LS, the dark olive-colored *lower peaty muck* (LM) contains a piece of spruce wood (*Picea* sp.) and conifer pollen. The climate must have been warming and coniferous forest, which later turned into a mixed hardwood forest of birch, oak, and alder, stabilized the slopes, reduced erosion, and allowed the accumulation of the sticky, slightly calcareous peaty muck.

About 8000 years B.P. a second material stratum, the black *Upper silt* (US), a weakly calcareous silty clay loam, or a mucky clay, began to cover the peat to a thickness of 200 cm, gradually becoming enriched in pollen of grasses and herbs. Walker estimates a hillside erosion of 1 cm in a century, triggered by a rapid climatic shift to dryness, and he contends that the 50 cm of erosional lowering of the upper slopes obliterated in what is now Prairie soil (Mollisol) all vestiges of forest horizons such as gray coatings on ped surfaces.

Three millennia B.P., grass pollen assumed prominence, slope wash subdued, and basin organic layers resumed growth, becoming *Upper muck* (UM). Today, a change toward humid climate seems in the making because oaks are invading the prairie.

C. Toposequences on Slopes

In the "closed basin" of the previous section the evolution of topography transformed the rough, bare surface of the ancient Cary till into a smooth undulating prairie landscape, its slopes descending gradually from watershed rim to bottom muck and harboring in Iowa and Minnesota the till-toposequence of soil series:

Clarion	→	Nicollet	→	Webster
well drained		imperfectly drained		poorly drained

Since erosion slightly modified the original Cary topography, today's r has become in part a genesis-dependent variable.

Slope Accruement of Organic Matter

In an Iowan loess landscape with virgin prairie and slope gradients of 13%, Aandahl (1) finds a doubling of soil organic matter from summit to downhill sites (Fig. 11.8). Highly significant correlations of % N with length of slope suggest small but continued surface-erosional transport of humus from higher to lower positions. The accumulations are corroborated by the N depth functions (inset of Fig. 11.8) which decline rapidly in summit (s) profiles and slowly at near-bottom (b) sites, leading there to thick, humus-rich B and C horizons. Riecken and Poetsch (43) document a progressive decrease of sand particles from crest to vale, the soils becoming finer textured and also darker in color. The foot-slopes acquire accruement profiles.

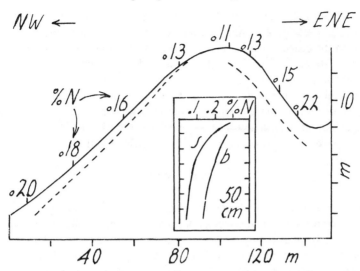

Fig. 11.8. Mean nitrogen percentages to 76 cm depth along slopes of a loess hill in Iowa, after Aandahl (1). Dashed line indicates depth to $CaCO_3$. In inset, N depth functions at summit (s) and near-bottom (b) position

On a loess mantle of 4.5 m thickness in Knox County, Illinois, Alexander et al. (3) differentiate well-drained *Tama* soil on 5% gradient, somewhat poorly drained *Muscatine* soil on gentler slopes, and poorly drained *Sable* soil on level and depressional areas (Fig. 11.9). All are Mollisols (Prairie soils). The calcareous montmorillonite-rich C horizons at 125 cm are near-identical heavy silt loams; the B horizons are clay loams with up to 37% clay, blocky structures, clay skins, mottles, and iron concretions. A1 horizons are 28-cm-thick silt loams in Tama and Muscatine, and 36-cm-thick silty clay loams in Sable. Leaching out of $CaCO_3$ contracts initial tesseras by as much as 50 cm (38).

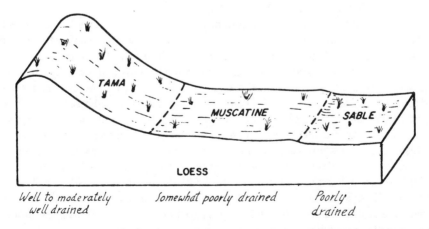

Fig. 11.9. Segment of an undulating loess landscape in the prairie belt of Illinois (3)

Weights of organic matter to a depth of 152 cm are 36 kg/m² for humus and 1.92 kg/m² for roots in Muscatine soil and slightly less for the other two soil types. Plant opals in the coarse silt fraction (20-50 μm) of the A horizons are 0.65% in Tama, 1.11% in Muscatine, and 0.46% in Sable, again pointing to best growth·on the middle portion of the slope.

Slow natural erosion is believed to balance the Tama soil by adding soil material from higher up and displacing equal amounts downslope. Muscatine is thought to receive slightly more than it loses. In Sable soils erosion is nil and wash-in during millennia might have been substantial and inflating the fine fraction of texture.

Redox Flush

In Blume's (9) German spruce-beech forest on clayey, calcareous moraine of the last glaciation, Brown Earth (an Inceptisol) occupies the crests, Pseudogley the low-slope portions. Water of maP 80 cm moves downslope principally as unsaturated flow accompanied by reduced Fe and Mn ions. Under rapid changes of redox potentials, resulting from evapotranspiration alternating with rains, small concretions form; under slow *Eh* changes broad rust spots and Lepidocrocite appear. Exchangeable aluminum decreases from ridge to trough.

Slope flush of Mn and Fe in the summer-dry Mediterranean region is demonstrated by Yaalon et al. (58).

Subtropical and Tropical Slope Sequences

In summer-dry coastal San Diego County (maP 40 cm, maT 15.6°C) near the Mexican border, rolling upland landscapes are carved into *tonolite rock,* which is a quartz diorite. Along a south-facing slope traverse 150 m long and 20 m in elevation difference, Nettleton et al. (37) recognize a toposequence in the grus comprising the soils Vista on smooth hill tops and upper slopes covered by xerophytic shrubs and rodent populations, Fallbrook on middle slopes, and Bonsall on lower and foot slopes with former grassland vegetation and scattered live oaks (*Quercus agrifolia*); Bosanko clay, on subdued relief, is outside of the transect proper.

Monthly measurements of *soil moisture* and calculations of water balances record highest humidities in downslope profiles. In an average year the Vista profile dries out in May, Bosanko in June, and Fallbrook and Bonsall in July. The soils remain at PWP and below for at least 60 consecutive days. In winter months that are wetter than average, water seeps into C horizon and the grus below. The latter has an available water capacity (AWC) of 0.08 cm of H_2O/cm depth and the calculated mean leaching value Li of 18 cm could wet uniform, dry C material to 225 cm depth, and much deeper if grus moisture were at FC. During rains water flows laterally in A2 of Bonsall soil.

Clay accumulates downslope and changes in kind as revealed by the divergence of the clay and CEC depth functions in Figure 11.10. In the consistently brown and reddish-brown B horizons vermiculite and kaolinite dominate in Vista, kaolinite and illite in Fallbrook, and kaolinite and montmorillonite in Bonsall. Its B horizon is sufficiently impregnated with Na derived from weathering to qualify as natric horizon. Gray Bosanko clay, a Vertisol, is montmorillonitic throughout.

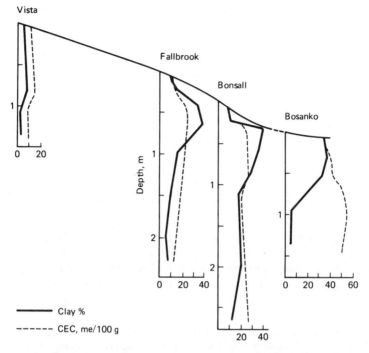

Fig. 11.10. Depth functions of clay and cation exchange capacities (CEC) of semiarid toposequence in tonolite grus. After Nettleton et al. (37)

In Llano County, Texas (maP 71 cm, maT 19°C), a gentle *granite slope* is covered with native savanna. At the two sites selected by Goss and Allen (20) the rock is coarse grained and has near-identical mineral modes, and the heavy mineral counts evince a high degree of initial uniformity among all horizons.

On the *upper-slope* (3.5%), brown Click sandy loam (a Typic Haplustalf) has a depth of 135 cm to bedrock and clay accumulations in B horizons (B2t, B3t). Illite and kaolinite exceed montmorillonitic clay.

On the *base-slope* (< 0.5%), gray Bauman sandy loam (a Typic Natraqualf) measures 160 cm to disintegrated granite, has $CaCO_3$ in B, and is enriched in fine (< 0.2 μm) montmorillonite clay. When the soil is moist, the dark gray color suggests high humus content, but, in Table 11.2, the organic C percentage is quite low in spite of clay and clay loam textures. In thin sections the metallic luster of illuviation cutans suggests staining of ped surfaces with dark-colored ferromanganese compounds.

The authors attribute the fine-textured foot-slope profile to downhill migration of weathering products, presumably within the soil mass, and possibly including ultrafine montmorillonite clay.

The brown and clay toposequences have many counterparts, e.g., in Israel (58) and in Bulgaria (30) where the dark-gray to black toe-slope phase is known as smolnitza. On the Deccan Plateau in India (maP 60-90 cm, maT 23-25°C) contrasting broad red and black soil belts are conspicuous from the air. The downslope black apron member

Table 11.2. Selected Properties of a Texan Savanna Brown-Gray Toposequence on Granite[a]

	Upper slope (Click sandy loam) Typic Haplustalf	Lower slope (Baumann sandy loam) Typic Natraqualf
pH	6.2-6.7	6.3-7.7
$CaCO_3$ horizon	None	Strong in B23, none in C
Percentage organic C	0.2-0.4	0.2-0.7
Color of profile	Brown	Gray
In B horizons		
Percentage clay	14-17	25-36
CEC (me/100 g)	10.2	26-35
Percentage exchange Na	2	11-28

[a] Goss and Allen (20).

belongs to the many types of fine-textured Black Cotton soils, or Regur, low in humus. The color shift from red to black had been interpreted long ago as slope-conditioned by A. Sen and other Indian soil scientists (42). Kaolinite prevails in the red variant and shrink-swell beidellite in the black one, but percentages of N and C are indiscriminate.

Tardy and his French colleagues (50) set *tropical* toposequences on granite into a climatic frame that orders gibbsite (G), kaolinite (K), montmorillonite (M), and sodium silicate (SS) as seen in Figure 11.11.

On the upper slopes clay genesis is accompanied by lateral eluviation and on the lower slopes by lateral illuviation. Fe and Mn concretions appear in seasonally alternating climatic regimes. Paraphrasing the authors, the upslope minerals foreshadow those of the downslope in a more humid climate, and vice versa, the downslope minerals predict those of a climatically more arid upslope.

On high level, tropical lands *laterite crusts* are residual accumulations of sesquioxides, as presented in Chapter 7. On slopes ferrous *iron* may flow downhill internally, rise to the surface at lower elevations, and oxidize, thereby cementing sand grains and pebbles to a "low level" ferruginous crust, as sketched by Sherman (46) for Hawaii in Figure 11.12. Whether iron may also wander along slopes as organic complexes remains to be demonstrated. Microscopic studies suggest to Hamilton (22) that iron migrates as colloidal oxide protected by silica. Sherman's drawing further postulates a Si flow that may generate a silcrete surface rock. Why low-level laterite crusts, or their precursors, do not extend northward to the southern United States remains to be explained.

During extended geologic periods natural erosion may denude adjacent softer strata and thereby "elevate" the crusts to prominent landscape features as seen on Apricum and Manzanita Hills near Ione, California.

Fig. 11.11. Clay minerals in tropical toposequences. After Tardy et al. (50). M, Montmorillonite; K, kaolinite; G, gibbsite; SS, sodium silicate

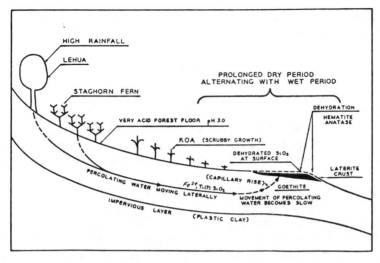

Fig. 11.12. Ferruginous laterite crust caused by lateral movement of iron in Hawaii (46). Courtesy University Press of Hawaii, Honolulu, HI

D. Salt-Affected Soils

The majority of humanity lives on the drier lands and Attila's historic threat of "sowing salt" is still a real fear.

Source of Salts in Arid Regions

Granites and basalts are low in chloride (Cl^-) and sulfate (SO_4^{2-}) anions. During weathering Ca^{2+} and Mg^{2+} become the but slightly soluble earth carbonates and tend to remain at the site whereas $NaHCO_3$, Na_2CO_3, and to a lesser extent $KHCO_3$ and K_2CO_3 follow the water courses. When flow ends in arid depressions evaporation of water leaves the ions behind as "alkali" or carbonate salts of pH 10 and higher.

> Sea water has an osmotic pressure of 24 atm and contains nearly 35 g of ions per liter of which 55.5% are Cl, 7.7% SO_4, and 31.7% Na. The remaining 5.1% are mostly Ca and Mg. The neutral salts NaCl (table salt) and Na_2SO_4 are highly soluble, 357 and 48 g/liter, respectively, at $0°C$. During evaporation sulfates precipitate much sooner than chlorides, which facilitates salt separation.

When an arid basin is inundated by sea water and then the sea retreats, evaporation incorporates the ocean salts in sediments and soils and Na_2SO_4 may be enriched relative to NaCl (31). The reaction (pH) of both salts is neutral but in the presence of soil part of Na becomes an exchangeable cation that promotes alkalinity by way of the hydrolysis equation of Chapter 3.

Soils rich in neutral salts are variously called *saline,* or white alkali, or solonchak, and soils with Na_2CO_3 and exchangeable Na are *sodic,* or black alkali soils, or types of solonetz. In essence, Kelley (28) points out, all are alkali soils and "saline" and "sodic" are convenient, simplifying distinctions. Sehgal et al. (45) suggest improved alkali criteria for the new Soil Taxonomy.

Plant Responses

In the 1880s Hilgard (27) pronounced the native succulent Iodine bush (*Allenrolfea occidentalis*) and the perennial herb pickleweed (*Salicornia subterminalis*) the most tolerant of white alkali, but not of black alkali. Greasewood shrub (*Sarcobatus* spp.) endures both. In those early days of exploration of natural alkali lands the native species were extensively calibrated as indicators of a soil's salt status. Now, field apparatus take their place.

Halophyte plants are able to withstand high osmotic pressures in tissue fluids (150 atm for *Atriplex* sp.) and the cellular enzyme systems endure high Na concentrations. Domestic plants react unfavorably (8) to the negative water potentials generated by sulfates and chlorides. As soil water is removed by evapotranspiration the resulting matric and osmotic potentials are additive and water stress becomes intolerable (Fig. 2.8).

At equal concentrations Na_2CO_3 is more toxic to plants than either NaCl or Na_2SO_4. Exchangeable sodium as Na clay in contact with roots disturbs cell wall stability and metabolic action.

Alkali Hydrosequence

The slightly sloping (0.4%) Panoche colluvial fan near hot, summer-dry Coalinga, California, consists of outwash from soft, calcareous and gypsiferous sandstones and shales, and the arid-type water table, drawn at flood stage in Figure 11.13, descends in proportion to distance from the river, being over 6 m deep on the left side of the outline. There is no water table during summer time. Seasonal capillary rise from the saline ground water divides the clayey fan into brownish-gray Panoche, Oxalis, and Levis soils of increasing salinization, measured as grams salt/100 g soil. The three series string together as a saline hydrosequence (26) with no diagnostic horizons beyond lime and gypsum impregnation and sporadic gley mottling in Oxalis subsoil.

Rise of the Alkali and Its Reclamation

Freshwater rivers emerge from the granite mountains of the Sierra Nevada, traverse the arid San Joaquin Valley, and enter the Pacific Ocean. Two of the rivers are marked A and B at opposite ends in the valley profile of Figure 11.14. In 1873, at the time of white man's agricultural explorations, the naturally salty ground water table was at 20 m depth, as shown by the lowest dashed line. The salts, mainly NaCl and Na_2SO_4, derived from weathering of salt-impregnated sedimentary rocks.

The pioneers diverted mountain waters into a network of canals to irrigate the fields. For years bounteous crops were grown. Since the conduits were unlined, excessive seepage slowly raised the water table with its dissolved salts. In 1888 water stood at 0.6-0.9 m below the land surface. Capillary rise and evaporation brought salts to the surface and caused extensive salinization, salt efflorescence, and crop damage. This deterioration of soil is a form of "desertification."

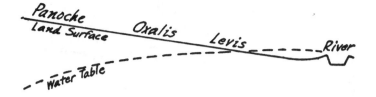

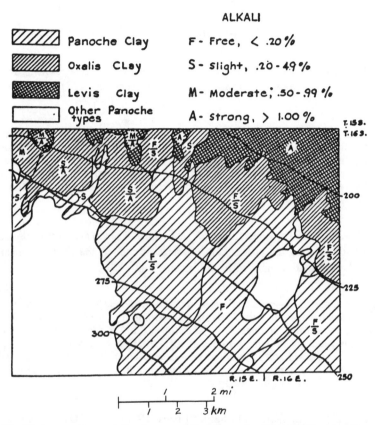

Fig. 11.13. Hydrosequence at the lower end of Panoche Fan, California. Above: outline of surface configuration and water table. Below: soil map with elevations (200 ft = 61 m, 300 ft = 91 m). Single letters indicate salt condition through the profile; quotients indicate salinity in surface soil (numerator) and in subsoil (denominator). (Courtesy The Williams and Wilkins Co., Baltimore, MD)

The rise of alkali initiates the following ion exchange reaction:

$$\begin{array}{c} \boxed{\begin{array}{c}\text{Clay} \\[4pt] \text{Humus}\end{array}} \begin{array}{c}\text{Ca}^{2+} \\[6pt] \text{Mg}^{2+}\end{array} + n\text{Na}_2\text{SO}_4 \rightleftharpoons \boxed{\begin{array}{c}\text{Clay} \\[4pt] \text{Humus}\end{array}} \begin{array}{c}\text{Na}^+ \\ \text{Na}^+ \\ \text{Na}^+ \\ \text{Na}^+\end{array} + \begin{array}{c}\text{CaSO}_4 \\[6pt] \text{MgSO}_4\end{array} + (n-2)\text{NaSO}_4 \end{array}$$

$$\qquad\qquad\qquad\quad \underset{\substack{\text{excess} \\ \text{salt}}}{} \qquad\qquad\qquad\qquad\qquad\qquad\qquad\qquad\qquad\qquad \underset{\substack{\text{excess} \\ \text{salt}}}{}$$

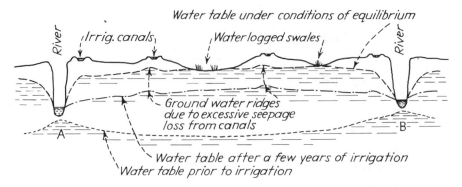

Fig. 11.14. "Rise of alkali" caused by rising ground water table resulting from leaky canals and overirrigation. After Etcheverry (25)

The original Ca, Mg, K ion swarm on the clay particles of the well-drained, arid soil is transformed into one of Na preponderance. Salt excess keeps clay and humus colloids flocculated and soil reaction neutral by repressing hydrolysis. The salt barrens are *white alkali* lands because of the light-gray color and the whitish crusts. Man-induced "rise of the alkali" still occurs in poorly designed irrigation projects here and abroad.

In 1920 Kelley (28) lowered the water table in the Fresno area by continual electric pumping and diverting the effluents in drainage ditches to the river. To leach out the excess salt he flooded the land with fresh river water.

In doing so, the flocculated soil reverts to dispersed Na clay and Na humus and the pH rises because of hydrolysis of Na colloids that yields NaOH, $NaHCO_3$ and Na_2CO_3, as outlined in Chapter 3. The hot sun bakes the clay to hard crusts, and Na humus blackens soil in blotches. White alkali soil converts to *black alkali* or *sodic soil*. Dispersed soil is puddled and structureless. Water infiltration is poor, as demonstrated in Chapter 6. Low gas diffusion induces oxygen deficiency that curtails water and nutrient uptake by roots.

Hilgard (27) counteracted the harmful $NaHCO_3$ and Na_2CO_3 in the soil solution with applications of gypsum, but as he was unaware of ionic exchange reactions and CEC he underestimated the amount of Ca needed to replace exchangeable Na and he achieved only partial success. Kelley augmented the dosage manyfold (30 tons/ha) and successfully converted Na clay into desired Ca clay (see equation in Chapter 6). He continued leaching for years, and gradually the soil became granular and mellow and the crop-producing power returned. After a decade the soil was practically free of salt and exchangeable Na, and annual hay yields of irrigated alfalfa (*Medicago sativa*) climbed to a high 9.8 tons/acre or 2.2 kg/m² , air-dry.

In newer reclamation techniques the fields are broadcast with powdered sulfur (S) which oxidizes to sulfuric acid, or land is doused directly with commercial acid that replaces Na^+ with H^+.

Acid Sulfate Soils and Jarosite Clays

Along sea shores with tidal marshes the littoral soils are either reduced sulfide muds or oxidized acid jarosite clays (cat clays) that conform to the sulfur redox pair:

$$\text{Sulfates} \quad \underset{\text{oxidation}}{\overset{\text{reduction}}{\rightleftharpoons}} \quad \text{Sulfides}$$

in which S is +6 on the left and −2 on the right. The ensuing elaboration relies on research done in The Netherlands and elsewhere (15) and it focuses on a nature preserve along San Pablo Bay near San Francisco (24,34) that has wide significance.

Reduction Regime. Bay mud flats with 52-64% clay and $< 1\%$ sand are populated by species-poor, salt-tolerant marsh vegetation that provides feeding and nesting for migrant and resident water birds. Tules (*Scirpus* spp.) and reed cord grass (*Spartina*) grow in shallow, brackish waters, while fleshy, salty, green- and red-colored pickleweed (*Salicornia*) and spiked salt grass (*Distichlis*) sharply segregate in zones of less frequent tidal flooding. Under tule the humic subsoils are permeated with root rhizomes and are dark bluish-gray or jet black from insoluble iron sulfides (FeS, FeS_2). The unpleasant odor of toxic hydrogen sulfide gas (H_2S) is evident or will appear upon addition of hydrochloric acid (HCl).

The sulfides originate from reduction of sea salts in the thick muds of pH 7-8, accomplished predominantly by microbes, e.g., *Sporovibrio* spp.:

$$\left.\begin{array}{c} Na_2SO_4 \\ \text{or} \\ CaSO_4 \end{array}\right\} + 8H \quad \xrightarrow{\text{reduction}} \quad \left.\begin{array}{c} Na_2S \\ \text{or} \\ CaS \end{array}\right\} + 4H_2O + 4CO_2$$

dissolved	from	soluble
colorless	organic	colorless
sulfates	matter	sulfides

Organic matter is oxidized (viz. dehydrogenated), accompanied by CO_2 evolution, and energy is made available. To the reaction $SO_4^{2-} + 4H_2 \rightarrow S^{2-} + 4H_2O$ Simon-Sylvestre (47) assigns an energy change of −60,000 cal. Supply of organic matter is viewed as the rate-limiting step of the reduction (7).

Subsequently, the soluble sulfides may turn into H_2S gas and the iron ore pyrite:

(a) generating H_2S:

$$CaS + H_2CO_3 \rightarrow CaCO_3 + H_2S$$

calcium	from	calcium	hydrogen
sulfide	organic	carbonate	sulfide gas
	matter		(slightly acidic)
	oxidation		

(b) generating FeS:

$$2Fe(OH)_3 \quad + \ 3H_2S \ \rightarrow \ 2FeS \quad + \ S \quad + \ 6H_2O$$

brown iron	blue-black	yellow
hydroxide	iron sulfide	sulfur
(insoluble)	(insoluble)	(solid)

In the presence of free sulfur, FeS becomes the deep-black, insoluble ferrous disulfide, FeS_2, or pyrite ore: $2FeS + S = FeS_2 + FeS$.

As seen, gypsum ($CaSO_4$) is ultimately converted to calcium carbonate, and Hardan (15) explains the lime accumulation in the soils of the Mesopotamian plain by the same biological sulfate reduction. In many Aridisols quantities of $CaSO_4$ and $CaCO_3$ appear inversely correlated.

The conversion of Na_2SO_4 to Na_2S may end in sodium carbonate (Na_2CO_3) and alkali soil, as demonstrated by Egyptian (2) and American (34) workers. The latter also infer synthesis of chlorite clay.

Oxidation Regime. When the mud flats rise above the tides, either by tectonic uplift, retreating sea level, or diking by man, rainfall and irrigation water flush out the alkalinity, and oxidation of sulfides, catalyzed enzymatically by the acid-tolerant sulfur bacteria (e.g., *Thiobacilli*), proceeds along various pathways. For H_2S the reaction is

$$H_2S \quad + \ 1/2\,O_2 \quad \rightarrow \quad S \quad + \quad H_2O \qquad \Delta G = -41,000 \text{ cal}$$

| hydrogen | solid |
| sulfide | sulfur |

The sulfur is seen microscopically as yellow granules inside the colorless bacteria. S oxidizes in air:

$$S + 1\tfrac{1}{2}\,O_2 + H_2O \quad \rightarrow \quad H_2SO_4 \qquad \Delta G = -118,500 \text{ cal}$$

sulfuric
acid

The free energy changes (ΔG) are those of Baas-Becking and Parks (6) and refer to natural conditions. Diffusion of O_2 is the rate-limiting step.

Oxidation of pyrite instead of H_2S may be written summarily (54):

$$4FeS_2 \ + \ 15O_2 \ + \ 14H_2O \quad \rightarrow \quad 4Fe(OH)_3 \ + \ 8H_2SO_4$$

pyrite

iron
hydroxide

The reaction does not go to completion and insoluble basic iron sulfate $KFe_3(SO_4)_2(OH)_6$ crystallizes as *jarosite*. It is a yellow gel that coats ped surfaces and rings root channels. Yellow mottling begins within 30 cm of the surface and may extend to 60 cm. Below, pyrite prevails. Red mottles and pipes of amorphous iron hydroxides or oxides accompany the yellows and generally extend to a greater depth. The color design and its substrate is named katteklei or cat clay. Van Breemen and Harmsen (54) document a chronosequence of oxidation. In the absence of carbonates the strong sulfuric acid of these "acid sulfate soils" maintains pH values of 2-4 that persist for decades. The acidity liberates K from mica, deteriorates chlorite, converts clays to stable Al clay aggregates, and imparts to the soil solution Al^{3+} concentrations that exceed the tolerance limit of crops. Successful reclamation is accomplished by liming.

E. Soil Erosion

The mountain landscapes along the Pacific and Atlantic Coasts are showplaces of man's interaction with nature because virgin forest ecosystems are still left for comparison.

Natural Erosion

Unlike grains in a dry sand pile, particles in the soil are bonded to humus and sesquioxide gels that impart shear strength and resistance to movement. Stability is aided by trees and shrub roots extending through the soil mantle into crevasses of bedrock, functioning as anchors. The shorter, fibrous grass roots are less effective. The forest floor cushions the impact of raindrops and the network of open-packed structures sucks in rain water to be dissipated by evapotranspiration and deep seepage.

In well-stocked woodlands with forest floors several centimeters thick surface erosion is less than 1 mm in a century, according to the C-factor of 0.1% in the soil-loss equation described later. In a pine-hardwood forest in Mississippi (maP 132 cm) a 2-year erosion wash of 4.5 g/m^2 from a small watershed (53) corresponds to an even lesser soil reduction (0.2 mm/century). Such minute surface losses may be matched by the rate of rock weathering below and the balance tends to maintain soil depth. Stability of slopes of over 30% is affirmed by their profiles of Ultisols and Oxisols with well-defined horizons.

When weathering rates exceed surface erosion rates the soil mantle thickens and the gravitational pull tends to drag the soil downslope. Something must give.

A brick resting on an inclined rock plane (Fig. 11.15, left) is acted upon by the force of gravity, mg, which is equal to the brick's weight W. In the right-hand sketch W is resolved into the pressing force N and the downslope pulling force P. The latter is the greater the larger W and the steeper the gradient α, written $P = W \sin\alpha$. The brick would slide were it not for the passive, frictional force F which opposes P and which connotes the interaction forces between brick and rock plane.

In contrast to a brick, wet soil is readily deformed and shear-strength characteristics come into play (21). Clay particles hydrate and swell, which diminishes cohesion and resistance to shearing. The interface of soil horizons suddenly may become a plane of sliding, more so with montmorillonite than kaolinite clays (40). Deep C and D horizons of thixotropic nature are especially prone and earthquakes may trigger slides.

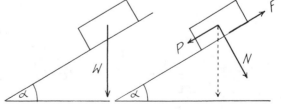

Fig. 11.15. To explain downward drag force P of sloping soil, brick of weight W, in left sketch, rests on inclined plane of gradient α. On the right the gravitational force W is resolved into the forces P and N

At many sites signals of infinitesimal soil creep become visible. Roots are exposed and isolated trees grow in curvature of tilt. On steep pastures fence posts lean downhill because surface soil glides more easily than lower horizons. Deep soils on slippery rock materials, like the Yorkville series, may be moving continually, becoming hummocky and carrying grassland in the midst of forest country because tree and shrub roots are torn apart. Drying and wetting and freezing and thawing further aid soil flow; on Arctic slopes and on barren ultrabasic materials anywhere "solifluction" transports stones, pebbles, and fines from crests to troughs.

When a downpour is severe and lasting, as in a once-in-a-century storm, the rate of rainfall exceeds the rate of infiltration and regardless of vegetation the runoff aggravates and disastrous floods ensue. The river waters are muddy because of bed trenching and collapse of creek and river banks. Undercutting of slopes by swollen rivers initiates landslides. Natural erosion is now superseded by man-made, accelerated erosion of unsurpassed intensity.

Accelerated Erosion on Forest Lands

Forests are a renewable resource, but the silvicultural aim of sustained yield is negated unless the soil is preserved.

Cutting trees curtails *transpiration*. The unspent water feeds runoff and seepage (Fig. 11.16). It raised water tables in Denmark and elsewhere. In Oregon, at maP 230 cm with summer moisture deficits, virgin, old-growth Douglas fir forest (*Pseudotsuga*

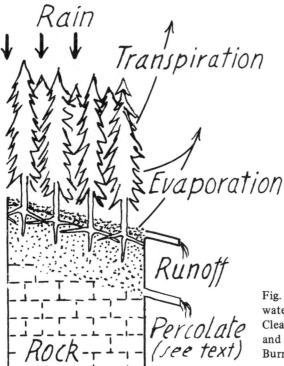

Fig. 11.16. Schematic sketch of water ways in a forest ecosystem. Clearcut eliminates transpiration and evaporation of intercepted rain. Burning of forest floor often enhances runoff

menziesii) on steep slopes provides an annual stream flow of 145 cm per unit area of watershed. Removal of the trees bolsters it by 46 cm (44).

When forests in New Hampshire, dominated by yellow birch (*Betula alleghaniensis*) and sugar maple (*Acer saccharum*) and growing on well-drained sandy loam Podzols (Haplorthods), are clearcut commercially, in 2 succeeding years (41) the *creek waters* carry away substantial quantities of Ca and 95 kg/ha of nitrate-nitrogen, mineralized from plant debris and humus (Table 11.3).

Clearcutting and burning of old growth western larch (*Larix occidentals*), Douglas fir (*Pseudotsuga menziesii*), and Engelmann spruce (*Picea engelmannii*) on glacial till soils in Montana (maP 64 cm) deliver in the first year's *runoff* (13) 14.2 kg/ha of nutrients (Ca, Mg, K, Na, and P) from burned sites but only 0.48 kg/ha from untreated controls. In the fourth year the loss is reduced to 3.00 kg/ha. The 4-year sum is 3.04 kg/ha for the controls and 31.9 kg/ha for the logged-burned plots.

In Idaho (36) 6-year sedimentation in small catchment reservoirs is 0.062 m^3/ha/year from undisturbed land and 0.097 m^3/ha/year from areas logged by the skyline method, which is less soil destructive than pulling logs with tractors. Harvesting methods that require substantial road construction contribute 13.4 m^3/ha of surface erosion and 33.4 m^3/ha of mass flow from disturbed lands annually. Road failures cause many small landslides in Oregon (16). Deforestation in Japan (19) increases landsliding 10-fold beginning 5-8 years after cutting because of advancing decay of soil-upholding roots, a tenet that is now being tested experimentally (55).

Stream flow multiplied by degree of muddiness gauges suspended *sediment discharge* (SED), expressed in weights per area of watershed. Hydrologists partition it statistically among causative site variables to ascertain erosion patterns in the source area, which is an indirect approach compared to the direct plot technique in the watershed.

In the forest-clad, high rainfall mountains of the Pacific Coast Anderson and collaborators (e.g., 4) find that man's activities (roads, logging, fire) elevate the average suspended sediment load 17 times. Discharge mounts as the square of stream flow which, itself, relates to the square of precipitation, hence SED varies approximately as the fourth power of maP, named *climatic stress factor* of denudation. Multiplied by soil erodibility and slope it forms an erosion hazard rating for soil protection in forest practices.

The plight of the *Redwood National Park* in northern California is attracting nationwide attention because the world's tallest trees (112 m) on the river flats are

Table 11.3. Net Nutrient Losses from Commercially Clearcut Forests, Gale River, New Hampshire[a]

Forest treatments	NO_3-N per ha/year		Ca per ha/year	
	kg	equiv	kg	equiv
Undisturbed forest	-2[b]	143	7	351
First year after clearcut	38	2,714	41	2,050
Second year after clearcut	57	4,071	48	2,400

[a] After Pierce et al. (41).
[b] Gain from rain.

threatened by soil flow from the privately owned, clearcut areas on the valley slopes of 30-70% inclinations. The tall upslope trees are valued at thousands of dollars each. Huge bulldozers prepare earth beds for the trees to fall upon to prevent shattering of wood. A routinely harvested clearcut tract on 40% slope looks (in 1974) like a bombed battlefield. The soil mantle is disrupted to over 1 m depth, turned upside down in many places, with gray C horizon placed on humus-rich reddish-brown topsoil, and with soil structures and horizon sequences destroyed. Incipient rills and gullies form and countless little ponds a few feet across are lined with films of washed-in clay, already mobilized by one winter's rain.

The slopes are replanted by seedlings and if the rains are gentle reforestation is successful, though probably delayed by lack of nitrogen. When torrential downpours propel the loosened soil mass downhill, the harm done below is blamed on the exceptional storm by one group and on soil disturbance by another.

Soil *losses* are often belittled and nature's healing power is praised. The forest is put on a par with an equilibrium system and its principle of virtual change: after a minor disturbance the system reverts to its former position of rest and all parts are stabilized again. More appropriate to an open system may be the analogy to biological radiation inputs in which small dosages cause accumulating alterations and eventual system deterioration, especially when erosion rates exceed weathering rates.

Complete severance of soil and forest from bedrock in historic times is visible in southern European and in Middle East countries. In the New World, in Colombia's upper Magdalena Valley (maP 120 cm) bare "desert slopes" appear to be recent if we may believe Friar Géronimo Descobar's sixteenth-century diary (14) that praises the region's luxuriant tropical vegetation. In many parts of the world mountains are becoming barren and depopulated and the fertile valleys inundated and crowded with highland refugees (17).

Accelerated Soil Erosion of Agricultural Lands

In storm intensities of 1-10 cm/hr raindrops have mean diameters of 2-3 mm and reach terminal velocities of 800 cm/sec or nearly 20 miles/hr (57). The kinetic energy of the billions of impingements on a hectare of bare, moist soil during a 30-min storm breaks up soil particles into mud and the splashes of raindrops lift fine particles centimeters and decimeters into the air. In a few minutes compaction and plugging of pores by dispersed clay lower infiltration rates and water begins to flow along the soil surface as a thin sheet, which is runoff or overland flow. The rough soil surface incites turbulent flow which is augmented by raindrops hitting the water layer from above. Turbulence drags clay, silt, sand, and gravel particles downhill as *sheet erosion.* The scouring action creates small downslope channels or *rills* which may merge to deep *gullies.* The loamy loess soils are especially prone to gullying (Fig. 11.17). The deep cuts interfere with soil management. As fertile topsoil is being carried away by sheet erosion, crop production suffers (Fig. 11.20) and farming becomes precarious (Fig. 11.18).

Decades of field-plot experimentations on runoff and erosion involving some 10,000 controlled plot-years permitted formulation of the *soil-loss equation* (57) by statistical, multiple regression analysis:

Fig. 11.17. Erosion in 12 years in loess soil (Howard Co., MO). Note man standing in deep gully. Courtesy Missouri Agricultural Experiment Station

Fig. 11.18. Sheet erosion has bred poverty (United States). The house points to a prosperous past. Photo by Post, courtesy F.S.A.; from the photo collection of the late Prof. W. A. Albrecht, Columbia, MO

A	=	R	·	K	·	$L \cdot S$	·	$C \cdot P$
soil loss		climatic factor		soil factor		topography factor		biotic factor

All variables are calibrated to a benchmark that is not the expected soil-preserving forest or grassland but its opposite, the *bare soil*, a standard plot 22.1 m long, of 9% slope, and kept free of vegetation by tilling up and down the slope. For this *unit* plot *L, S, C,* and *P* have assigned values of 1.0 each. The symbols mean the following:

A = *weight of loss of soil* per year per unit area, caused by sheet and rill erosion,

R = *rainfall factor*, the interaction product of total kinetic energy (E) of a storm and its maximum 30-minute intensity (I).

R is the long-time average yearly total of the storm products EI and is an expression of the *erosive maP*. East of the Rocky Mountains R fluctuates between 50 in the northern Great Plains area and 600 at the Gulf Coast, the values having dimensions of tons/acre/year or, multiplied by 0.224, kilograms/square meter/year.

K = *soil erodibility*, determined from *unit* plots by dividing measured losses A by R, hence it is the amount of erosion for R equal to 1. Dunkirk silt loam in the state of New York has the high erodibility K of 0.69, Tifton loamy sand in Georgia, the low one of 0.10. Soils that are high in silt and low in clay and organic matter are the most erodible.

$L \cdot S$ = *topography factor*, which is the ratio of soil loss from a given bare field to that of the *unit* plot. LS is read off from the curves in Figure 11.19, which follow the empirical equation

$$L \cdot S = \sqrt{\lambda} \cdot (0.0076 + 0.0053s + 0.00076s^2)$$

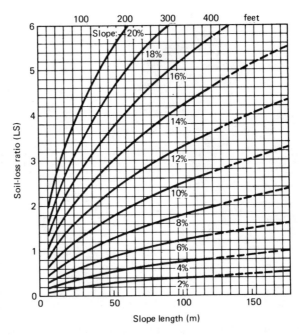

Fig. 11.19. Topography factor ($L \cdot S$) of soil-loss equation. For unit plot, $LS = 1$; for length 400 ft and a slope of 10%, $LS = 2.74$

in which λ is slope length in feet and s is the number of the slope percentage. In a field of variable slopes the mean of s would underestimate A, hence the steepest slope is inserted in the equation.

$C \cdot P = biotic\ factor$, consisting of species planted in a certain way (C) and of erosion control practices P (terracing, strip cropping) applied. A growing cotton crop without winter cover crop seeding has a high factor of 80%, a meadow has a low one of 0.4% and a forest a still lower one of 0.1%, the slight losses caused in part by faunal action, e.g., worm casts.

To get an idea of absolute soil losses, select a *unit* plot in eastern North Dakota having the low rainfall factor R equal 50. Since L, S, C, and P are 1, for a soil of average erodibility (K of about 0.3) the loss A is $50 \cdot 0.3 \cdot 1 \cdot 1 \cdot 1 \cdot 1 = 15$ tons/ acre/year or 3.4 kg/m^2/year. For meadow land C is 0.004 and $A = 50 \cdot 0.3 \cdot 1 \cdot 1 \cdot 0.004 \cdot 1 = 0.06$ tons/acre/year or 13.4 g/m^2/year. The removal of this fine earth lowers 1 m^2 of the surface of a soil of 1.3 g/cm^3 of bulk density by only one-hundredth of a millimeter a year. A growing crop of corn (*Zea mays*) with C of 0.38 evokes intermediate losses, A being 5.7 tons/acre/year or 1.28 kg/m^2/year.

Along the Gulf Coast the high R of 600 induces on bare land of *unit* plot and average K (0.3) the severe loss of $600 \cdot 0.3$, or 180 tons/acre/year, or 40.3 kg/m^2/ year. It lowers the soil surface by about 3 cm a year. A corn crop reduces erosion to 68 tons/acre/year and a forest to 0.04 kg/m^2/year. All calculations deal with the average conditions in an average year (57).

The A horizon is usually the most productive part of a soil profile. As it is being eroded crop yields suffer, illustrated in R. E. Uhland's Figure 11.20 for corn yields (*Zea mays*) on deep prairie soils (Mollisols) in Missouri (48). Mineral fertilizers (N, P, K, etc.) applied to the truncated profiles substantially restore biomass production at considerable cost and expenditure of energy, yet without reconstituting the soil.

Geographer Carl O. Sauer contends that soil erosion is rampant on lands newly converted from virgin grasslands and forests in systems of exploitive, pioneer agriculture and forestry. Based on soil profile truncation, Trimble (52) estimates for 34 areas in the southern Piedmont Plateau a gross erosion of 18 cm since European settlement 200 years ago. Only 5% of it left the system as suspended river sediment, the

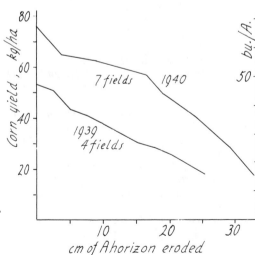

Fig. 11.20. Erosion of surface soil lowers corn (maize) production, in absence of fertilizer additions. After Uhland (48)

rest filled channels and valleys, literally burying old bridges. In spite of decades of educational efforts costing billions of dollars and the operation of over 3000 local soil and water conservation districts, barely 50% of American farmers practice the erosion control measures advocated (11).

North-European landscapes display fewer signs of active erosional devastation and even steep slopes appear stabilized. Possibly, their erosion period was maximal during earlier centuries, for the writer was shown valley fills of washed-in loess material many meters deep containing buried medieval agricultural tools, swords, and helmets, now in museums.

F. Review of Chapter

Broadly, topography modulates the regional climate by speeding up heating and evapotranspiration on south-facing slopes (in the northern hemisphere) and slowing them on north exposures where soils are deeper, more humic, and more thickly vegetated. Troughs and basins harbor salty playas in deserts and mucks and bogs in rain country.

On near-level surfaces weathering may extend soil depths to tens of meters. On slopes runoff waters carry clay and humus particles as overland flow and as runin on lower sites. When rate of surface erosion balances the rate of weathering at the base of a profile the thickness of the soil mantle is maintained.

High temperatures stimulate rock weathering but curtail runoff because of higher evapotranspiration. It helps to explain why hillsides tend to support deeper soils near the equator than away from it, other state factors being equal.

References

1. Aandahl, A. R. 1948. *Soil Sci. Soc. Am. Proc.* 13: 449-454.
2. Abd-el-Malek, Y., and S. G. Rizk. 1963. *J. Appl. Bacteriol.* 26: 14-26.
3. Alexander, J. D., J. B. Fehrenbacher, and B. W. Ray. 1968. In *Symposium on Prairie and Prairie Restoration,* P. Schramm, ed., pp. 34-38. Knox College, Galesburg, IL.
4. Anderson, H. W. 1975. In *Am. Soc. Civ. Eng. Symp: Watershed management,* pp. 347-376. Logan, UT.
5. Arkley, R. J., and H. C. Brown. 1954. *Soil Sci. Soc. Am. Proc.* 18: 195-199.
6. Baas-Becking, L. G. M., and G. S. Parks. 1927. *Physiol. Rev.* 7: 85-106.
7. Berner, R. A. 1971. *Principles of Chemical Sedimentology.* McGraw-Hill, New York.
8. Bernstein, L. 1975. *Ann. Rev. Phytopathol.* 13: 295-312.
9. Blume, H. P. 1968. *Z. Pflanz. Düng. Bod.* 119: 124-134.
10. Borst, G. 1975. *Soil Surv. Hor.* 16: 20-24.
11. Carter, L. J. 1977. *Science* 196: 409-411.
12. Coaldrake, J. E. 1961. C.S.I.R.O. Bull. 283. Melbourne, Austr.
13. De Byle, N. V., and P. E. Parker. 1972. In *Watersheds in Transition,* S. C. Csallany, T. C. McLaughlin, and W. D. Striffler, eds., pp. 296-307. Symp. Am. Water Res. Assoc.
14. Descobar, G. 1884. *Doc. Inéditos Arch. Indias* 41: 479-480, Madrid.
15. Dost, H. (ed.). 1973. *Acid Sulphate Soils,* 2 vols. Inst. Landreclam, P. O. Box 45, Wageningen.

16. Dyrness, C. T. 1967. U.S. Forest Serv. Res. Pap. PNW-42.
17. Eckholm, E. P. 1975. *Science* 189: 764-770.
18. Elgabaly, M. M. 1953. In *Unesco Symposium Land Use in Arid Regions*, pp. 125-152, Cairo.
19. Fujiwara, K. 1970. *Hokkaido Univ. Forest Res. Bull.* 27: 297-345.
20. Goss, D. W., and B. L. Allen. 1968. *Soil Sci. Soc. Am. Proc.* 32: 409-413.
21. Gray, D. H. 1970. *Assoc. Eng. Geol. Bull.* 7: 45-66.
22. Hamilton, R. 1964. In *Soil Micromorphology*, A. Jongerius, ed., pp. 269-276. Elsevier, Amsterdam.
23. Hayakawa, M. (ed.). 1976. *Fremontia* 4, Oct. issue, Berkeley, CA.
24. Janitzky, P., and L. D. Whittig. 1964. *J. Soil Sci.* 15: 145-157.
25. Jenny, H. 1941. *Factors of Soil Formation.* McGraw-Hill, New York.
26. Jenny, H. 1946. *Soil Sci.* 61: 375-391.
27. Jenny, H. 1961. *E. W. Hilgard and the Birth of Modern Soil Science.* Agrochimica, Pisa.
28. Kelley, W. P. 1951. *Alkali Soils.* Reinhold, New York.
29. Klemmedson, J. O. 1964. In *Forage Plant Physiology and Soil-Range Relationships*, M. Stelley, ed., pp. 176-189. Am. Soc. Agron. Sp. Publ. 5, Madison, WI.
30. Kojnov, V. 1964. *Pédologie* 14: 179-204.
31. Kovda, V. A. 1965. *Symp. Sodic Soils, Agrok. Talajtan* 14: 15-48.
32. Lorio, P. L., and J. D. Hodges. 1971. *Soil Sci. Soc. Am. Proc.* 35: 795-800.
33. Lotspeich, F. B., and H. W. Smith. 1953. *Soil Sci.* 76: 467-480.
34. Lynn, W. C., and L. D. Whittig. 1966. *Clays Clay Miner.* 14: 241-248.
35. McGinnies, W. J. 1960. *J. Range Mgnt.* 13: 231-234.
36. Megahan, W. F. 1975. In *Techn. Sediment Yield and Sources*, pp. 74-82. U.S.D.A., ARS-S-40.
37. Nettleton, W. D., K. W. Flach, and G. Borst. 1968. *A Toposequence of Soils in Tonalite Grus in the Southern California Peninsular Range.* U.S.D.A., S.C.S. Report 21.
38. Nielsen, G. A., and F. D. Hole. 1963. *Wisc. Acad. Sci., Arts. Lett. Trans.* 52: 213-227.
39. Nikiforoff, C. C. 1941. *Hardpan and Microrelief in Certain Soil Complexes of California.* U.S.D.A. Techn. Bull. 745.
40. Paeth, R. C., M. E. Harward, E. G. Knox, and C. T. Dyrness. 1971. *Soil Sci. Soc. Am. Proc.* 35: 943-947.
41. Pierce, R. S., C. W. Martin, C. C. Reeves, G. F. Likens, and F. H. Bormann. 1972. In *Watersheds in Transition*, S. C. Csallany, et al. (eds.), pp. 285-295. Symp. Am. Water Res. Assoc.
42. Raychaudhuri, S. P., M. Suleiman, and A. B. Bhuiyan. 1943. *Ind. J. Agr. Sci.* 13: 264-272.
43. Riecken, F. F., and E. Poetsch. 1960. *Iowa Acad. Sci.* 67: 268-276.
44. Rothacher, J. 1970. *Water Resour. Res.* 6: 653-658.
45. Sehgal, J. L., G. F. Hall, and G. P. Bhargava. 1975. *Geoderma* 14: 75-91.
46. Sherman, G. D. 1950. *Pacif. Sci.* 4: 315-322.
47. Simon-Sylvestre, G. 1960. *Ann. Agr.* 11: 309-330.
48. Stallings, J. H. 1957. *Soil Conservation.* Prentice-Hall, Englewood Cliffs, New Jersey.
49. Sukatschew, W. N. 1932. In. *Handbook of Biological Work Methods*, E. Abderhalden, ed., Sect. 11, Vol. 6, pp. 191-250. Berlin.
50. Tardy, Y., G. Bocquier, H. Paquet, and G. Millot. 1973. *Geoderma* 10: 271-284.
51. Tomanek, G. W. 1964. In *Forage Plant Physiology and Soil-Range Relationships*, M. Stelley, ed., pp. 158-164. Am. Soc. Agron. Sp. Publ. 5, Madison, WI.
52. Trimble, S. W. 1975. *Science* 188: 1207-1208.
53. Ursic, S. J., and F. E. Dendy. 1965. In *Federal Inter-Agency Sedimentology Conference* (1963), pp. 47-52. U.S.D.A. Misc. Publ. 970.

54. van Breemen, N., and K. Harmsen. 1975. *Soil Sci. Soc. Am. Proc.* 39: 1140-1148.
55. Waldron, L. J. 1977. *Soil Sci. Soc. Am. J.* 41: 843-849.
56. Walker, P. H. 1966. *Iowa St. Univ. Res. Bull.* 549: 839-875.
57. Wischmeier, W. H., and D. D. Smith. 1965. *Predicting Rainfall-Erosion Losses from Cropland East of the Rocky Mountains.* U.S.D.A., A.R.S. Agr. Handb. 282.
58. Yaalon, D. H., H. Nathan, H. Koyumdjinsky, and J. Dan. 1966. *Proc. Int. Clay Conf.* 1: 187-198.

12. State Factor Climate

To begin with a historical note, the documented recognition of climate as a molder of soil is but a century old, though earlier de Saussure had suggested in his *Voyages dans les Alpes* (1796) that climate is responsible for variations in contents of soil organic matter.

Eugene W. Hilgard was 3 years old when the lawyer family shunned Germany and settled in 1835 on a farm in Illinois. When free of chores youthful Hilgard was an avid plant collector under the tutelage of his famed botanist cousin, G. Engelmann. Later, he studied geology with Escher in Zürich and earned a Ph.D. in chemistry with Bunsen in Heidelberg (1853). For nearly 20 years he was State Geologist of Mississippi and Professor of Chemistry and developed chemical analyses to evaluate soil profiles and their agricultural potentials. When called to California in 1874 to become Professor of Agriculture and of Botany he soon discovered that arid lands are richer in acid-soluble constituents such as potassium and calcium than the humid, subtropical soils of the southern United States, even when developed on the same geologic stratum. In the late 1870s he announced a lixivation or leaching theory and in 1892 published a voluminous documentation of over 1000 analyses on the Relations of Soil to Climate, from which the summary in Table 12.1 is taken.

Meanwhile, in 1883 V. V. Dokuchaev proclaimed that he considered climate a cause of the distribution of humic black soils (Chernozems) in Russia. In the mid-1890s Sibirtzev integrated Hilgard's and Dokuchaev's works (26) and advanced the doctrine of zonal soils which dominated European and American soil classifications for many decades. It is fair to say that the climatic principle of soil genesis has its roots independently in the United States and in Russia (18).

A. Climatic Indices

The internal climates cl' of vert and soil spaces vary from point to point and change as soil and vegetation develop. The air climate cl above the ecosystem is the preferred state factor for sequences of entire systems.

Table 12.1. Hilgard's Analyses of Soils from Arid and Humid Regions[a,b]

Region	Number of analyses	Total soluble material (%)	Soluble SiO$_2$ (%)	Al$_2$O$_3$ (%)	Fe$_2$O$_3$ (%)	CaO (%)	MgO (%)	K$_2$O (%)	Na$_2$O (%)
Arid	573	30.84	6.71	7.21	5.47	1.43	1.27	0.67	0.35
Humid	696	15.83	4.04	3.66	3.88	0.13	0.29	0.21	0.14

[a] From Hilgard (15).
[b] Five-day hydrochloric acid digestion, specific gravity 1.115.

The P-T Grid or Field

For orientation, a map of the United States showing mean annual values of precipitations (maP) and temperatures (maT) is assembled in Figure 12.1. A climatic field or grid may be constructed by arranging vectors of maP and maT at right angles to each other, as illustrated in Figure 12.2 for the meteorological stations of India where the warm places are located in the plains and the cool places in the Himalayan foothills (e.g., Darjeeling, 2300 m) and on southern mountains (e.g., Kodaikanal, 2343 m).

To solve the climosequence

$$l,v,a,s = f(cl)_{\phi,r,p,t,...} \quad \text{or} \quad l,v,a,s = f(cl, \phi,r,p,t,...)$$

soil and vegetation locations are selected in a grid along isotherms from left to right with humidity rising and temperature being constant, and from top to bottom along isomoisture lines and increasing heat input, written

$$l,v,a,s = f(maP, maT, \phi,r,p,t,...) \quad \text{Moisture sequence}$$

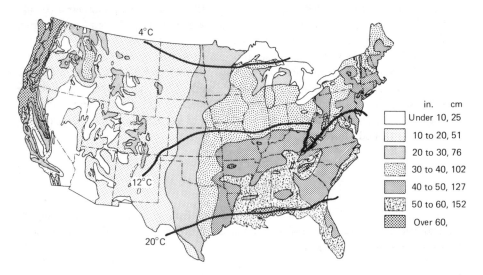

Fig. 12.1. Distribution of mean annual precipitation (maP) and temperature (maT) in the United States

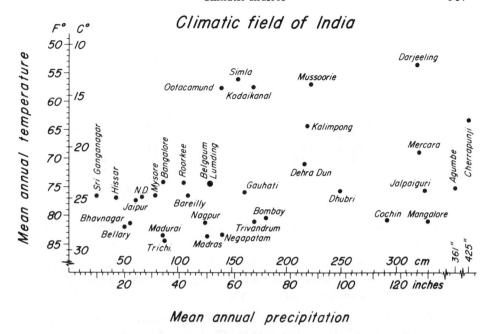

Fig. 12.2. Climatic field of India with maP and maT coordinates. Each point corresponds to a meteorological station

$$l,v,a,s = f(\mathbf{maT}, maP, \phi, r, p, t, \ldots) \quad \text{Temperature sequence}$$

Success depends on degree of control of the subordinate factors over a wide range of dominant maP or maT.

Along a climatic vector such as maP the flora of influx is ϕ and is considered an independent variable. The orderly alignment of species actually growing, as seen in Figure 12.6, is the active flora ϕ', which is a dependent variable. The independence of ϕ is ascertained by demonstrating that bluestem seeds arrive at the dry end of the transect, or that sprinkling them into the buffalo grass sod will not displace it.

Seasonal Patterns

Precipitation falls in even monthly portions (Eastern type), or congregates in summer (Plains type) or in winter (Pacific or Mediterranean type), as patterned in Figure 12.3. In mid-latitudes the monthly air temperatures march from summer highs to winter lows, the amplitudes lessening in the subtropics and practically disappearing near the equator, even at high altitudes. The role of this latitudinal T-dampening in soil genesis is still not known.

Evapotranspiration

Inputs of heat stimulate evaporation, transpiration, and rates of chemical reactions. At St. Paul, Minnesota, maP is 70 cm and the native vegetation belongs to the semihumid

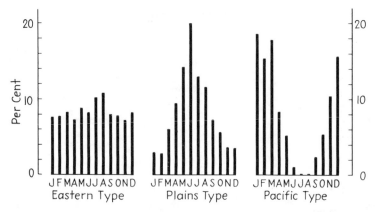

Fig. 12.3. Types of seasonal distribution of precipitation in the United States (17)

grassland-forest transition zone. Warm San Antonio, Texas, registers the same maP, yet vegetation and soil mirror semiaridity because the higher temperature (maT 20.5 versus 6.8°C) materially enhances moisture dissipation.

Transeau (17) substituted for the unknown evapotranspiration of an ecosystem the amount of water (E) that evaporates from an open pan. For St. Paul the annual ratio of precipitation to evaporation, P/E, is 1.02, for San Antonio 0.51. Meyer (17) divided precipitation by the absolute saturation deficit of the air, which is temperature controlled, and designated the quotient NSQ. At the two cities chosen the annual NSQs are 329 and 120. Clearly, both P/E and NSQ discriminate the two stations whereas maP does not. Within the United States the two moisture indices are linearly correlated but are limited geographically because few meteorological stations measure E or relative humidity. Maps of P/E and NSQ have been published previously (17).

Thornthwaite Indices

The moisture criteria of Chapter 2 are applied in Figure 12.4 to an *arid* → *humid transect* of the Pacific Coast region, detailed later in this chapter. Precipitation (maP) extends from 8 to 200 cm and temperature (maT) is confined to the narrow belt of 10-16°C.

The temperature-sensitive potential evapotranspiration (Pet) declines slightly with rising maP (r = -0.27) because it responds to trends in latitude and altitude. Actual evapotranspiration (Aet) for an assumed water capacity of 15.2 cm approaches a maximum that is substantially below Pet because of prolonged dry summers. Below maP 50 cm, leaching factor Li is less than Aet and leaching is nonexistent; at high rainfall Li is proportional to maP.

To explore the paths of moisture from *north* → *south* at variable P and T in the central part of the United States, Dr. R. J. Arkley kindly furnished calculations for a traverse starting in cold, dry North Dakota, passing over Iowa and Missouri, and ending in warm, humid Louisiana. In Figure 12.5, maT quintuples, maP trebles, and Pet doubles. South of North Dakota Aet always matches Pet. Li peaks in Arkansas, and at the highest precipitation (maP 144 cm) is less than half that of the cooler western transect.

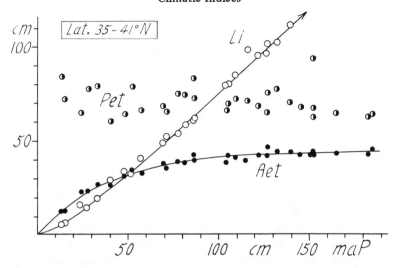

Fig. 12.4. Moisture indices of West Coast arid-humid transect. *Pet,* a measure of temperature and radiation input, lies above *Aet,* which is an indicator of seasonal water stress and biomass production on fertile soils. All values of Pet, Aet, Li, and maP are given in centimeters

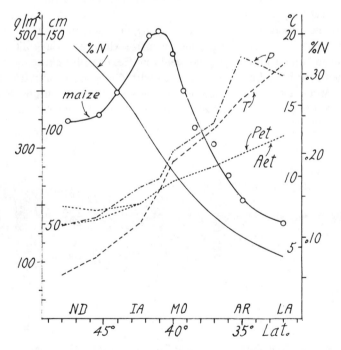

Fig. 12.5. Latitude transect from North Dakota to Iowa, Missouri, Arkansas, and Louisiana. P, maP (centimeters); T, maT (°C); Pet (centimeters), potential-evapotranspiration; Aet (centimeters), actual evapotranspiration (courtesy R. J. Arkley); also plotted are average maize yields (grams/square meters) of the 1930s and percentage total soil nitrogen (17)

Climate and Biomass Production

Plants thrive on warmth and water and many attempts have been made to deduce the function: biomass = f (climate). Solutions are unrealistic unless soil conditions are explicitly taken into account.

In an *arid* → *humid* direction the grasslands of aeolian soils of the Great Plains respond to moisture in the manner sketched pictorially by Shantz (28) in Figure 12.6. The rise in biomass is documented in Figure 12.7 by the grass clippings of Clements and Weaver (17) in the 1920s and by the cuts of the workers of the Grassland Biome (8) of 1970. Maize yields in Kansas (17), converted to above-ground dry matter, support the same upward trend. Biomass culminates in the deciduous forest farther east for which Olson (27) quotes 1400-2100 g/m² of annual production.

From *cold* → *hot* regions, or north to south, the annual hay yields of natural *tall grass* prairies, plotted in Figure 12.8 as county averages, exhibit two slight maxima, but the physiologically expected biomass stimulation at higher temperatures fails to materialize. Perhaps nutrients of the southern uplands are limiting, both total P and N being low.

A similarly uneventful north-to-south trend comes to light in biomass harvests of the ungrazed, drier *short grass* belt (Fig. 1.10) that extends from Canada to Texas. Noteworthy (Table 12.2) are the relatively high root weights secured by the Grassland Biome (8) in 1970. The depressed yield of the Pantex field is attuned to the very low waterholding capacity of its soil.

The latitudinal discrepancy between the rising, favorable trend of Aet and the nonresponding biomass is most conspicuous for maize which is a heavy nitrogen feeder. In prehybrid and prefertilizer days grain yield rose from North Dakota to central Iowa, then declined sharply to Louisiana in parallel with diminishing soil nitrogen contents (Fig. 12.5). Modern agronomic methods and heavy N fertilization double and treble yields in the South and have narrowed the span from crest to trough.

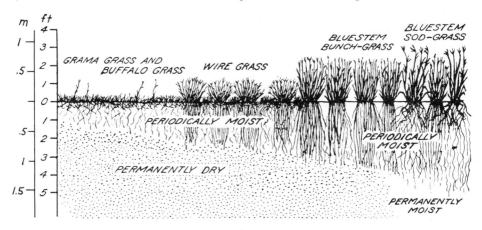

Fig. 12.6. Sketch by Shantz (28) showing plant communities and depth of lime (CaCO₃) horizon across the Great Plains area from Colorado through Kansas to Missouri; maP = 40-97 cm, maT = 11°C

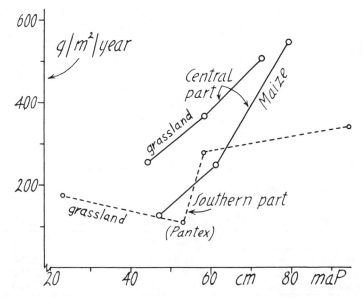

Fig. 12.7. Dry matter production (above-ground) of never-cultivated grasslands in the central and southern parts of the *Great Plains,* and maize yields of 1926 in the central part (Kansas)

In the forested humid region plant dry weights in grams/square meter/year, based on tabulations of Olson (27) and given in abbreviated form in Table 12.3, are more in line with expectations of abundance of gross production in warm climates.

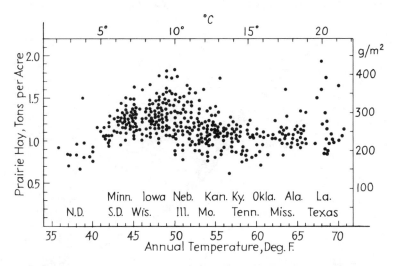

Fig. 12.8. North-south transect of hay yields of tall grass prairies. Each point is a county average

Table 12.2. Annual Production of Biomass along a N-S Transect in the Ungrazed Short-Grass Sections of the Great Plains[a]

Locality	Climate		Soil water storage (cm)	Dry matter (g/m^2/year)		
	maT (°C)	maP (cm)		Above ground	Below ground	Total
Dickinson, ND	4.8	41	15	328	391	719
Bridger, MT	7.7	61	12	179	372	551
Pawnee, CO	8.3	30	16	142	458	600
Cottonwood, SD	8.3	38	27	197	269	466
Hays, KS	12.2	58	8	273	288	561
Pantex, TX[b]	13.9	53	3	107	177	284

[a]After Caldwell (8).
[b]Soil has a very low water-holding capacity.

Table 12.3. Annual Production of Dry Weights (g/m^2) of Forests of Humid Regions[a]

	Net production	Gross production
Cold region	480-1,280	2,250- 4,880
Temperate region	1,400-2,100	2,360- 4,660
Tropical region	743-2,100	3,400-12,730

[a]From tabulations by Olson (27).

B. Altitude Sequences

With increasing elevation moisture tends to rise and temperature to fall and contrasts between ecosystems become formidable. Viewing a mountain through the frame of state factors focuses attention on sites having similar azimuthal exposures, degrees of slope, and kinds of parent material (Fig. 12.9). Under control of altitudinal p and r, plant and soil populations appear and fade out gradually.

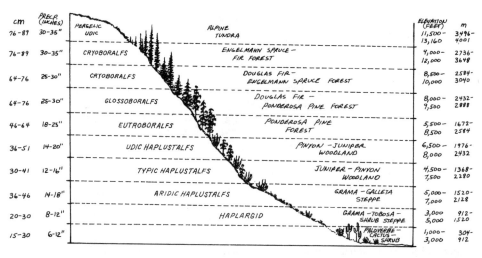

Fig. 12.9. Altitude sequence of soils and vegetation in New Mexico. After Carleton et al. (9), U.S.D.A. Forest Service

Fresno Altitude Transect

From near Fresno toward Kaiser Pass, California, factors are documented as follows:

cl: Wet winters and dry summers, maP 43-91 cm, maT 5.8-16.3°C, given in Figure 12.10.

ϕ: Grasses, native (*Poa* ssp., *Stipa* ssp.) and introduced (*Avena* ssp., *Bromus* ssp.); Blue oak (*Quercus douglasii*), Live oak (*Q. wislizenii*), Black oak (*Q. Kellogii*); Digger pine (*Pinus sabiniana*), Ponderosa pine (*P. ponderosa*), Jeffrey pine (*P. jeffreyi*); White fir (*Abies concolor*), Red fir (*A. magnifica*).

r: South-east facing slopes sampled between 16 and 32%.

p: Quartz-diorite bedrock rich in plagioclase, hornblende, and biotite.

t: Very old soils that have never been glaciated.

From base to peak, the foothill grasslands carrying cattle give way to oak-grass savanna which is followed by oak-pine woodlands and by dense pine-fir forests that have been logged selectively. The sequence of v and s properties on the chosen r and p sites is considered climatic, although some observers pronounce the grasslands as artifacts of Indian burning.

The C/N Drift. Soils (0-20 cm) collected in SE direction 1.8 m from tree boles furnish the widening C/N trend of Figure 12.10. It is the consequence of a trebling of organic carbon, from about 0.8% at 300 m to 2.2% at 2400 m, coupled with a lagging rise of nitrogen. Fir sites deliver wider quotients than grass sites but the difference cannot be assigned to species control alone because the shift in climate modifies the decomposer

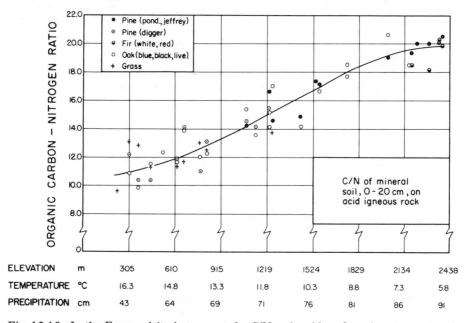

ELEVATION	m	305	610	915	1219	1524	1829	2134	2438
TEMPERATURE	°C	16.3	14.8	13.3	11.8	10.3	8.8	7.3	5.8
PRECIPITATION	cm	43	64	69	71	76	81	86	91

Fig. 12.10. In the Fresno altitude transect the C/N ratio widens from base to crest and the vegetation becomes more coniferous

populations and their activities. On related Arizona mountains none of the soils provides a suitable environment for N-fixing Azotobacter, and in the acid forest soils at high altitudes nitrifiers are absent, according to Martin and Fletcher (23).

Clay and Red Colors. The elevation sequence comprises the six soil series:

Vista Ahwahnee	→	Musick	→	Holland Shaver	→	Corbett
at low altitudes				at high altitudes		

All have brownish (10 YR) A1 horizons (0-20-28 cm) of sandy loam texture, and their pH values descend from near 7 to 6 to 5.7. Subsoil colors intensify from light-yellow brown of Xerochrepts (Vista, Ahwahnee) to red (2.5 YR 5/8, dry) and reddish yellow of the deeper Musick soils (Ultic Haploxeralfs), and they fade to reddish brown of Holland of the same order and to pale brown of Shaver, an Ultic Haploxeroll. Shallow Corbett is a Psamment.

In Huntington's study (16) of eight representative soil tesseras, clay contents (grams/square centimeter/d) to bedrock achieve a maximum in Musick soils because of their deeper profiles (d) and higher clay percentages in all horizons (Fig. 12.11). The structure of the Musick B is subangular-blocky compared to structureless-massive in the other soil series. The strong profile development is aided by the overall gentle topography of the Musick surface which sponsors water infiltration and enhances the optimal weathering climate of mid-elevations.

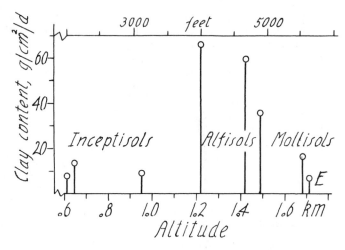

Fig. 12.11. Clay contents to bedrock of Fresno altitude sequence on granodiorite, according to Huntington (16). Alfisols have profile depths (d) of 180-250 cm compared to the 60-135 cm of the drier Inceptisols and the cooler-forested Mollisols and Entisols (E). Horizontal distance is about 13 km

C. The Humus-Climate Paradigm

In long-standing debates on soil organic matter, some assert that tropical soils are high in humus because of lavish vegetation, and others that they are low because of rapid oxidation of organic residues. In 1913, E. C. J. Mohr (24), doyen of tropical soil sciences, modeled production and decay but lacked quantitative records to arrive at realistic predictions.

Rate of Decomposition of Organic Matter

Crushed litter is put into a series of beakers and wetted to different degrees of moisture. Replicates are placed in constant temperature rooms, conveniently at 5, 10, 20, 30, and 40°C. Rates of organic decay as weight loss or CO_2 production are monitored at regular intervals, as shown in the inset of Figure 12.12 for alfalfa leaves decomposing 100 days. When moisture is low, high temperatures hamper soil microbe activity; excessive wetness does the same, at any temperature.

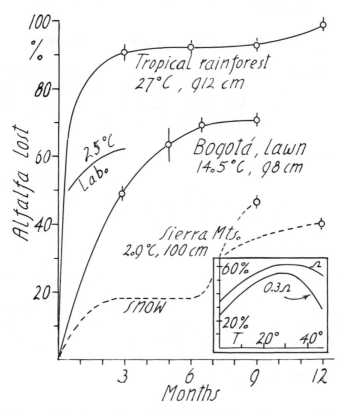

Fig. 12.12. Rates of decomposition of alfalfa buried in soil in equatorial and temperate regions (Sierra Nevada). Inset: laboratory decompositions (100 days) at various temperatures (T) and at optimum (Ω) and lower (0.3 Ω) moisture contents

In Gessel's (19) laboratory simulation of autumnal versus continuous litterfall decomposition of 20 g of alfalfa in bulk and in 8 weekly installments reaches about 60% in both sets. A cyclic temperature regimen rising from 5 to 31°C and returning back imposes as much loss as continued incubation at the mean of 19°C.

Ground alfalfa leaves in metal cans with perforated tops and bottoms are buried vertically into soil. In the 8-fold replications the upper can lids match the soil surface. Duplicate cans are removed and analyzed every 3 months. In Figure 12.12 the upper curves portray the fast decay rates in the tropical rain forest of Calima near sea level, and at 2640 m high, cool Bogotá, both in Colombia, S.A. Compared to oxidation in laboratory tests at 25°C and at optimum moisture, Calima's loss rates are higher, possibly because of invertebrate activity and downward leaching of degradation products.

Microclimate site variations appear ineffectual, for decay rates of alfalfa (3.5% N) are the same whether cans are dug into soil under large trees or in lawns, in the shade or in sunny openings, in dense pine stands or in chaparral (20).

In the Fresno altitude sequence of Figure 12.10 the biologically most active soil climate exists at about 1000 m where decay reaches 74% compared to 40-50% at 3000 m and 64% in the drier foothills.

N and C Climate Functions, Illustrated for India

When Indian scientists matched their soils with those of the fertile Middle West (e.g., Iowa) they were appalled by the low N and C levels of their lands and attributed the gap to exhaustion from centuries-long use. As realized later, the comparison is confounded with the humus-climate connections (19).

Moisture Functions. The climatic field of India of Figure 12.2 reaches from deserts to perhumid Cherrapunji of maP 1080 cm. Along the 24°C isotherm relatively undisturbed forests grow in former domains of Maharajas, in holy sanctuaries, Reserve Forests, and in regions of high rainfall. At low precipitation native vegetation is scarce and consists of thorny, leguminous shrubs, succulents, and clumps of bunch grasses here and there. No typical grasslands with Mollisols are met; instead, in semihumid Gundlupet-Bandipur open stands of tall trees with dense brush and patches of grass teem with wild life, including elephants.

Carbon contents of 150 random samples of surface soils to 20 cm depth, exclusive of forest floor, and adjusted to loam texture, are averaged over suitable rainfall intervals and the dot size of the means in Figure 12.13 is proportional to the sample aggregate. Lengths of vertical bars denote the standard error. Cultivated soils, exclusive of paddy (rice), have C/N ratios of 9-11; perhumid forest soils average 11-13. The two lowest points of the curve originate outside of India, i.e., a mean of 0.14% C for upland soils in southern Egypt (maP $<$ 2 cm) and 0.20% C at maP 2-8 cm in the Mohave desert.

Clearings in the high-rainfall forests have been farmed for 100-150 years and the conversion lowered the contents of C to one-half and those of N to two-thirds. The sharp discontinuity at moderate rainfall is located near centers of population pressures and antiquity of civilizations (e.g., Delhi, Jaipur, Mysore). There, firewood has

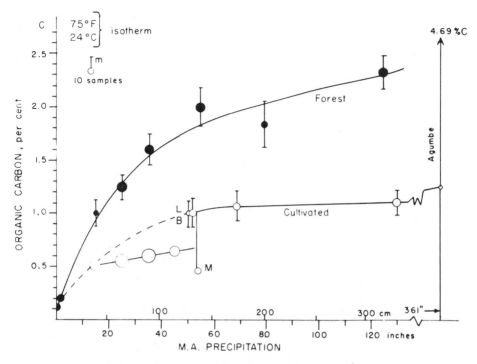

Fig. 12.13. Organic C dependency on rainfall (maP) along 24°C (maT) isotherm in India (19). Surface soils (0-20 m) from native vegetation and cultivated fields. L, Lumding; B, Belgaum; M, Mohar

become scarce and inhabitants must resort to cow dung for fuel. Organic returns to soil are minimal, C and N are depressed, and the crops are poor. Plant breeding and costly N fertilization are initiated, requiring industrialization, atomic energy, and changes in life styles.

Temperature Functions. In the perhumid sections of the lower Himalayas, of Coorg, and the Malabar Coast the monsoon pulsations deliver yearly 250-500 cm of water, and soil N and C decline exponentially with rising mean annual temperatures (Fig. 12.14). Logarithms convert the curvature to linearity, and least-square fitting through all individual samples elicits the equations for °C:

Forest soils—[48]: $\log(100N) = 2.1552 - 0.0371\,T\ (r = -0.79)$

Cultivated soils—[26]: $\log(100N) = 1.9477 - 0.0376\,T\ (r = -0.87)$

C/N ratios average 13.3 for forested and 12.7 for cultivated soils. Regressions, distances between curves, and coefficients are statistically well documented. The same is true for similar curve-pairs in the humid and semihumid regions of the subcontinent.

Temperature exerts a powerful influence on organic matter contents of Indian soils. For the most part the hot plains and plateaus have low nitrogen and humus levels, primarily because of climatic control and secondarily because of exploitation of land.

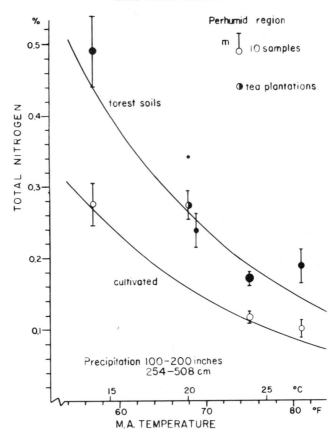

Fig. 12.14. Soil nitrogen-temperature dependency in perhumid India; size of circle is
proportional to number of samples; m = mean or standard error (19)

To return to the introductory paragraph of the Indian climate functions, Iowa's
agricultural soils at maP 76 cm and maT 8.9°C hold 0.21% N according to Figure
12.15. In India the same climatic coordinates are met at 3000 m of Himalayan altitudes where N and C contents are likely to exceed those of the Middle West.

A Nitrogen-Climate Surface

An early model of dependency of soil properties on moisture and temperature is based
on total organic nitrogen contents of cultivated loamy surface soils (0-20 cm) of former
grasslands of the Great Plains and points east. Topography is level to undulating and
parent material is confined to loam and silt loam A horizons of loesses and other sedimentary deposits. Soil age is unknown but much older than 1000 years, hence humus
content is possibly near steady state and practically time-independent, except for
changes brought about by white man's cultivation which heightens sample variability.

The observed N-temperature and N-moisture trends are similar in shape to those of
India. They are fused mathematically into the surface model of Figure 12.15 which is

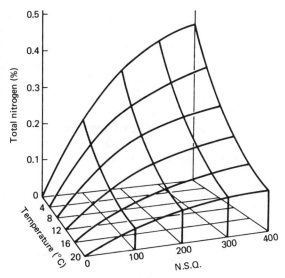

Fig. 12.15. Model of soil nitrogen variations (maT, NSQ) of the Great Plains area from Canada to Mexico (back to front curves) and from the desert to the humid regions (left to right curves)

$$N = 0.55e^{-0.08T}(1 - e^{-0.005m})$$

in which T denotes maT and m is maNSQ.

For a given isotherm, or zone of constant temperature, the equation reduces to

$$N = K_T(1 - e^{-k_1 m}), \quad k_1 = 0.005$$

and represents the curves rising from left to right, from the desert to the humid region. The rise of N, and also of C, is most pronounced at low T (Canadian border) and least at high T (Texas).

At chosen moisture indices m the surface formula becomes the family of lines

$$N = K_m e^{-k_2 T}, \quad k_2 = 0.08$$

which descends from the back (Canada) of the model southward to its front, which is near Mexico. Loamy soils of the southern United States contain much less N and humus than their relatives of comparable state factors along the northern border.

In trying to understand the climatic trends of soil organic matter it should be kept in mind that each point on a curve balances production of biomass, its consumption by decomposer organisms, and the soil's capacity of humus retention.

The N- and C-curves that ascend with rising rainfall imitate the positive trends of crop yields (Figs. 12.6 and 12.7); further, clay, soil acidity, and maP rise together and with them the exchangeable Al ions that bind humus molecules to clay (13). On the other hand, taking the similarity of hay yields from north to south in Figure 12.8 at face value and invoking the temperature stimulation of microbial metabolism puts the decomposer organisms in control of the humus-temperature curves.

Sequences with Maxima

In New Zealand, at about maT 8°C, graywacke sandstone extends through a rainfall belt of 30-380 cm, and slopes of 9-27% support tussock grassland and subalpine scrub tussock on medium-textured soils. As seen in Table 12.4, nitrogen culminates at maP 51 cm whereas carbon reaches a modest peak at 150-200 cm.

Phosphorus is Ca-bound in the dry sector and converts to organic form in the humid zone, up to 83% of total P, the remainder being tied to sesquioxides. Over the entire transect base saturation veers from 100% to an abysmal 8%. The overall humus pattern is repeated on mica schists and on basalt, the latter accumulating nearly twice as much N and C as graywacke.

On mountain slopes the rise of soil organic matter from valleys to high peaks is expected to reach a maximum and then decline as vegetation succumbs to the organic vicissitudes of frigid climates (17).

The Latitude Mode of Humus

With increasing altitudes, forested soils gain more humus near the equator than away from it. To observe a doubling of soil nitrogen one must climb an elevation interval of 2760 m in the Sierra Nevada, 1350 m in India, and only 890 m in Colombia. Putting it differently, doubling of soil nitrogen content requires a maT drop of 14.6°C in the Sierra, 7.6°C in India, and 5.0°C in Colombia.

In mineral soils of depth 0-20 cm, a latitude trend is visible in Table 12.5. For comparable state factors, including maP and maT, mean C and N contents rise sharply from temperate to the equatorial regions. To rephrase the mode, all N-temperature curves (over 20) decline as maT rises, regardless of latitude; but, the absolute position of the curves is staggered, those from near the equator lying above those from mid-latitudes. In consequence, the temperate N-T curves in Figure 12.15 do not extrapolate to the tropics.

> As causes one might infer a relatively high nitrogen fixation in the tropical forests, often legume-rich, and cite the high production of litterfall which decays rapidly, enabling rains to wash copious amounts of humus into the mineral soil below. Volcanic ash soils are especially prone to accumulate N and C because their allophane clays strongly attract humus substances (10).

Table 12.4. Soil Properties—Rainfall Function on Graywacke in New Zealand[a]

Soil number	maP (cm)	Soil depth[b] to bedrock (cm)	N in 0-20 cm (%)	C (g/m²)	C/N	Total P (g/m²)
1	31-38	49	0.111	4,040	8.8	313
2	38-46	47	0.163	6,390	8.5	349
3	51	38	0.279	10,240	13.1	259
4	140-152	54	0.242	11,210	16.9	313
5	152-203	60	0.242	13,790	18.9	191
6	381	39	0.219	11,960	21.5	126

Column heading "Entire tessera to bedrock" spans C (g/m²), C/N, and Total P (g/m²).

[a] After Walker and Adams (34).
[b] Exclusive of 0 horizon.

Table 12.5. Latitude Trend of Soil Nitrogen in A1 Horizons (0-20 cm Depth) of Forest Soils Having Similar r, p, maT, and maP for *n* Sampling Sites

Region	Latitude north	*n*	maT (°C)	maP (cm)	Total N (%)	C/N
United States						
West (California)	37-40°	16	12.8-15.0	127-206	0.134±0.005	16-20
East (Tennessee)	36-37°	17	14.5-16.7	122-130	0.123±0.013	—
India						
North (Mussourie)	26°	15	13.3-17.2	216-224	0.271±0.022	14-18
South[a] (Ooty, Kodaikanal)	10°	6	14.5-18.3	140-158	0.499±0.051	15-18
Colombia[b]	3-6°	8	12.2-18.9	150-203	0.809±0.118	12.6

[a] Parent rock is charnokite.
[b] Parent rock is volcanic ash.

D. Precipitation, Base Status, and Carbonate Regimes

Hilgard's classical Table 12.1 of HCl extracts of soils *not* derived from calcareous rock formations assigns much higher quantities of CaO, MgO, K_2O, and Na_2O to soils of arid than to those of forested humid regions, and Coffey's (17) later compilation puts the transitional grassland belt between the two.

A Moisture Transect on Igneous Rocks

In the foothills of the Sierra Nevada and Cascade Mountains of California a strip 1000 km long and about 50 km wide offers Mediterranean-type climates of maP 8-200 cm and maT 10-16°C, as outlined in Figure 12.4. To illustrate factor selection the transect in Figure 12.16 is cut into blocks of SE-facing slopes of 5-25% that are

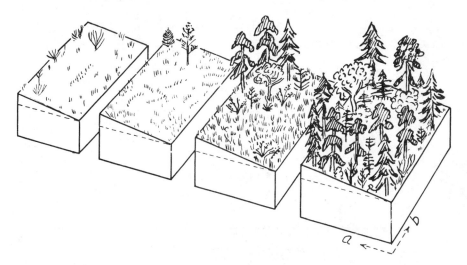

Fig. 12.16. Segments of the Sierran moisture transect from the desert to the humid region. All blocks have the same parent rock (either Ai or Bi), slope range and SE exposure. On the second block from the right, which receives 60-80 cm maP, grassland and tree clusters coexist

underlain by either Ai rocks (granites, granodiorites) or Bi rocks (basalts, andesites). Rainfall and the 56 Ai and 41 Bi rock sites are uncorrelated ($r = -0.08$), meaning that parent rock and climate are independent of each other. The vegetation mantel displays shrub species in the Mohave desert and majestic pine-fir forests at high maP; between the two, extensive grasslands reign. Since tree-covered mountains tower above the deserts and semihumid foothills, seed distribution faces no obstacles when given sufficient time, and the strip is assumed to be blanketed by a common biotic factor ϕ. The distribution of the growing ·species is cl-dependent. Except for grazing, selective logging, and occasional fires the soils have suffered little from white man's interference.

In the desert the soils are shallow and rocky. With increasing precipitation they deepen, develop clayey B horizons, and assume a reddish-brown color as Ultisols. In the northern humid portion geologists map Tertiary and Pleistocene lava flows and both carry the deep, reddish soils of the Cohassett series.

Within the frame chosen, soil sampling by F. F. Harradine, J. O. Klemmedson, D. A. Cappanini, A. E. Salem, and the writer is randomized throughout the transect. Under Ponderosa pines, site selection is restricted to the SE side 1.8 m from the trunk surface. The samples reported here are from the A1 horizon to a depth of 20 cm, one sample per site.

Soil Reaction and Exchange Acidity. As seen in Figure 12.17, below maP 15 cm, both Ai and Bi soils give *pH readings* above 7 and fizz with HCl, which denotes the presence of carbonates. At maP 13-15 cm Mehlich exchange acidity, aci, begins appearing in Ai

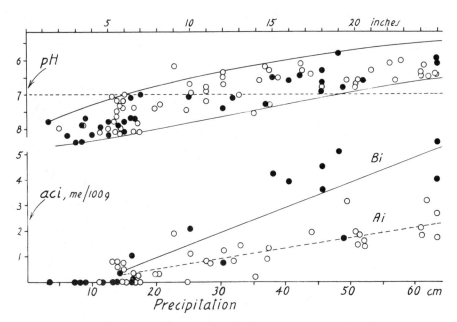

Fig. 12.17. Changes of soil pH and exchange acidity (aci) with maP in a narrow temperature belt. Soils derived from acid igneous (Ai) are indicated by open circles, those of basic igneous rocks (Bi) are indicated by filled circles

soils, even at pH above 7 because of hydrolysis of clay and humus. At higher rainfall aci becomes prominent, more so with Bi than Ai soils. Up to maP 65 cm the collinearity (r) of aci with maP is 0.824 for Ai and 0.877 for Bi soils.

At a given precipitation the scatter of aci values is closely linked to variation in percentage of organic carbon (C) in the soils. For instance, at maP 180-200 cm an aci-C dependency among 19 Ai soils is calculated by A. E. Salem as aci = 1.37 + 3.52% C ($r = 0.88$).

Degree of Unsaturation. Acidity, aci, and the sum of exchangeable bases, sba, constitute the soil exchange capacity CEC. Division of aci by CEC, multiplied by 100, yields *percentage acidity, pac,* or degree of unsaturation which controls availability of bases to plants and the stability of clay aggregates. For 41 Bi soils, pac and maP (in centimeters) regress in Figure 12.18 as the dashed line

$$pac = 0.701 + 0.443 \, maP \ (r = 0.913)$$

Over 83% (100 r^2) of the pac variability is attributable to the spread of maP. The remaining 17% reside mainly in the assembling of the growing biotic factors grassland species, desert shrubs, and Ponderosa pine to an overall ϕ. Germules of pines fail to survive at low maP and those of grasses and desert shrubs disappear at high maP (Fig. 12.18, inset). The equations for the two linear segments are

$$grass, Gr: \ pac = 0.070 + 0.337 \, maP \ (r = 0.817)$$

$$pine, Pi: \ pac = 16.411 + 0.337 \, maP \ (r = 0.797)$$

where Gr denotes the aggregate of grass and desert shrub species. Incorporating Gr and Pi as dichotomous variables (Gr = 0, Pi = 1) provides the trivariate formula

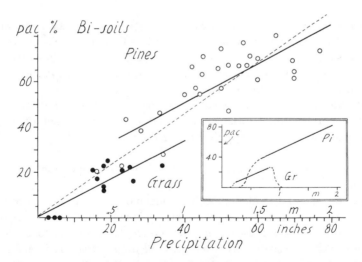

Fig. 12.18. Percentage acidity (pac) of 41 Bi soils. The dashed line, fitted to all points, reports the overall trend; the two solid lines pertain to grasslands and desert shrubs and to Ponderosa pines. Inset: visualizing the overlap of pines and grass

$$pac = 0.070 + 0.337\,maP + 16.341\,\phi$$

which improves the correlation coefficient to 0.938, corrected for degrees of freedom. Inserting the values 0 or 1 for ϕ converts the equation to those of grasses and pine above.

Web of Soil Properties. All of the 29 s properties analyzed at each of the 97 soil sites of the moisture transect may be regressed individually with rainfall. Some correlate well (C, pH, exch. Ca), others poorly (CEC, quartz, exch. Na). Not only are the soil properties linked in various degrees to maP they also engage in affinities and disparities among themselves, coded by 406 r values.

Among the 176 pairs of s that undercut the low r of 0.20 are N X K, Ca X aci, and C X sba. On the list of 13 high pairs ($r > 0.80$) are sba X Ca ($r = 0.984$), C X aci ($r = 0.945$), C X N ($r = 0.919$), N X aci ($r = 0.853$), and pac X pH ($r = 0.839$).

Problems of correlation among s properties are basic to soil genesis and classification but remain largely unsolved. Why should, as shown, N be highly covariant with aci but barely with K, or why should C relate positively to halloysite and negatively to montmorillonite? The extensive correlation matrix of the transect samples points to precipitation as an overall coordinator of rs, for two properties that are highly responsive to maP enter into close alliance with each other. Hence, state factors deserve being included in s correlations.

Pairs of properties joined by high rs are twins or dyads that may approach the $r = 1$ of identity properties like pH X pOH or ash X loss of ignition. N X C X aci is a triplet blended by principal component analysis. Its statistical "eigenvalue" accounts for 93.5% of the combined variance of the three soil attributes. These clusters or amalgams of s properties may themselves be regressed with maP, bringing about parsimony of functions but not necessarily a fostering of understanding.

Distribution of $CaCO_3$ and $MgCO_3$

In the rainy, eastern United States carbonate rocks and the calcareous variants of alluvium and glacial drift are leached (Fig. 12.19). Contrarily, soils in the arid West retain carbonates, often as subsoil nodules, "white gravel," and white seams that the pioneer farmers from humid Europe and the Atlantic Seaboard viewed with suspicion. The genesis of this "lime horizon" has been elaborated in Chapter 7.

In the stylized sketch (Fig. 12.6) of Shantz the carbonate horizon disappears in the humid region. In the central, calcareous loess belt of Nebraska continuity of Ca-decline is proven quantitatively in Alway's Figure 12.20.

Depth of Carbonate Horizon Related to Precipitation

In Kansas along the 11°C isotherm the lime horizon on broad ridges descends in Figure 12.21 proportionally to maP. The great scatter of points is caused by lack of control of parent material and soil age. Near maP 100 cm, large isolated concretions rather than continuous carbonate sheets characterize the deeper strata. The trend of depth D (centimeters) to carbonates is estimated as

$$D = 2.5\,(maP - 30.5)$$

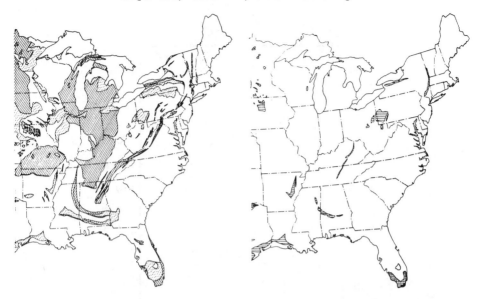

Fig. 12.19. Approximate distribution of calcareous parent materials (left side map) and calcareous A horizons (right side map) of the eastern United States, compiled by R. Richter

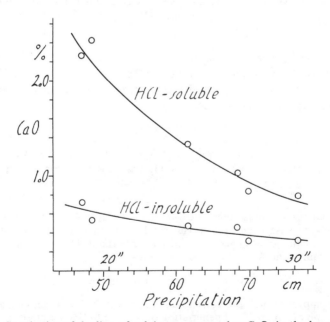

Fig. 12.20. Continuity of decline of calcium, expressed as CaO, in the loess belt of the arid-humid transition zone of Nebraska, after Alway (1). Soils to depth of 183 cm were extracted with hydrochloric acid

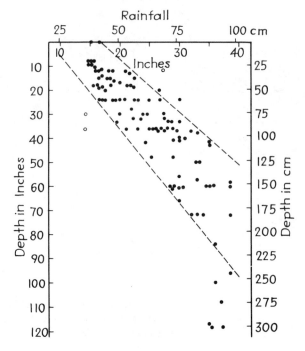

Fig. 12.21. Depth to carbonate horizon, or thickness of carbonate-free soil, related to maP in the central Great Plains area (17). Courtesy The Williams and Wilkins Co., Baltimore, MD

and, on the average, negligible carbonate leaching is predicted below maP 30.5 cm.

On old mesas, terraces, and fans of the arid West, Arkley (2) selected 27 calcareous soils that are well-drained, lack mottling and strong alkali, and are not Vertisolic. The solid line in Figure 12.22 is described as

$$D = 1.39 \, (\text{maP} + 4.41), \, (r = 0.761)$$

Compared to Figure 12.21, the position of D suggests that the cold California winter rains are more effective displacers of $CaCO_3$ than the warm summer showers in the western Great Plains.

The four black points above the solid line are clay soils and they are leached but moderately because 1 cm of rain will wet dry clay to a lesser depth than dry sandy loam. If the equation is expanded to include texture of the A horizon, quantified as moisture equivalent (ME), the trivariate expression

$$D = 8.28 + 1.62 \, \text{maP} - 0.45 \, \text{ME} \, (R = 0.872)$$

yields an improved multiple correlation coefficient R. The negative slope attached to ME pinpoints the reduced penetration of water into fine-textured soils. Selection of 11 sandy loams (ME 8-17%) augments r to 0.966, shown in Figure 12.22 as open circles and a dashed line that begins near the origin.

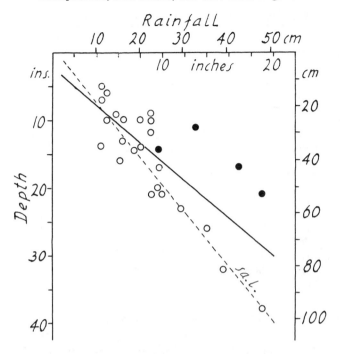

Fig. 12.22. Depth to carbonate horizon of alluvial terraces in Nevada and California (2). Sandy loams (open circles) and clay soils (filled circles)

Carbonate Horizon and Past Climates

Since the carbonate horizon relates to present-day maP, what happened to the impact of past pluvial periods?

A point of departure is offered by ancient *Lake Lahontan* in the Carson desert area of Nevada, adjacent to the Sierra Nevada mountains of California. This vast Quaternary lake—now dry—had depths of over 180 m, as judged by well-preserved shorelines. The geologist Morrison (25) distinguishes several oscillations of lake levels during and subsequent to late Pleistocene times. He correctly surmised that exposed and buried soils might record the intensities of the various pluvial periods responsible for the lake's enormous expansions and contractions. Springer's (31) investigation of the soils of an alluvial basalt-gravel terrace of 1% slope, situated on highlands above the highest lake level, is summarized in Table 12.6. Mean annual precipitation of 13.5 cm occurs mainly during the winter, and shad-scale (*Atriplex confortifolia*) and greasewood (*Sacrobatus baileyi*) dominate the sparse shrub vegetation. The humus content is extremely low, and C/N fluctuates between 6.3 and 8.1. Soil reaction varies from 7.3 to 8.6. In this typical "pre-Lahontan soil profile" (Fig. 7.14) the clay B horizon, which is relatively rich in montmorillonite and lacking in kaolinite species, lies above the carbonate horizon. It coincides with the zone of spring moisture according to Springer's observations and water balance calculations. The upper horizons (0-23 cm)

Table 12.6. Pre-Lahontan Soil Profile on Pleistocene Desert Alluvium[a]

Depth (cm)	Remarks	Munsell color	N (%) (whole soil)	CO_2[b] from carbonates (whole soil)	Percentage stones > 2 mm
-2.5-0	Desert gravel pavement, 75% cover	Brown (desert varnish)	—	—	64.7
0-6.4	Vesicular layer, fine sandy loam	Pink-gray	0.022	0.2	11.7
6.4-18	Clay loam	Reddish-brown	0.027	0.1	2.3
18-23	Clay loam, soft concretions	Light reddish-brown	0.028	11.7	53.2
23-38	Gravelly loam, lime-cemented	Pink	0.011	16.4	71.3
38-61	Gravelly loam, lime-cemented	Pink	0.007	14.6	66.1
61-76	Gravelly loam, partly cemented	Pink	0.0016	6.1	40.3

[a] After Springer (31).
[b] To convert CO_2 to $CaCO_3$ multiply by 2.275.

are gravel-free; Springer attributes this to upward movement of stones, as elaborated in Chapter 7. All features are typical of eight profiles chosen from alluvium, rhyolite tuff, and basalt and andesite lava table lands of Pliocene and early Pleistocene age.

Rather than being truncated erosion relicts, these soils portray full-fledged horizonation in a desert environment. The depth of carbonate horizon is consistently at 20-35 cm, depending on soil texture, and correlates very well with Arkley's climosequence in Figure 12.22; in fact, the five points at lowest rainfall are taken from Springer's set. No significant pluvial periods are reflected in these profiles, a point stressed by Springer.

Possibly, each pluvial period of a lake rise might have leached out previously formed carbonate horizons and subsequently a new $CaCO_3$ layer accumulated. But dense and thick carbonate sheets are slow in forming, especially below a clay layer. Further, the adjacent younger profiles of mid- and post-Lahontan and Holocene age have but weakly developed lime horizons or none at all. And how would the depleted Ca be renewed? Rock weathering is time consuming, and significant aeolian Ca deposition is improbable in view of the paucity of limestone rock formations in the Carson desert. Either the soil profiles have been insensitive to climatic shift or the lake levels were not linked to the local desert rainfall regime that formed the profiles. As a possibility, the lake may have been fed by melting ice of the Sierra Nevada or lake water accrued in response to climatic cooling (7). More field work needs to be done on fixing the lower boundary of the carbonate horizon to learn how far down $CaCO_3$ has moved and whether today's climate and soil water balance can account for such depths.

E. Climate and Clay Features

It used to be taught that "no harmonious relationship exists between soil textures and environment, that they occur in a wholly hit or miss way" in any kind of landscape. A different story is told when texture components such as clay are allocated to state factors.

Clay Temperature Sequences in the Eastern United States

In an evenly humid climate of maP 100-150 cm igneous and metamorphic rocks stretch from Maine to Alabama, accompanied by the wide maT span of $6\text{-}19°C$. Clay contents of 21 soils to a depth of about 1 m and underlain by Bi rocks, mainly gabbro and diorite, surround the straight line as shown in Figure 12.23.

Percentage clay = 4.94 maT – 37.4, having r of 0.814. Scatter of points is substantial because rock mineralogy is diverse, land surfaces have unequal ages, and erosion is spotty. For seven other rock provinces r ranges from 0.49 to 0.86, the regressions demonstrating clay gains with increasing maT, or, decreasing latitude from 45 to 30° N. When percentage clay is adjusted to a constant moisture index, e.g., NSQ 400, the curves rise exponentially, for Bi soils as log of percentage clay = 0.10 + 0.147 (r = 0.918).

Discarding the soils of the young glaciated areas in the North ($< 8°C$), the mean clay contents in percentage and their standard errors are 25.6 ± 1.24 at maT 10-13°C and 38.0 ± 1.30 at 16-19°C, the sample sizes being 38 and 33, respectively. When the comparison is restricted to subsoils the clay means are 24.8 ± 1.32 and 44.8 ± 1.42, the difference of 20.0 being highly significant (17).

In the same humid region Baver and Scarseth's (4) plot of SiO_2/Al_2O_3 or sa of clays in surface soils declines linearly from about 4 in the State of New York to < 2 in southern Alabama where the ratio is independent of rock material (17). Contrastingly, in the semiarid region of the West, sa remains at 3.4-5.3 regardless of maT.

Clay Mineralogy in the Warm-Temperate Region

In their sand fraction dry-land soils average 63% quartz and 37% nonquartz minerals and the latter drop to 8% in high rainfall regions. Similar proportions are found in the silt fraction (17). The vanishing feldspars, micas, and pyroxenes transform in part to clays, but the belief that clay content of profiles and maP rise together (17) is challenged by Birkeland (6).

Barshad (3) finds clay mineralogy attuned to state factors, and he has been kind enough to put the 97 X-ray interpretations of the aforementioned California moisture transect at the writer's disposal. The percentage sum of silicate clays, quartz ($< 2\,\mu m$), and gibbsite, subtracted from 100%, is assigned to iron oxyhydroxy minerals. For convenience of interpretation the species rich in OH ions, namely, kaolinite, halloysite, gibbsite, and iron oxides, are summed up as "OH clays" and the intergrading crystals are allocated to their montmorillonitic or vermiculite fractions. For Bi soils the four groups—OH-clays, illite, vermiculite, and montmorillonite—are joined to maP in Figure

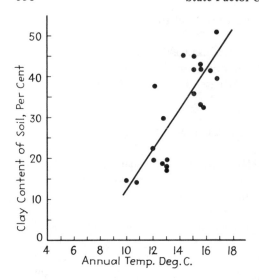

Fig. 12.23. Latitude or maT trend of average clay percentage to 1 m depth of soils derived from basic igneous rocks along the East Coast. Each circle denotes a profile

12.24, and the strong scatter of points is again charged to dissimilarities of mineral assemblages among the Bi rocks.

Above maP 100 cm, OH clays and vermiculites achieve dominance. Ai soils, not shown, accumulate kaolinite in preference to halloysite, 51 to 1, and so do Bi grassland soils, 12 to 1. All 28 Bi pine soils contain halloysite instead of kaolinite. No A1 horizon carries both minerals simultaneously.

At low rainfall montmorillonite and illite assume importance, most prominently in the desert. In Bi soils neither occurs above maP 80 cm. Because of the presence of micas in granites, the Ai soils possess a broader illite field than the Bi soils.

Principal component analysis for the 97 soil samples transforms their 10 clay types into 10 independent blends or components that decrease in statistical importance from the first to the tenth. If the first principal component, which is heavily loaded with halloysite, illite, kaolinite, and iron minerals, is regressed with precipitation, the correlation coefficient of all samples is 0.520; if temperature is included, the multiple correlation coefficient R rises to 0.625; and if parent rocks Ai and Bi are also entered, R becomes 0.859. In substance, the first mineral composite by itself responds significantly to the state factors climate and parent rock.

Weathering of a mineral grain takes place in a small soil subsystem, often a microsystem, and the process is governed by its own localized set of state factors. Insofar as these are functions of the factors of the macroecosystem, the clays will correlate with macroclimate, as illustrated by the graphs.

The mineral suite of a soil horizon is not necessarily an image of genesis of clay. It is the balance of formation, alteration, and destruction of clay, and of clay influx from whatever source and efflux to other horizons. Thus, at high rainfall the absence of illite and montmorillonite in A1 may reflect instability and translocation, not necessarily lack of synthesis.

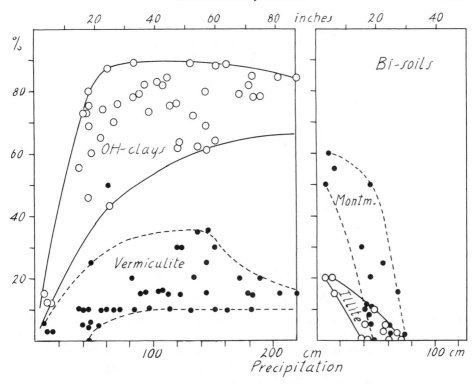

Fig. 12.24. Clay species of 45 surface soils (0-20 cm) derived from basic igneous rocks (Bi soils) plotted on a maP axis. To avoid crowding of points, illite and montmorillonite fields are graphed separately (Barshad's analyses)

The climatic suite of clays guides response to fertilizers. When 400-g aliquots of the 45 Bi soils are put into small pots and provided with the high nutrient treatment of $N_3P_3S_{0.5}$, 1 unit corresponding to 20 mg of element, and planted to barley, the yields diminish as maP of the collection site rises (Fig. 12.25). The decline is traceable to *phosphate fixation* by the red clays that accumulate in the humid sector. Trebling the P dosage to P_9 partially overcomes the limitations set by clay mineralogy (22).

Moisture Sequences in Hawaii

Interiors of unweathered basalt rocks at the base of C horizons, analyzed by Sherman and Ikawa (29), are rather uniform and have the following SiO_2/Al_2O_3 ratios (sa): 4.84 at maP 114 cm, 5.00 at maP 152 cm, 4.81 at maP 229 cm, and 5.75 at maP 305 cm. In the saprolite, 5 cm outward from the rock surface, the corresponding ratios are 1.80, 1.45, 0.98, and 0.10, hence the "C" horizon material is itself a function of climate, silica having been leached differentially.

In basaltic *Pahala ash* of a 10,000- to 17,000-year-old *Kilauea eruption*, Hay and Jones (14) find in ash-soil samples at 25-150 cm depth the course of silica leaching as graphed in Figure 12.26. Unweathered ash has a SiO_2/Al_2O_3 ratio of 6.66, and the

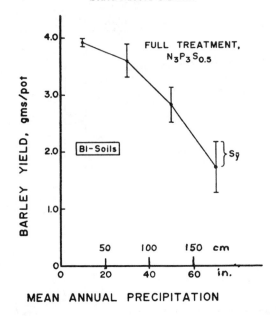

Fig. 12.25. Barley yields of pots filled with transect soils and fertilized with N, P, S are plotted against maP of soil sampling sites (22). Decline is caused by P fixation. $S_{\bar{y}}$ = standard error. Courtesy The Williams and Wilkins Co., Baltimore, MD

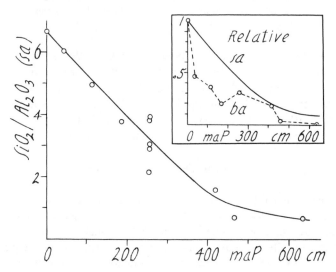

Fig. 12.26. Narrowing of silica-alumina ratio of soils with increasing maP in Hawaii (14). Inset: relative to initial state, the bases, ba, as $(Na_2O + MgO + CaO)/Al_2O_3$, are leached more readily than silica

bases ratio $(Na_2O + MgO + CaO)/Al_2O_3$ is 3.92. Using these as units the trends of the curves in the inset of Figure 12.26 show the bases Na, Mg, and Ca being leached more strongly than silica, leading to nutrient-impoverished soils. At low rainfall montmorillonite and calcite line root channels and fuse into 1 cm seams, and at high rainfall gibbsite (12-25 μm diameter) fills cavities and small veins and appears as pseudomoph; otherwise, the fines are hydrated and amorphous.

Tanada's (33) compositions of clays from *various islands* keep pace inversely with rising maP (40-580 cm): SiO_2/Al_2O_3 narrows from 2.01 to 0.76, SiO_2/Fe_2O_3 from 12.5 to 0.47, and SiO_2/TiO_2 from 48.7 to 1.46, with correlation coefficients of -0.76, -0.91, and -0.86.

The weathering of basalt on *Kauai Island* (29) creates montmorillonite below 50 cm of maP. At high rainfall kaolinite takes over. It becomes abundant at about 90 cm of maP and disappears at 100-110 cm (Table 12.7), provided free drainage prevails. The silicate clay reappears in copious amounts in the perhumid zone because the poor internal drainage of the permanently swollen, clay-rich soils either arrests its dissolution or encourages resilication of gibbsite to kaolinite.

Near 200 cm of maP iron oxides accumulate to nearly half of the soil's weight. The dry months foster dehydration of amorphous iron hydroxide gels to colloidal crystals among which the red color-inducing hematites (32) are heavy (density of 5.3) and relatively large. The soils are ferruginous latosols (Gibbsihumox) that may form laterite crusts. In the perhumid, fern forest-covered Hydrol Humic Latosols (Tropaquepts) iron content is low and humus high, 15-40%. Possibly, chelation and reduction enhance iron lixiviation (32).

On *eastern Kauai* the geomorphic surface Kumukumu belongs to a basaltic lava flow of age 0.5-1 million years. Over a distance of 8 km and an elevation interval of 200 m climate rises from warm-humid at the lower end to warm-perhumid at the upper. A climosequence of reddish Oxisols has been mapped (5):

Table 12.7. Moisture Sequence of Dominant (X) Clay Minerals in Percentage of Soil Weight on Hawaiian Basalts[a]

Precipitation (maP, cm)	Number of Dry months	Number of Wet months	Montmorillonite (%)	Kaolinite (%)	Halloysite (%)	Gibbsite (%)	Fe-oxides[b] (%)
50	–	–	X				
50-80	–	–	X	X			
80-100	–	–		X			
100-115	8-9	3-4			70	–	20
130-180	4-5	4-5			35	10	40
180-250	2-5	4-6			20	25	45
250-381	2-4	5-8			30	30	10
380+ (perhumid)	0-2	8-12		50		15	10

[a] After Sherman and Ikawa (29).
[b] Goethite + hematite + magnetite.

$$100 \text{ cm} \xrightarrow{\text{maP}} 300 \text{ cm}$$

Eutrustox	→	Euthrorthox	→	Gibbsiorthox	→	Gibbsihumox
kaolinite		oxidic clay		gibbsite-rich		gibbsite-rich
clay				gravels		gravels
						$> 16 \text{ kg C/m}^2/\text{m}$

On the near-level portions of the geomorphic surface Oxisols reside; on the steeper flanks Ultisols with clay skins and Bt horizons develop.

Other Tropical Moisture Sequences

Along an arid to humid transect on the basaltic island of *Mauritius* (maT 23°C) the clay fraction of surface soils decreases and silts and sands augment because, as stated by Craig and Halais (11), advanced weathering induces cementation of colloidal particles to silt and sand-sized aggregates. Simultaneously, exchangeable bases diminish from 24.0 to 4.0 meq/100 g soil and SiO_2/Al_2O_3 from 1.68 to 0.37.

North Queensland, Australia (lat. 17° S) has reddish soils on Late Pleistocene basalt flows that extend over a rainfall range of 90-370 cm and a temperature interval of 20 to 24°C. Below maP 100 cm the native vegetation is Eucalyptus savanna woodland, above 130 cm rain forests dominated by Indo-Malaysian species.

Nine profiles collected by Simonett and Bauleke (30) are high in clay, 60 to 70% in the upper horizons of the climatically extreme sites of Mareeba and Topaz selected in Table 12.8. Acidity straddles pH 5.9-6.3 in the humid and pH 4.3-4.7 in the perhumid profile. Weight percentage of kaolinite in the clay fraction of the 0- to 30-cm layer decreases with increasing rainfall as

$$\log \% \text{ kaolinite} = 2.302 - 0.270 \log \text{maP} \ (r = -0.96)$$

The equation is of the same form as Tanada's, but Simonett's slope coefficients are larger, suggesting that the same amount of rainfall might be more effective in Hawaii than in Queensland. Molecular ratios of clays also correlate well with rainfall.

Table 12.8. Percentage Composition of Clay Fraction ($< 2 \mu$m) in Well-Drained End Members of an Australian Climosequence[a]

Depth (cm)	Kaolinite	Gibbsite	Hematite[b]	Magnetite[b]
Mareeba (maP 90 cm)				
0-15	77	1	9	6
15-46	74	1	6	7
46-152	79	3	8	4
Topaz (maP 370 cm)				
0-10	52	19	12	8
10-76	60	13	11	7
91-102	68	9	6	7
183-239	67	7	5	8

[a]After Simonett and Bauleke (30).
[b]Goethite occurs only in less than well-drained profiles.

Soils as Indicators of Past Climates

The question of soils mirroring climatic shifts is touched upon in a preceding section on carbonate horizons. The following criteria deserve to be kept in mind:

1. All extrapolations into the past rest on known correlations of soil properties with present-day climates, and these are relatively scarce.
2. Most of the correlation coefficients between s properties and climate stay below 0.90 and only drastic climatic shifts become discernible.
3. Many properties (e.g., humus, pH, CEC, etc.) are reversible and partly readjust to environmental change.
4. The competing roles of climate and length of soil genesis are not readily separated. What will today's young Inceptisols and Mollisols of temperate regions look like half a million years from now, climate remaining unchanged?
5. Over long periods of time, not the mean precipitation but the number of wet years dictates depth of weathering and leaching.
6. Within the same climate coarse-textured parent materials are leached to greater depth than fine ones, and an acid matrix may lead to acid soils in arid regions.
7. Different climates may produce the same soil order (e.g., Spodosols in Scandinavia, California, Brazil).

The points listed question whether pedology is an easy diagnostic tool for assessing climatic change.

F. Review of Chapter

A century ago Hilgard discovered climatic imprints in mineral soils and shortly afterward Dokuchaev recognized them in the humus economy. In the face of questioning, both observations have been abundantly verified and recast as climofunctions and as separate maP and maT sequences, provided control of the other state factors is affirmed.

Not all soil properties respond to climatic stimuli to the same degree. Quartz and zircon minerals are resistant, $CaCO_3$ and clays are pliable. Excluding soil moisture and temperature, the organic matter may be one of the most sensitive of soil attributes. This yielding, however, denies a property the status of being a key witness of past climatic changes.

References

1. Alway, D. F. 1916. *Soil Sci.* 1: 197-258, 299-316.
2. Arkley, R. J. 1963. *Soil Sci.* 96: 239-248.
3. Barshad, I. 1966. *Proc. Int. Clay Conf.* 167-173, Jerusalem.
4. Baver, L. D., and G. D. Scarseth. 1931. *Soil Res. (Int. Soc. Soil Sci.)* 2: 288-307.
5. Beinroth, F. H., G. Uehara, and H. Ikawa. 1974. *Soil Sci. Soc. Am. Proc.* 38: 128-131.
6. Birkeland, P. W. 1974. *Pedology, Weathering and Geomorphological Research.* Oxford Univ. Press, New York.
7. Brackenridge, G. R. 1978. *Quat. Res.* 9: 22-40.
8. Caldwell, M. 1975. In *Photosynthesis and Productivity in Different Environments,* J. P. Cooper, ed., pp. 41-73. Cambridge Univ. Press, New York.

9. Carleton, O., L. Young, and C. Taylor. 1974. *Climosequence Study of the Mountainous Soils Adjacent to Santa Fe, New Mexico.* U.S.D.A., Forest Serv. S. W. Region.

10. Cortes, A., and D. P. Franzmeier. 1972. *Soil Sci. Soc. Am. Proc.* 36: 653-659.

11. Craig, N., and P. Halais. 1934. *Empire J. Exp. Agr.* 2: 349-358.

12. Dokuchaev, V. V. 1883. Russian chernozem. *Selected Works,* Vol. 1, Israel Progr. Sci. Transl., U.S.D.A., 1967.

13. Harradine, F. F., and H. Jenny. 1958. *Soil Sci.* 85: 235-243.

14. Hay, R. L., and B. F. Jones. 1972. *Geol. Soc. Am. Bull.* 83: 317-332.

15. Hilgard, E. W. 1892. *U.S.D.A. Agr. Weather Bur. Bull.* 3: 1-59.

16. Huntington, G. L. 1954. The effect of vertical zonality on clay content in residual granitic soils of the Sierra Nevada Mountains. M.A. Thesis, Univ. of California, Berkeley.

17. Jenny, H. 1941. *Factors of Soil Formation.* McGraw-Hill, New York.

18. Jenny, H. 1961. *E. W. Hilgard and the Birth of Modern Soil Science.* Agrochim. 3, Pisa.

19. Jenny, H., and S. P. Raychaudhuri. 1960. *Effect of Climate and Cultivation on Nitrogen and Organic Matter Reserves in Indian Soils.* Ind. Council Agr. Res., New Delhi.

20. Jenny, H., S. P. Gessel, and F. T. Bingham. 1949. *Soil Sci.* 68: 419-432.

21. Jenny, H., A. E. Salem, and J. R. Wallis. 1968. In *Soil Organic Matter and Soil Fertility,* P. Salviucci, ed., pp. 5-37. Pont. Acad. Sci. Scripta var. 32, Vatican.

22. Klemmedson, J. O., and H. Jenny. 1966. *Soil Sci.* 102: 215-222.

23. Martin, W. P., and J. E. Fletcher. 1943. *Vertical Zonation of Great Soil Groups on Mt. Graham, Arizona, as Correlated with Climate, Vegetation and Profile Characteristics.* Ariz. Agr. Exp. Sta. Bull. 99.

24. Mohr, E. C. J. 1930. *De grond van Java en Sumatra,* 2d ed. De Bussy, Amsterdam.

25. Morrison, R. B. 1964. *Lake Lahontan: Geology of Southern Carson Desert, Nevada.* U.S. Geol. Surv. Prof. Paper 401.

26. Neustruev, S. S. 1927. *Genesis of Soils.* Russ. Pedol. Invest. U.S.S.R. Acad. Sci., p. 5.

27. Olson, J. S. 1975. In *Productivity of World Ecosystems,* D. R. Reichle, J. F. Franklin, and D. W. Goodale, eds., pp. 33-43. Nat. Acad. Sci., Washington, D.C.

28. Shantz, H. L. 1923. *Ann. Assoc. Am. Geogr.* 8: 81-107.

29. Sherman, G. D., and H. Ikawa. 1968. *Pacif. Sci.* 22: 458-464.

30. Simonett, D. S., and M. P. Bauleke. 1963. *Soil Sci. Soc. Am. Proc.* 27: 205-212.

31. Springer, M. E. 1953. Soil formation in the desert of the Lahontan Basin, Nevada. Ph.D. Thesis, Univ. of California, Berkeley.

32. Tamura, T., M. L. Jackson, and G. D. Sherman. 1953. *Soil Sci. Soc. Am. Proc.* 17: 343-346.

33. Tanada, T. 1951. *J. Soil Sci.* 2: 83-96.

34. Walker, T. W., and A. F. R. Adams. 1959. *Soil Sci.* 87: 1-10.

13. Biotic Factor of System Genesis

Planet Earth houses millions of species of plants and animals and all may perform as biotic factors. The old truism that sun and rains impinge on plant and soil, that soil affects the plants and plants affect the soil is often written:

climate

vegetation ⇌ soil

v-properties   s-properties

The v and s components also temper the air environment above the ecosystem but wind gusts and turbulent diffusion soon dissipate the incursion. The plant's substrate environment instead bears long-time testimony to organism occupation, which is the main subject of the present chapter.

A. Formulation of Biotic Factor Sequences

We shall probe the intricate soil-plant contract from the vantage point of state factor analysis.

Soil-Plant Interactions

The chronosequence of Chapter 9:

$$l, v, a, s = f(t, cl, \phi, r, p, \ldots)$$

recognizes that two soil properties, s_1 and s_2, that belong to the same time sequence are also correlated with each other as $s_1 = f(s_2)$, in Figure 9.11 as $pH = f(CaCO_3)$. Collinearity extends to other ecosystem properties at other state factor combinations, hence

$$s = f(v), \text{ and } v = f(s)$$

provided s and v are responding to the same constellation of cl, ϕ, r, p, t, ... The story of the redwood seedling growing on a sand bank, told in Chapter 1, and the carbon curves of Figure 13.2 in this chapter are cases in point.

The accord fails to divulge what is direct action and what is feedback. Is soil pH the cause or effect of vegetation? In the case of roots bathing in a flowing clay suspension of a given pH the substrate remains constant and its steerage of metabolism is demonstrable, but in soils plants soon alter the initial pH and interlocking sets in.

Biotic Factor Equation

The factor function scheme is written as:

$$l,v,a,s = f(\phi)_{cl,r,p,t,...} \quad or \quad l,v,a,s = f(\phi,cl,r,p,t,...)$$

$$\text{ideal case} \qquad\qquad\qquad\qquad\qquad \text{actual case}$$

in which ϕ are the biotic factors as genotypes and v the phenotype vegetation properties such as size and health. Written in words:

$$\left.\begin{array}{l} \text{vegetation properties} \\ \\ \text{soil properties} \end{array}\right\} = F(\underbrace{\text{genotypes}}_{\phi}, cl,r,p,t,...)$$

where t is either specimen age, soil age, or ecosystem age. An organism's genotype is independent of the ecosystem because it is acquired prior to its growth within it. Vegetation is not a biotic factor, but flora is.

To learn how biota mold soils, an organism is introduced into an abode of substrate, air and light, as in a phytotron or a plowed field. The act sets time zero for the specimen and the induced soil alterations are observed at different periods of time, as has been traditional in soil-plant research.

In natural ecosystems a vascular plant is nurtured in company with many other species, including invertebrates, microflora, and symbionts such as rhizobium and mycorrhiza. In forest and prairie the individuals of a chosen species, together with their organismic cohorts, function as biotic definer variables.

B. Partitioning Biotic Factor and Environment

The interaction of organisms with their environments delineates the domain of ecology. Practical ecologists, such as agronomists and foresters, want to know—if only as a scientific challenge—how biota and their surroundings share in the production of biomass. Systems analysis provides an answer.

Conventional and Unconventional Pot Experiments

Ulrich (45) fills large cylindrical crocks ("unit pot") of 613 cm^2 cross-sectional area (A) and 33 cm depth with vermiculite flakes and nutrient solution, the latter being replaced every other day. As biotic factor ϕ he plants seedlings of a sugar beet variety in numbers (n) of one, two, and four per pot and replicates the treatments 20 times to assure a mean genotype of the seed population. (Hindsight suggests prismatic containers of square cross-section that touch each other to minimize border effects.)

From an unpublished Ulrich experiment three mean fresh weights (Y) of entire beet plants as kilogram/pot are plotted in Figure 13.1 (inset) and a linear regression is fitted to the three means. This graphic procedure is common but has its shortcomings. The most reliable datum point, zero yield at zero seedling, is arbitrarily omitted and though the horizontal axis is discontinuous, the regression line and the two segments spell continuity.

In the smooth drawing of Figure 13.1 some of the queries are answered, but the idea of a fractional seed (one-half) giving rise to a valid yield makes no sense biologically; it does, however, ecologically.

On some farms a 20-acre barley field delivers twice as much grain as a 10-acre field, and a 30-acre field delivers three times as much. If this proportionality of yield and soil area holds for beets growing in vermiculite, then planting one seed in a double pot (2×613 cm^2 area and constant 33 cm depth) corresponds to one-half seed in unit pot. Experimentally, the yield of one beet plant in a double pot is 5.396 kg, and half this value is plotted in Figure 13.1 as a white-black circle. Whereas the conventional experiment varies the number of plants, n, in a unit pot of area A and establishes $Y_n = f(n)_A$,

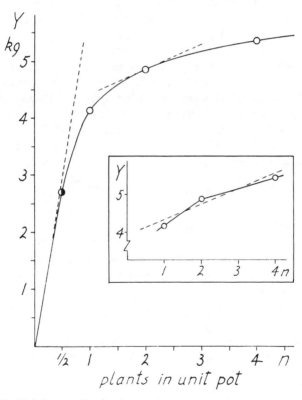

Fig. 13.1. Yields Y (kilogram/pot) of sugar beet plants plotted against number (n) of plants per pot. Dashed lines are slopes, initial and at $n = 2$. Inset: customary mode of data plotting, and regression line

the unconventional trial keeps the number of plants constant and varies pot area A while preserving geometry and structure of the abode. It forges the curve for $Y_A = f(A)_n$ which is a *continuous function*. Intuitively, we suspect that the two response curves are interrelated and that yields of fractional plants may indeed be computed from tests employing variable sizes of pots.

Nature of Initial Slopes

In Figure 13.1 the dashed line starting at the origin is the initial slope of the smooth, freehand curve $Y_n = f(n)_A$. To interpret, it is the yield of an imaginary, small fraction of a germule in unit pot, adjusted to the whole plant; or, expressed more realistically, it is the yield of one plant growing alone in a very large pot of standard depth (33 cm). The computed biomass of 6.00 kg at $n = 1$ represents the highest phenotypic fulfillment of the genotype. It is a plant's maximum productivity under the conditions of the chosen environment of air, light, and substrate. Ecologically, the experiment alludes to the colonization of bare land by a plant pioneer. Its genetic make-up is the limiting factor.

The initial slope of the companion curve $Y_A = f(A)_n$, which is not shown, is the yield of one plant in a tiny abode (a narrow tube of 33 cm depth) and referred to unit pot size. More visibly, Y is the biomass produced by very many plants growing in a single unit pot, extrapolated in Figure 13.1 as 5.65 kg. Here, the surrounding of the plants is the limiting factor and the yield is a measure of the maximum productivity of the chosen combination of air, light, and substrate. Agronomically, this case underpins the popular Neubauer test of soil fertility in which K and P uptake by 100 crowded rye seedlings growing on 100 g of soil is determined.

When the experiments are conducted with different soils in different climates, as was done in part by Borden (25) in Hawaii, the separate productivities of air-light and soil are amenable to quantification.

The Euler Ecosystem

Systems for which a doubling or trebling of the area, subject to preserving geometry and structure of the abode, doubles or trebles the biomass or any other extensive property conform to Euler's homogeneous functions of the first degree. In such systems the two curves $Y_n = f(n)_A$ and $Y_A = f(A)_n$ are linked mathematically by the Gibbs-Duhem equation, and further (34)

$$Y = Y'_n \cdot n + Y'_A \cdot A$$

in which Y'_n are the slopes of the curve $Y_n = f(n)_A$, and Y'_A those of $Y_A = f(A)_n$. These slopes express the effectiveness of biotic factor and of environment at any combination.

Although the beet experiment of Figure 13.1 is not strictly Eulerian because of leaves drooping over the pot walls, it satisfies the *equation* fairly well. Thus, at the point marking two plants in unit pot the slope Y'_n shown is 0.43 kg/plant and the computed slope Y'_A is 4.13 kg/pot. The predicted yield Y is $0.43 \cdot 2 + 4.13 \cdot 1 =$

4.99 kg whereas the observed production of the ecosystem is 4.86 kg. Similar concordances satisfy all other points of the curve. Effectiveness or productivity of a factor, defined as slope, gradient, or derivative, means that the introduction of one germule into a large beet ecosystem of composition, say, $n = 2, A = 1$ is much less productive than enlarging the environment by a unit of pot size.

The partition equation is strictly operational system analysis and no plant physiological, biochemical, or pedologic principles are utilized. In the above inquiry the analysis succeeds in logically connecting the disparate events of colonization of bare land and bioassay of soil fertility. Phenomenologically, it gives an answer to the eternal, ecological question: what is more effective, biotic factor or environment?

C. Plant-Controlled Initiation of Ecosystem Genesis

The spectrum of germules received by bare ground is a multiple biotic factor of plants, animals, and microbes. When seeds encounter suitable microhabitats they survive and become, in Crocker's words (12), *effective* biotic factors that contribute to system genesis.

Glacial Drift and Dune Pioneers

At Glacier Bay, Alaska, till material to 5 cm depth contains 7-10% $CaCO_3$, as observed by Crocker and Major (13). On bare sites the carbonate persists during two decades, but is reduced to 6% under willow (*Salix barclayi*), to 5% under poplar (*Populus trichocarpa*), to 3-4% under *Dryas* ssp. and to 0.3% under alder (*Alnus crispa*). In the same thin soil stratum organic nitrogen (grams/square meter) accumulates to 5.5 under *Dryas,* 7.0 under poplar, 13.9 under willow, and 23.0 under alder, and organic carbon adapts to the same biosequence.

Below a lone, 45-year-old shore pine (*Pinus cortorta*) on a young sand dune, Zinke (51) observes a maximum of 0.029% N in 6 cm of surface soil, which is 11 times higher than N in bare sand. Tree influence extends beyond the tree crown periphery and lines up with wind direction.

Purshia and Pine

Among the andesitic mudflow sequences of Mt. Shasta, described in Chapter 9, A-flow descended in 1924 and 40 years later Glauser (23) collected tesseras of four bare sites, of three isolated antelope bushes (*Purshia tridentata*), and of five ponderosa pines. What follows is based on his analyses but calculations and interpretations go beyond his thesis.

Bare Areas. It is assumed that the soil material at 61-91 cm depth comes closest to the original composition of the mudflow deposit. For 13 bare and vegetated sites, analyzed in 1964, the means of C horizons are $0.00230 \pm 4 \times 10^{-4}\%$ for N and $0.0311 \pm 3 \times 10^{-3}\%$ for C, and for statistical computations more digits are carried than are analytically justified. A fresh tessera of 0-91 cm depth would have had, in 1924, 27.72 g/m^2

of N and 374.8 g/m² of C. In 1964 the bare surface layers (0-7.6 cm) had 0.0048% N and 0.1088% C which signifies a net gain of 2.18 g/m² of N and 68 g/m² of C in 40 years, valid at the 5% level. The increase is attributed to deer droppings and to decay of windblown leaves and feeble herbs here and there.

Purshia-22 and Pine-22. Glauser decapitated a 22-year-old, isolated *Purshia* shrub (91 cm tall) surrounded by bare soil and dug up four soil tesseras from near bush center to periphery. Then he excavated the remaining roots to the crown periphery. In Table 13.1 the ecosystem gain in 22 years is 2443 g/m² of C and 63.9 g/m² of N or, yearly, 2.90 g/m² of N. Purshia roots are profusely covered with nodular, collaroid growth which is suspect of harboring a N-fixing mechanism.

Since none of the five young pines harvested matches the age of *Purshia* a hypothetical 22-year-old pine ecosystem is calculated in Table 13.1 by interpolation of isolated, 18- and 27-year-old pine ecotesseras. The annual N gain is 1.99 g/m² but the mode of acquisition remains hidden, because no legumes grow near or under the trees and the roots lack visible mycorrhiza.

Purshia-22 weighs 9.306 kg compared to the pine's 5.381 kg and it covers the larger surface area. Pine-22 is taller. *Purshia* mobilizes six times more Ca than the pine and the superiority prevails in absolute (me) and in relative (me/m²) amounts. Compared to the conifer the shrub endows the ecosystem with four times the amount of N and it puts twice as much N/m² into the mineral soil. As to soil C, the two species behave similarly, and pine maintains higher C/N ratios than *Purshia*.

Table 13.1. Comparison of 22-Year-Old Ecosystems of Purshia (Area = 2.625 m²) and of Pine (Area = 0.9213 m²)[a,b]

System components	Nitrogen				C/N		Calcium	
	Purshia		Pine		Purshia	Pine	Purshia	Pine
	(%)	(g)	(%)	(g)			(%)	(%)
Vert space	0.633	42.58	0.575	24.47	79.64	91.42	0.498	0.220
Roots	1.054	27.19	0.208	2.34	47.93	247.48	0.939	0.167
Forest floor	2.512	47.98	0.802	5.54	20.95	68.43	1.762	0.524
Soil[c] 1964		139.68 ± 3.00%		39.57	15.50	19.72		
Soil 1942[d]		89.75		31.50	16.17	16.17		
Soil gain		49.93		8.07	14.30	33.56		
Ecosystem gain		167.68		40.42	38.25	85.75		
Ecosystem gain/m²		63.88		43.87	38.25	85.76		

[a] From Glauser's (23) analyses.
[b] Soil depth is 91.4 cm; oven-dry materials; A.D. 1964.
[c] For C the standard error is ± 1.38%; bulk density of pine soil tessera adjusted to Purshia's (1.25).
[d] Intrapolation of 1924 and 1964 values.

Percentage of Ca in the forest floor is species-characteristic, the values being 1.48, 1.76, and 1.81 for Purshias, and 0.48, 0.51, 0.62, and 0.64 for pines. The values parallel those of Ca in fresh leaves which average 0.86% in Purshias and 0.28% in pines.

Contract of Plant and Soil. The premise that plant and soil evolve together is shown quantitatively in Figure 13.2. Carbon of three shrubs and four pines, including their roots, is plotted against carbon gains in litter layer and A1 horizon combined. *Purshia* is the greater humus builder, possibly because its ecosystem has a higher N input and mineral metabolism than pine.

D. Trees in Forests

In an established forest a seed germinates in a ready-made soil that becomes the organism's initial state S_0.

Giant Sequoia

The 3-millennia-old sequoias, *Sequoiadendron giganteum* of the Sierra Nevada (maP 109 cm, maT 8.3°C), maximize the role played by a single plant specimen. Zinke and Crocker (52) find a tree's trunk surrounded by a litter ring of acidic (pH 4.4) bark

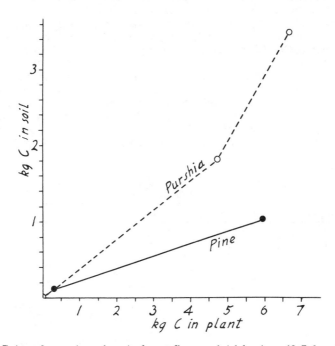

Fig. 13.2. Gains of organic carbon in forest floor and A1 horizon (0-7.6 cm) related to C in standing biomass, inclusive of roots. The partial circle at the origin comprises a Purshia and two pines

humus 150 cm wide and 90 cm thick. Toward the projection of the crown periphery, leaf and branch humus is less than 10 cm deep, and outside of the tree the ground is bare, neutral to slightly acid, and has in its surface a lower degree of base saturation than under the canopy.

During growth the sequoia's colossal trunk of over 8 m diameter must have pushed, heaved, and warped the adjacent soil radially over a distance of 4 m, destroying the original horizon sequence.

Deciduous Trees and Conifers

In a forest two fully grown species are pedologically comparable if they germinated in similar soil and if interaction, such as pine needles being blown under the deciduous trees, is minimal. Only differences between species can be distinguished because the initial soil state S_0 at germination time is unknown.

Soil tesseras under 200- to 300-year-old individuals of oak (*Q. kelloggii*) and pine (*P. ponderosa*) growing side by side on old granitic soils contrast strongly (Table 13.2) in forest floor but little, if any, in N and C in mineral soil, an observation reported frequently. The N reservoir in the mineral soil and the clay loam B horizons of the Ultic Haploxeralf had been formed long before the two trees started growing and, probably, oaks and pines had alternated on the sites.

In Japan, Yamaya (50) adds 250 g of air-dry leaves of alder (*Alnus inokumae*) to a pot holding about 2 kg of a loamy volcanic soil of bulk density 0.72 and 0.07% of N. A parallel, identical pot receives needles of Japanese cedar (*Cryptomeria japonica*). The alder leaves contain 3.24% N (C/N = 15) and 3.20% Ca; the cedar needles only 0.86% N (C/N = 60) and 1.51% Ca. The pots, covered with wire net and shown in Figure 13.3, are partially buried in a red pine forest and left there from May to November, receiving 84 cm of rain. The alder leaves lose 60.4% of their weight, the cedar needles only 40.4%. Humus from cedar forms a thin layer on the soil surface, containing 0.19% N, and infiltrates the lower pot section as B, with 0.11% N. Alder produces a thin, crumb-like A layer (0.43% N) and 9 cm of dark, humus-colored "B1" (0.22% N) and "B2" (0.13% N). The alder pot acquires a rich soil fauna of springtails, centipedes, and earthworms, gains a small amount of exchangeable Ca,

Table 13.2. Comparison of Impact of Oak and Pine on Soil in a Forest, *Quercus Kelloggii* and *Pinus ponderosa*

System property	Oak	Pine
Litterfall/year		
Weight (g/m^2)	155	314
N (g/m^2)	1.27	1.54
C/N	58.7	106.4
Forest floor		
Weight (g/m^2)	4,381	18,837
N (g/m^2)	31.1	117
C/N	39.4	55.3
k' for N	3.42%	1.30%
Mineral soil (0-127 cm)		
N (g/m^2)	602	533
C/N	17.3	20.4

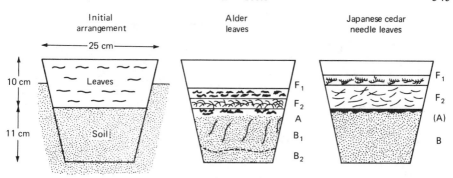

| Initial arrangement | Alder leaves | Japanese cedar needle leaves |

Fig. 13.3. Yamaya's (50) experiment on leaf decay which is fast for alder and slow for cedar

and dispatches 63% of the Ca it receives from the rotting leaves. Cedar soil loses none. Alder leaves induce high gains of hydrolytic acidity (Table 13.3) presumably through infiltration of unsaturated humus molecules.

Crown Drip and Stemflow

Raindrops impinge in random fashion upon a forest ecosystem. Once the drops enter the tree canopy they are intercepted and diverted, running along twigs, branches, and trunk as stemflow. In 1896 E. Hoppe claimed that in beech forests up to 15% of all rains reach the soil by way of stemflow, in an annulus area of about 30 cm width around the bole, as determined later by Voigt (46).

The measurements of Franklin et al. (18) of radioactive cesium (^{137}Cs) from fallout in Ohio are shown in relation to distance from stem in Figure 13.4. Quantities of Cs in forest floor and mineral soil to 7.5 cm depth are highest near the trunk. Radioactivity decreases in the species-sequence:

$$\text{beech} \quad > \quad \text{maple} \quad > \quad \text{oak} \quad > \quad \text{no tree}$$

(*Fagus grandifolia*)   (*Acer saccharum*)   (*Quercus alba*)

The same order is observed (27) in stemflow, which is conditioned by smoothness, roughness, and sponginess of the bark.

During summer rains stemflows from a beech contain the following mean concentrations in ppm: carbon 133, calcium 13, potassium 12, magnesium 2.2, phosphorus

Table 13.3. Soil Acidification by Alder Leaves Compared to Fir Needles[a]

	pH	KCl acidity	Ca-acetate acidity	Calcium
Initial soil	6.40	0.6	13.8	16.50
Cedar soil (A + B)	6.90	0.6	12.9	18.50
Alder soil (A + B1)	5.80	1.3	37.7	20.50

[a] After Yamaya's pot experiment (me/100 g oven-dry).

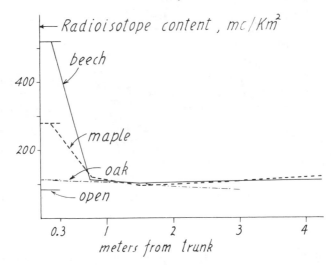

Fig. 13.4. Cesium content of surface soils at 0-30, 75, 150, and 450 cm from trunk perimeter (18). Courtesy Soil Science Society America, Madison, WI

1.6. Gersper and Holowaychuk (21,22) estimate for a 30-cm annulus around a tree of 80 cm diameter (at breast height) an area-concentration of carbon flow of 1037 g/m^2.

Under a 24-m-tall beech tree on Miami loam, a Typic Hapludalf, the A horizons along the vertical stem-surface projection are thicker and the B2 horizons have more subdivision and mottling than the corresponding horizons 50, 100, and 200 cm outward from the trunk. For A1 horizons at stem surface (± 5 cm) and at a distance of 200 cm, the following pairings deserve notice: *organic carbon,* 18.6 versus 5.3%, accompanied by increased *cation exchange capacity* (CEC) of 48.1 versus 26.4 me/100 g, which reflects the richness of stemflow in organic substances; *exchangeable K* increases in the annulus (1.23 versus 0.56 me/100 g), presumably from potassium leaching of leaves and bark, and *exchange acidity* mounts drastically from 10.6 to 35.3 me/100 g and is nearly equivalent to the gain in CEC. Its depth functions are shown in Figure 13.5.

E. Prairie-Forest Biosequences

For centuries the coexistence of natural grasslands, prairies, and steppes with woodlands and forests in the same regional climate has fascinated botanists and pedologists alike.

The thought once prevailed that the American pioneers had entered an unspoiled, primeval continent in which Indians lived in innocent harmony with nature. More and more we are being made aware of a long-time residency of the aborigines and of their *burning practices* that may have created part of the "natural vegetation," but how much of it is still controversial (47). In debating the origin of grasslands simplistic explanations are avoided by writing

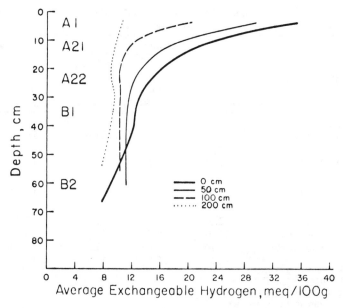

Fig. 13.5. Soil depth functions of exchange acidity at various distances from the stem of a beech tree (21)

$$\frac{\text{grassland,}}{\text{forest}} = f(cl, \phi, r, p, t, \text{fire}, ..)$$

The two biomes may be expected to coexist under a variety of factor combinations, with or without burning.

Sierran Transition Zone

In the arid humid transect of Figure 12.16, pines and grasslands overlap in the interval of maP 60-80 cm. Clusters and groves of ponderosa pines alternate with patches and expanses of grasslands that seem randomized and that may be viewed as a biosequence. The setting is one of long standing because soil criteria harmonize with vegetative cover.

In Table 13.4 the rows of maP and maT document the climatic similarity of the fringe zone. The sums of bases (sba) differentiate parent rocks but not vegetation, whereas % C and C/N respond to grass and pine but not to rocks; aci and pac react to both. All differences are statistically significant at the 1% level.

It should be noted that in opposition to the prairies of the Middle West these interior foothill grassland soils are not dark-colored Mollisols but lighter-colored Mollic Haploxeralfs and Typic Xerochrepts. However, their dark companions on granodiorites along the cooler, foggy coast qualify as such (4).

The column of Table 13.4, "Isolated pines in grass," pertains to tesseras 1.8 m from the trunks of lone, 100- to 200-year-old digger pines (*P. sabiniana*). The soils are

Table 13.4. Soils of Grasslands and Pine Groves in the Arid-Humid Transition Zone of California[a,b]

System properties	Ai soils[c]			Bi soils[d]	
	Grass $(n = 10)$	Pine $(n = 8)$	Isolated pines in grass $(n = 6)$	Grass $(n = 4)$	Pine $(n = 5)$
Annual precipitation (cm)	58-76	66-76	64-66	58-76	58-71
Annual temperature (°C)	13-15	11-12	12-15	13-15	11-16
Sum of bases (sba) (me/100 g)	6.28	7.19	7.63	16.19	18.27
Acidity (aci) (me/100 g)	2.55	8.77	2.49	4.45	13.00
Percentage acidity[e] (pac)	28.9	54.9	24.6	20.8	41.6
Percentage organic carbon	1.03	2.98	1.11	1.31	3.20
C/N	12.8	25.8	14.6	12.5	27.9

[a] Jenny et al. (26).
[b] n = number of sampling sites, which is equal to number of soil samples.
[c] Soils derived from acid igneous rocks.
[d] Soils derived from basic igneous rocks.
[e] Percentage acidity = 100 aci/(sba + aci) = pac.

still the grassland type. The lesson is conveyed that in this Mediterranean-type climate a single pine towering above grass does not make the darker pine soil. Because build-up of humus to steady state is slow, a permanent clump of trees is needed that imprints its mark during generations.

The C/N ratio deserves further comment. For the 52 Ai soils of the transect, C/N widens from Aridisols to the humid Ultisols along the solid, regression line ($r = 0.662$) of Figure 13.6. The tempting inference that decomposer organisms respond to mean annual moisture stress is premature because it overlooks the species alignment with maP. Separating C/N of pine soils from those of grasslands and desert brush splits the regression into two distinct, nearly horizontal lines that spell immunity of C/N to maP variations.

Complementary to the equation of C/N with maP (solid line) is the trivariate equation

$$C/N = 13.50 - 0.00685 \text{ maP} + 17.20\phi \quad (r = 0.902)$$

Insertion of the dichotomous variables Gr = 0 and Pi = 1 produces the two dashed lines in Figure 13.6. In controlling C/N in soils of these acid igneous rocks the plant species are more forceful than either decomposers or leaching.

Conversion of Grassland Soil to Forest Soil

In cool southwestern Alberta the lower slopes of Porcupine Hill consist of sandstone veneered with glacial till. In 1880 they were covered with prairies of rough fescue association (*Festuca scabrella*). Trees have since descended from higher elevations and in-

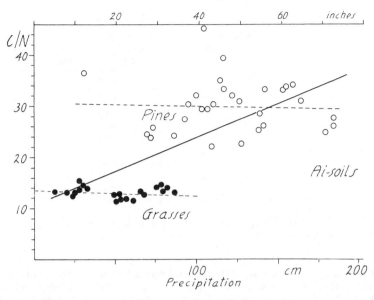

Fig. 13.6. Alignment of soil C/N ratios with precipitation (maP) and flora along the California moisture transect

vaded the original Black Chernozem soil, a dark, humus-rich body with CaCO₃ in the subsoil. Domaar and Lutwick (15) recognize several soil stages induced by tree species:

1. Grass-covered Black Chernozem soil of indeterminate age, chosen as initial state.
2. Dark Gray Chernozem with 45-year-old aspen trees, *Populus tremuloides.*
3. Eluviated Dark Gray Chernozem with 85-year-old balsam poplars, *Populus balsamifera L.*
4. Degraded Brown Wooded soil with a 150-year-old Douglas fir, *Pseudotsuga menziesii.*

Variations in the topography factor are subordinate and their effectiveness is negligible as judged by profile features. Although time and tree species are confounded the authors credit the biotic factor with dominance and designate the four profiles a biosequence.

Under trees the massive, 10- to 25-cm-thick Chernozem Ah horizon alters to light colored, weakly platy Ahe and Aej subhorizons, accompanied by a progressive reduction in total and organic phosphorus. In the lower horizons the sum of exchangeable bases (sba) in Table 13.5 remains invariant throughout the entire sequence whereas in the surficial stratum it declines to one-third of the amount in the initial state. Exchange acidity as pac rises threefold. The Ca/Mg ratio narrows as trees keep replacing grass and the organic matrix virtually collapses, but the clay regime is barely affected. In extracted humus the infrared spectral patterns, which are signposts of molecular configurations, are systematically altered.

For 2 years Domaar (personal communication) leached columns of Chernozem Ah horizon with distilled water, with aqueous extracts of either aspen leaves or

Table 13.5. Grass-Forest Conversion in Southwestern Alberta, Canada[a]

Soils	Sum of bases (sba) (me/100 g)		Percentage acidity (pac)[b]		Ca/ Mg	Percentage N[c]	
	0-20	20-80	0-20	20-80	0-20	0-20	20-80
Black Chernozem	33.2	19.4	12.2	8.5	6.7	0.683	0.148
			(4.6)	(1.8)		(11.7)	(9.6)
Dark gray Chernozem, 45 years	26.7	20.3	17.6	14.7	6.4	0.469	0.135
			(5.7)	(3.5)		(11.4)	(9.5)
Illuviated dark gray Chernozem, 85 years	15.7	20.4	20.7	15.4	5.4	0.204	0.092
			(4.1)	(3.7)		(10.9)	(9.3)
Degraded brown wooded, 150 years	10.7	19.4	31.4	18.5	3.8	0.127	0.110
			(4.9)	(4.4)		(12.9)	(12.3)

[a] Dormaar and Lutwick (15).
[b] pac = 100 · (CEC − sba)/CEC; figures in parentheses are CEC− sba, me/100 g.
[c] Figures in parentheses are C/N ratios.

balsam poplar. Water and aspen extracts barely affect soil pH but balsam leaf extract lowers it from 5.6 to below 4.5. The exchangeable base content (sba) of 28 me/ 100 g is slightly depleted by water, a little more by aspen extract, and decidedly, down to 16-18 me/100 g, by balsam solution.

To dwell on soil genesis, when the fine, fibrous grass roots are replaced by the woody, long-living tree roots the starved invertebrates and microbes deplete the humus reservoir at rates of 0.6-1.4% per year, according to Table 13.5. Concomitantly, CEC of soil is lowered and, aided by acid litter leachates, the liberated bases are washed away.

Stabilized Biotic Contact

St. Arnaud and Whiteside (42) perceive an accomplished biosequence on the gently sloping Touchwood Hills (Saskatchewan) that rise 122 m above the surrounding till plain of late Wisconsinan age. Orthic Black soil, a grassland Chernozem, confronts aspen-covered Gray Wooded soil that occupies slightly higher elevations. Although the small, local climatic divergence favors trees over grass the cl differential in itself is not considered effective in setting the two soils apart; rather the separation resides with grass versus aspen of long standing.

At each of the four sites chosen three sequential C horizons possess remarkably uniform loam textures and carbonate contents that vary from 19 to 23%. Within 80 cm profile depth leaching of $CaCO_3$ rises from 55% under grass to 67-78% under trees, and Ca/Mg in the 0-20 cm surface layer declines from 6.2 to 3.5. As in Alberta, humus is highest under grass and lowest under trees by a two- to threefold margin, the direct opposite of the Sierran foothill situation previously recounted.

The dominant clay minerals are montmorillonite and illite, in that order, and since the latter proportionally increases from C to B to A horizons St. Arnaud and Mortland (41) postulate illitization, a conversion of montmorillonite to illite by way of K-fixation that is favored in the grassland surface soil and is least conspicuous in the densely forested end member of the biosequence.

In the fine earth of the densely timbered Orthic Gray Wooded soils division of non-quartz by quartz minerals gives 0.748 in C, 0.735 in grayish Ae (2.5-13 cm), and 0.648 in clayey Bt (18-46 cm) horizons. Slight weathering of feldspars, micas and horn-blende, and production of silt induced by freezing and thawing, are inferred. Clay/quartz ratios suggest a clay loss of 71% in Ae and a 36% gain in Bt. All wooded profiles indicate clay migration which is corroborated by oriented clay skins in root channels and other voids. Eluviation of clay from A leaves behind banded and platy fabrics, 1.5 mm or less in thickness, that may relate to ice lenses in frozen soil.

This contrast of grassland and forest soils confirms the earlier observations in Minnesota and Illinois (25).

Biosequences in Iowa

North America's "Middle West" has long been the focal area of grass-forest transition research, a claim well substantiated by the 1950 monograph "Prairie Soils of the Upper Mississippi Valley" by Smith et al. (40). Prairie, a luxurious grassland where big bluestem (*Andropogon gerardi*) once commonly reached heights of 180 to 240 cm, coexists with deciduous forest of the genera *Quercus* (oak), *Ulmus* (elm), *Fraxinus* (ash), *Tilia* (basswood), etc., on sites near-identical in undulating topography, parent material (drift and loess), and macroclimate.

At maP of 75-85 cm and maT of 8-10°C, Riecken and collaborators (10,36) arrange over 30 soil series on upland divides into numerous bio- and toposequences, exemplified in Table 13.6. Each slope or drainage class has its own biosequence with transition profiles being younger than the end members. No analytical rigor is applied in proving exact identity of parent materials but the scheme is persuasive because A and B horizons are consistent over large segments of landscapes.

A well-drained biosequence on loess proceeds as

$$\text{Tama} \longrightarrow \text{Downs} \longrightarrow \text{Fayette}$$
$$\text{(prairie)} \qquad \text{(transition)} \qquad \text{(forest)}$$

Soil organic matter is higher under grass than under forest and the C/N ratio is 10-12 in the A horizons and 8-10 in the B horizons. The loss in humus is accompanied by a graying of the lower A horizon. Eventually it becomes A2.

The forest soils have less clay in the A horizons and more in the B horizons than their grassland associates. In a profile the ratio

$$\frac{\text{maximum percentage of clay in a B horizon}}{\text{minimum percentage of clay in an A horizon}}$$

about doubles for the change in the biotic factor (Table 13.6). Trees reduce water permeability and aeration in B which encourages formation of Fe^{2+} and Mn^{2+} cations and concomitant mottling discoloration.

From grass to forest the depth to $CaCO_3$ horizon increases and the exchangeable bases (sba) are lowered. In the poorly drained biosequence in Table 13.6, Webster clay loam of the prairie has 30-40 me/100 g of bases in the As, forested Ames silt loam has only 9-10. The loss in Ca, Mg, and K is coupled with a large gain in exchangeable Al (35), and pac rises from 7-12% in the As of Webster to 33-40% in those of Ames. When

Table 13.6. Arrangement of Soil Types According to Bio- and Toposequences on Glacial Drift in Iowa[a,b]

State factors	Biosequences		
	Prairie	Transition	Forest
Toposequences			Haydon silt
Slopes of 2-8%	Clarion loam	Lester loam	loam
(well drained)	1.04	1.33	2.00
Slopes of 1-3%	Nicollet clay	LeSueur silt	Luther silt
(moderately drained)	loam	loam	loam
	1.02	1.54	1.96
Slopes of 0-1%	Webster clay	Dundas silt	Ames silt
(poorly drained)	loam	loam	loam
	1.01	1.59	2.22

[a] Cardoso (10).
[b] The figures below the names of soil types are ratios of clay in B to clay in A horizons.

tested for plant-available P the forest B horizons exceed their prairie counterparts as much as fivefold (37).

Today's forests respond to the profile features floristically and in growth rates. On the Webster-Ames biosequence in Table 13.6, dominant trees reach heights of 17-18 m in Ames forests and 20-23 m on Dundas intergrades, and higher in planted groves on Webster prairie soil. McComb and Riecken (31) point out that plant and soil diversity evolve simultaneously and no shift in climate need be postulated, but what triggers the biosequence in the first place remains something of a mystery.

In the Trelease Woods Reserve in Illinois, former prairie grassland on loess has been forested for an estimated period of 400-600 years. Geis et al. (20) see transition profiles developing on the better drained sites. These Mollic Hapludalfs reveal appreciable reductions of Ca, Mg, CEC, pH, and organic carbon prior to visible alteration of the physical and morphological countenance.

Unanswered ecopedological queries remain. Not enough is known about evapotranspiration in grassland and adjacent forest to explain the accelerated lowering of the $CaCO_3$ boundary under trees; and, if it is true, as claimed by Ruhe and Scholtes (38), and elaborated in the models of Arnold (2), that the midwestern prairies are younger than 5000 years and that they had been preceded by forests of much longer duration, then a way has to be visualized to change a well-developed forest soil (Alfisol) into one of prairie (Mollisol) with reversal of Ca regime and extent of clay migration. If Walker's alternative of rapid sheet erosion of the ancient forest soil just prior to grass take-over is chosen, as outlined in Chapter 11, problems of slope stability under natural vegetation and soil losses on level ridges come to the fore. Interestingly enough, the pedologists do not invoke fire hypotheses. Yet charcoal is common in such soils.

F. Animal Activities and Impact by Man

In ecology, the small invertebrates and the microscopic fauna and flora are herded together as decomposers, and they have been described as such in Chapter 5. Some of the animals modify the soil in bulk, rivaling the much larger vertebrates. Owing to their mobility, animals are not readily arranged into biosequences, and the presence or abscence of a given species must provide the comparison.

Invertebrates

Crawfish (*Cambarus*) of the Crustaceans live in imperfectly drained soils of the claypan (Albaqualfs) types. To maintain contact with water tables they bury to depths of 3-4 m, penetrating claypan horizons. They push mud, including pebbles, above the ground as "chimneys," 20 cm tall and up to 10-20 cm in diameter, weighing as much as 1.5 kg, with areal densities of $0.1-0.2/m^2$. In Texas Thorp (44) counted up to 12.4 small chimneys per m^2.

On delta lands of West Borneo the mud lobster (*Thalissina anomala*) builds 1- to 1.5-m-high mounds out of sulfidic subsoil materials that oxidize to acid sulfate soil and threaten agricultural development, as reported by Andriesse and van Breemen (16).

Earthworms eat their way through the soil, and they are more common in humid than in arid regions. The famous *Lumbricus terrestris* came to North America from Europe and damaged Chernozem soils in western Canada (44). Worms also destroyed humic crumb structures and horizons in South Dakota (9). Bouché's (7) adult specimen of *L. terrestris* from Versailles has means of 2.69 cm^3 for volume and 2.97 g for fresh weight, 88% being water. On dry weight basis C is 49.0% and N is 10.6%. Satchell's detailed review (39) quotes for rich, grassy sites as many as 500 worms per m^2 with biomass of about 120 g/m^2.

In Austrian laboratories (19) daily consumption of dried hazelnut leaves amounts to 20.4 mg per worm, 52% reappearing in the excrements, partly as humus precursors. For a field density of 100 lumbricids/m^2 the quantity eaten in captivity exceeds twice a tree-litter crop of 300 g/m^2/year.

To a depth of 10 cm Stöckli (43) finds 7-50% of the soil's air capacity (noncapillary pore space) consisting of tunnels and worm cavities. In a Wisconsin (33) deciduous forest soil burrows extend nearly vertically to a depth of 1-2 m and the A1 horizons are lined with leaves pulled in. Casts ejected per year upon the soil surface by *Allolobophora* species amount to 0.2-8 kg/m^2 in Europe, where most of the worm studies are done, and a high of 21 kg/m^2 in the Cameron tropics (29). Many species cast unknown quantities inside the soil.

In some localities Gray-Brown Podsolic soils (Typic Hapludalfs) covered with oaks have slightly calcareous surfaces (pH 7.3) underlain by acid A12 (pH 4.8) and A2 (pH 5.3) horizons. Wiecek and Messenger (48) attribute the inversion to the $CaCO_3$ sacs of earthworm glands and estimate to a depth of 5 cm 5.6 g/m^2 of calcite gland casts.

In Aridisols of the western United States the nymphs of *Cicada insects* of the genera *Platypedia* and *Okanagana* burrow vertical tunnels of about 2 cm diameter to a depth of 1 m, according to Hugie and Passey (24). The burrows are filled with cylindrical cast segments up to 4 cm long, derived from the host horizon. The face of a pit exhibits an abundance of blocky, cylindrical peds and horizons may be saturated with filled passageways. Many peds are calcified.

Stories about *mound-building termites* make fascinating reading. According to pedologist P. H. Carroll (11) who "lived with" *Macrotermes natalensis* in the southwest region of the Ivory Coast, Africa, the soft bodied insects reside in colored mounds that stand 2-6 m in height (Fig. 13.7). Red mounds occupy the hill crests and upper slopes; gray, white, and yellow ones squat in the swales and brown ones, often blunted and rutted by erosion, are deployed along gentle gradients of the long side slopes.

The large castles are prominent in areas of shifting cultivation; deep in the primeval rain forest they are missing. Carroll thinks that the termites build mounds in response to forest clearing and thinning for agricultural use. Accordingly, if large mounds bearing the marks of erosion and old age are found in dense forests, they are suspect of being relics of prehistoric agricultural use.

> It is hard work to slice one of these edifices with its 15- to 22-cm-thick, dense, hard walls, made of selected subsoil material, and to unearth the basement of the fortress and the elaborate 1-m-deep tunneled runways used for foraging expeditions.
>
> To construct the termitaria, the workers haul the clayey portions of the subsoil in their gullets and the sand separates between their mandibles. Upstairs, the clay is regurgitated into neat piles and the insect masons trowel the clay onto the wall with their scimitar-like mandibles. The sand grains then are imbedded into the clay matrix and additional clay is smoothed over the top. When dried out, no pores remain and no cleavages develop.

The mounds are so reliable as indicators of subsoil color and sand/clay ratios of fine earth that in soil survey work they are used as indicators of subsoil textures and drainage conditions. Prospectors analyze them for nickel, molybdenum, or tungsten. Since particles larger than 2 mm are not moved, iron nodules, ironstone, gravel, and quartz fragments in the soil are enriched, gravel concentrations often reaching 80 or 90% by volume in the upper 50-100 cm of soil. It is manifest in "stonelines" in nearly all upland soils of the region.

The narrow subterranean runways and pores also are lined with clay coating, and at first sight they might be taken for illuvial clay films resulting from weathering and clay migration. Under magnification, however, the termitogenous films are uniformly thick and pebbly surfaced, whereas illuviated clay films on oblique pore walls are thickest at the lower end of the wall and exhibit a tallow-flow pattern.

The unrelated but equally busy *ants* are less spectacular mound builders but not less efficient movers of soil particles. In parts of the arid Great Plains area it is not uncommon (44) to observe 50 hills/ha, containing 0.38 kg/m^2 of soil material largely brought up from subsoil horizons (17) often as limy balls which rejuvenate the surface soil. Around each hill, over a radius of 1-3 m, all vegetation is cut. On aerial photos the clearings appear as light-colored dot patterns.

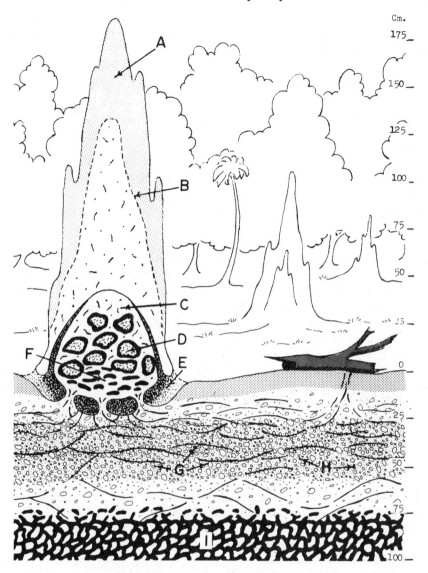

Fig. 13.7. Carroll's drawing of a termitarium and the soil profile below. (A) Outer wall of the earthen nest, (B) interior of mound exposed through sectioning along dashed line, (C) comb, (D) fungus gardens, (E) columned basement, (F) chambers grouped around the queen cell, (G) termite runways, (H) ironstone gravel, and (I) plinthite

Vertebrates

Fossorial species may spend the greater part of their lives underground, digging long runs and tunnels, throwing up mounds, and turning over the soil. Among the mammal burrowers are moles, gophers, ground squirrels, prairie dogs, rabbits, marmots, badgers, mice, and shrews, the armadillo in tropical America, the mole-rat in Africa, and the marsupial mole in Australia. The reptiles are represented by burrowing snakes, lizards, and skinks, and the amphibians by *Caecilia* of the tropics.

Gophers and Mima Mounds. In the grasslands of the western United States are extensive areas of Mima mounds and vernal pools, described in Chapter 11. The mound hills are populated by *gophers* (*Thomomys*), which, in itself, is no proof that the rodents built the mounds; in fact, many other agencies of earth displacements have been championed.

Circumstantial evidence relating to shallow depth of soil, hardpan and claypan, loamy texture, high water table during the wet season, and the animal's living and digging habits convince Arkley and Brown (1) and others (e.g., 8) that generations of gophers are the cause of mound microrelief.

Ground Squirrels and Crotovinas. Calcareous Shedd clay loam and noncalcareous Vista coarse sandy loam of grassy uplands of the Southwest expose empty and partially filled animal burrows, named crotovinas. In the light-colored subsoils they are readily recognized by their distinct granular and darker colored fill-material which resembles A horizons. The channels have diameters of about 10 cm and vary in length from 2 to 13 m. The shapes and dimensions of burrows and nests fit those of the ground squirrel (*Citellus beecheyi*), according to Borst (6). On the average the animals move 123 dm^3 of soil in a season, either to the surface or from new to abandoned burrows. Based on the observations of zoologists, Borst calculates that the squirrels are capable of mixing surface and subsoil horizons to a depth of 75 cm in about 360 years. Other observations provide estimates of 2000 years.

Although these soils mantle old, relatively stable landscapes they have the profile features of young soils, lacking the well-developed argillic horizons of similarly situated upland soils. Borst attributes the immature profile features to the intensity of burrowing and churning.

Prairie Dogs. In Colorado (44) prairie dogs (*Cynomys*) alter clayey subsoil horizons and heap fresh limy or salty material on the surface. On a loessial site the silt loam A horizon has been converted to a loam because the animals bring up sand and gravel from a depth of 2-3 m. A 2 ha field has 84 mounds and the estimated pile-up of earth is 7.28 kg/m^2.

Lemmings and Soil Thaw. The indirect effect of a vertebrate herbivore upon soil is documented by the lemming of the arctic tundra, briefly mentioned in Chapter 9. During the long winters the animal lives, builds nests, and breeds in the grass and sedge layer between frozen soil and cover of snow. In Alaska its preferred nourishment consists of *Dupontia fischeri* and species of *Poa*, *Carex*, and *Eriophorum*. During the brief summer period the snow evaporates and melts and the soil thaws, producing to a depth of about a decimeter an "active soil layer" that enables plants to resume growth. The crucial speed of soil thaw is aided by the foraging activities of the lemmings. At Point Barrow, Alaska, wire exclosures of 3 m^2 area prevent lemmings from consuming the grass biomass inside. The depth of thaw is greater outside than inside (A. M. Schultz, H. Jenny), 12-15 versus 5-8 cm, late in July 1961. (In an adjacent quadrate two holes enable entry of rodents. Inside, the depth readings to ice are 13, 13, 10, 15 cm, outside 13, 13, 15, 13 cm; that is, no significant difference.) In June 1970 the writer measured in fertilized plots 1.45 cm of thaw depth inside and 5.03 cm outside for 10 measurements each, the difference being highly significant. The dense mat of old dead grass inside an exclosure acts as an insulator that retards thawing, whereas outside lemming grazing increases active soil depth and encourages root growth.

Homo sapiens

The biotic factor ϕ pertains to the genetic constitution of an organism. The criterion is insufficient for humans, for our outlook on nature is also guided by the views and economics of the society in which we grow up.

While this book emphasizes soil formation in the absence of modern mankind—*in order to establish benchmarks for our activities*—numerous anthropic situations have been incorporated in preceding chapters, e.g., the losses of soil caused by agricultural and silvicultural erosion, the decline of humus and soil nitrogen during cultivation, the alkalinization of arid lands, and acidification by fertilization.

Soil fertility is being depleted by shipping mineral-rich grains to cities and dumping sewage into the seas. The latest threat is triggered by the energy crisis and proposes plant biomass conversion to fuel (e.g., alcohol). Unless a substantial portion of organic matter is returned to the soil its structure decays and wind and water erosion will soar.

In spite of creeping soil deterioration industrialized society has succeeded in greatly improving plant and animal productions. The outputs keep it well-fed. Nevertheless, injecting crop-stimulating fertilizers into soil does not recreate soil mass lost or restore natural soil structures and soil life.

Man has the power to modify state factors. He changes *climates* locally by altering the heat balance through harvesting of biomass and heating of cities, by aerosol enhancement of cloudiness, and by infringements of the ozone layer. He manipulates the *biotic factor* by cropping and plant breeding, he lowers the water table by drainage installations and raises it by watering, and he remodels entire landscape *topographies* with monsters of earth-moving hardware. New *initial states* arise by confining the sea with polders, by floods caused by dam failures, and by strip mining. *Influxes* toxify soils and plants with lead compounds from car exhausts, with herbicides and pesticides, and by industrial emissions that end up in "acid rain" (14,30).

Man suppresses natural ecosystems by covering the soil with the thick concrete of streets, highways, and airfields and by building endless grids of tract homes and bigger and better factories. Of course, these innovations benefit people in many important ways, some people more than others, but soil as a national resource is seldom taken into consideration.

Lands surrounding the Mediterranean Sea are replete with tragic cases of soil devastation and desertification that used to be attributed to climatic change. Naveh and Dan (32) put the blame squarely on early and late man. We in the United States are striving to do better but we have barely started on the comparable span of 2000 years.

The Global Carbon Issue. Though doomsday predictions are frowned upon, the atmospheric CO_2 *prognosis* is of pedological interest. Since 1860 the CO_2 content of the air has slowly risen from 295 to 330 ppm (0.0033% on a volume basis). The inflationary trend has exponential character and extrapolation predicts 600 ppm by the year 2030 A.D. Because CO_2 molecules absorb the infrared heat radiation that leaves the Earth's surface their multitudes will usher in a warmer climate that expands the deserts toward the poles, melts the ice caps, and raises the world-wide sea level by as much as 50 m (3). Among the causes held responsible for the CO_2 enrichment, highest ranks are accorded to the continuing burning of fossil fuels and the cutting of forests (49). The contributions of soil organic matter appear underestimated. Probably more CO_2 would become oxidized from debris, roots and humus for a number of years after cutting (to regenerated woods) or clearing (to pasture or crops) than would be released promptly by fire and immediate decay (3).

The ways trees and soil humus function as sinks and sources in the carbon cycle is indicated in Figure 13.8. During centuries of primary succession CO_2-fixation stores C in cellulose and lignin of tree boles and roots, and during millennia the soil organisms build up the humus capital. In global estimates the vert space averages (over all land and ice area) approximately 4 kg C/m^2 (3) to 5.6 kg C/m^2 (49) and a similar quantity was once assigned to humus until Bohn (5) proposed for soil space to 1 m depth a mean of 24.6 kg C/m^2. Whatever the figures, vert space and soil space together serve as a sink of global carbon.

After 1000 years of system growth the model of Figure 13.8 assumes a devastating forest fire transforming the stored chemical energy ΔH to entropy and the carbon pool to a source of CO_2. Later, the land is utilized agriculturally and annual field crops provide short-term sink and source sequences of C. Though the fire barely touched tree roots and humus reservoir (28), soil C and N decline in line with the Great Plains and Ohio (25) cultivation curves. Even in the absence of erosion, in less than a century soil space may volatilize more C as CO_2 than the burning of the forest (3). In what proportions fossil fuel burning, deforestation, and humus oxidation contribute to the rise in atmospheric CO_2 is being ardently explored. Determining how the loss fluxes and renewed inputs of carbon vary among stands, regions and times of global history can be clarified with the aid of the organic matter evaluations of Chapters 5 and 9. A major challenge remains: to relate those changes over time to the varied geographic conditions specified in terms of the other factors of climate, parent material, topography and biota (Chapters 10-13).

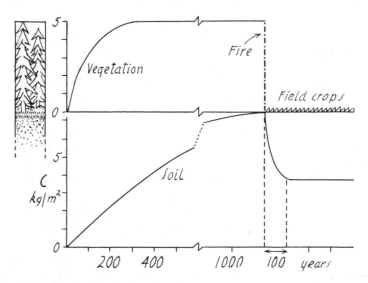

Fig. 13.8. Role of vegetation and soil as sinks and sources in the global carbon cycle

G. Review of Chapter

Trees are seen scattering leaves on the ground, upheaving soil by root penetration, and depleting the water reservoir by transpirational pumping; therefore, we espouse vegetation as a biological soil-forming agent. Since, however, the intensities of actions are soil-conditioned, *vide* rich and poor soils, the role of biota is properly accredited to the genotype features of organisms.

In a manner parallel to rainfall as a state variable the "species rain" is chosen as the biotic factor of an area. In turn, the germules that survive and participate in ecosystem genesis make up the "effective biotic factor" of Crocker (12). (Analogously, one might call runoff the effective rainfall of erosion and Li the effective rainfall of soil leaching.)

The pedogenic impact of organisms is inferred from comparisons of bare and vegetated soil, described for mudflow A of Mt. Shasta, or of soils covered by combinations of species—provided clorpt is satisfied—illustrated by the Iowan biosequences and the grassland-forest transition zones. Man, the willful converter of natural ecosystems to his preferred man-made settings, is lately preserving native fragments that will serve as benchmarks to coming generations of naturalists.

References

1. Arkley, R. J., and H. C. Brown. 1954. *Soil Sci. Soc. Am. Proc.* 18: 195-199.
2. Arnold, R. W. 1965. *Soil Sci. Soc. Am. Proc.* 29: 717-724.
3. Baes, C. F., Jr., H. E. Goeller, J. S. Olson, and R. M. Rotty. 1977. *Am. Sci.* 65: 310-320.
4. Barshad, I. 1946. *Soil Sci.* 61: 423-442.
5. Bohn, H. L. 1976. *Soil Sci. Soc. Am. J.* 41: 468-470.
6. Borst, G. 1968. *Trans. 9th Congr. Soil Sci.* 2: 19-27.
7. Bouché, M. B. 1967. In *Progress in Soil Biology,* O. Graff and J. E. Satchell, eds., pp. 595-600. F. Vieweg, Braunschweig.
8. Branson, F. A., R. F. Miller, and I. S. McQueen. 1965. *Ecology* 46: 311-319.
9. Buntley, G. J., and R. I. Papendick. 1960. *Soil Sci. Soc. Am. Proc.* 24: 128-132.
10. Cardoso, J. 1957. Sequence relationships of Clarion, Lester and Hayden soil catenas. Ph.D. Thesis, State Univ. of Iowa, Ames, IA.
11. Carroll, P. H. 1969. *Soil Surv. Horiz.* 10: 3-16.
12. Crocker, R. L. 1952. *Quat. Rev. Biol.* 27: 139-168.
13. Crocker, R. L., and J. Major. 1955. *J. Ecol.* 43: 427-448.
14. Cronan, C. S., and C. L. Schofield. 1979. *Science* 204: 304-305.
15. Dormaar, J. F., and L. E. Lutwick. 1966. *Can. J. Earth Sci.* 3: 457-471.
16. Dost, H. (ed.). 1973. *Acid Sulphate Soils,* 2 vols. Inst. Land reclam., P.O. Box 45, Wageningen, The Netherlands.
17. Forcella, F. 1977. *Soil Surv. Horiz.* 18: 3-8.
18. Franklin, R. E., P. L. Gersper, and N. Holowaychuk. 1967. *Soil Sci. Soc. Am. Proc.* 31: 43-45.
19. Franz, H. 1950. *Bodenzoologie.* Akad. Verl. Berlin.
20. Geis, J. W., W. R. Boggess, and J. D. Alexander. 1970. *Soil Sci. Soc. Am. Proc.* 34: 105-111.
21. Gersper, P. L., and N. Holowaychuk. 1970. *Soil Sci. Soc. Am. Proc.* 34: 786-794.
22. Gersper, P. L., and N. Holowaychuk. 1971. *Ecology* 52: 691-702.
23. Glauser, R. 1967. The ecosystem approach to the study of Mt. Shasta mudflows. Ph.D. Thesis, Univ. of California, Berkeley.

24. Hughie, V. K., and H. B. Passey. 1963. *Soil Sci. Soc. Am. Proc.* 27: 78-82.
25. Jenny, H. 1941. *Factors of Soil Formation.* McGraw-Hill, New York.
26. Jenny, H., A. E. Salem, and J. R. Wallis. 1968. In *Soil Organic Matter and Soil Fertility,* P. Salviucci, ed., pp. 5-37. Pont. Acad. Sci., Scripta var. 32, Vatican (Wiley (Interscience), New York).
27. Kittredge, J. 1948. *Forest Influences.* McGraw-Hill, New York.
28. Klemmedson, J. O., A. H. Schultz, H. Jenny, and H. H. Biswell. 1962. *Soil Sci. Soc. Am. Proc.* 26: 200-202.
29. Kollmannsperger, F. 1956. *Trans. 6th Int. Congr. Soil Sci.* C: 293-297.
30. McColl, J. G., and D. S. Bush. 1978. *J. Environ. Qual.* 7: 352-357.
31. McComb, A. L., and F. F. Riecken. 1961. *Recent Advances Botany* 1627-1631. University of Toronto Press, Toronto.
32. Naveh, Z., and J. Dan. 1973. In *Mediterranean Type Ecosystems,* F. di Castri and H. A. Mooney, eds., pp. 373-390. Springer-Verlag, New York.
33. Nielsen, G. A., and F. D. Hole. 1964. *Soil Sci. Soc. Am. Proc.* 28: 426-430.
34. Planck, M. 1945. *Treatise on Thermodynamics* (§ 201), 7th ed. Dover, New York.
35. Richardson, J. L., and F. F. Riecken. 1977. *Soil Sci. Soc. Am. J.* 41: 588-593.
36. Riecken, F. F. 1965. *Soil Sci.* 99: 58-64.
37. Riecken, F. F., and B. R. Tembhare. 1978. In *Fifth Midwest Prairie Conf. (1977),* D. C. Glenn-Lewin and R. Q. Landers, eds., pp. 46-50, State Univ., Ames, IA.
38. Ruhe, R. V., and W. H. Scholtes. 1956. *Soil Sci. Soc. Am. Proc.* 20: 264-273.
39. Satchell, J. E. 1967. In *Soil Biology,* A. Burges and F. Raw, eds., pp. 259-322. Academic Press, New York.
40. Smith, G. D., W. H. Allaway, and F. F. Riecken. 1950. *Adv. Agron.* 2: 157-205.
41. St. Arnaud, R. J., and M. M. Mortland. 1963. *Can. J. Agr. Sci.* 43: 336-349.
42. St. Arnaud, R. J., and E. P. Whiteside. 1964. *Can. J. Soil Sci.* 44: 88-99.
43. Stöckli, A. 1949. *Z. Pfl. Düng. Bod.* 45: 41-53.
44. Thorp, J. 1949. *Sci. Monthly* 68: 180-191.
45. Ulrich, A. 1959. *J. Am. Sugar Beet Tech.* 10: 448-458.
46. Voigt, G. K. 1960. *Forest Sci.* 6: 2-10.
47. Wells, P. V. 1970. *Science* 167: 1574-1582.
48. Wiecek, C. S., and A. S. Messenger. 1972. *Soil Sci. Soc. Am. Proc.* 36: 478-480.
49. Woodwell, G. M., R. H. Whittaker, W. A. Reiners, G. E. Likens, C. C. Delwiche, and D. B. Botkin. 1978. *Science* 199: 141-146.
50. Yamaya, K. 1968. In *Biology of Alder,* J. M. Trappe, et al., eds., pp. 197-207. Pac. N.W. For. Range Exp. Sta., Portland, OR.
51. Zinke, P. J. 1962. *Ecology* 43: 130-133.
52. Zinke, P. J., and R. L. Crocker. 1962. *Forest Sci.* 8: 2-11.

14. Integration of Factors and Overview of Book

In preceding chapters the "clorpt" formula was taken apart and *explicit* factor sequences were secured. Now, a case of *implicit* correlation is to be resolved by stratified resampling. Later, previous sequences are put together in multivariate synthesis.

A. Integration of Factors

From Scatter of Points to Collinearity

In a mountainous landscape with wide variations in cl, r, and p, 80 profile tesseras of the California Soil Vegetation Survey range from 4.8 to 6.8 in pH and from 5 to 97% in base saturation (bs), which is 100 sba/(sba + aci). Colloid chemical theory puts pH and bs into mutual contract, but plots for A and B horizons in Figure 14.1 produce

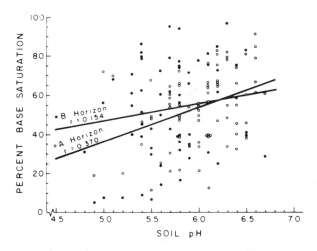

Fig. 14.1. Scatter diagram of percentage base saturation (which is 100 pac) of A and B horizons in 80 Soil Survey profiles having variable state factors. Courtesy Soil Science Society America, Madison, WI

but broad scattering and the very low correlation coefficients (*r*) of 0.370 for A and 0.154 for B that communicate a severe lack of accord.

Blosser (1) resampled the area by controlling state factors, confining topography to southwest exposures and 10-26% slopes, and confining parent rock to greenstone, a metamorphosed, fine-grained ferromagnesian rock. From low to high altitudes maP rises from 69 to 198 cm and maT falls from 17.6 to 10.7°C, and this shift in climate is accompanied by four soil series. Among them bs and pH correlate well, *r* being 0.917 for the 39 B horizons (Fig. 14.2). For the A horizons *r* is 0.836, and for 13 B horizons under grass and shrub vegetation *r* is 0.966. Stratified sampling turns chaos into order.

If 100 random soil samples from the entire United States were subjected to the pot test outlined in Chapter 10, a very low correlation between yield and soil nitrogen would be expected because of the jumble of factors and the multitude of contravening interactions. On the other hand, for the West Coast moisture transect the regression of the 97 relative barley yields (% RY) of Figure 10.2 furnishes % RY = 3.22 + 176.0 N with *r* = 0.79, and when the collection is further stratified to basic igneous rocks (Bi) alone *r* rises to 0.83. When regression is broadened to include four additional soil properties the multiple correlation coefficient for % RY becomes 0.84 for all soils, 0.87 for Bi soils, and 0.92 for forested Bi soils. It is astonishing that five soil properties should explain over 80% of the yield variation when the plant itself needs over a dozen different elements, not to mention suitable soil conditions for microbial activities. The explanation is to be sought in the high collinearity of many soil properties brought about by the stratified sampling procedure of the state factor approach.

Integrated Clorpt Formula

Ideally, the individual time and space successions are recombined and integrated to evaluate the nature of f in the original model l, v, a, s = f(cl, ϕ, r, p, t, . . .). In part it has been done in Chapter 12 in which the individual N-moisture and N-temperature functions of loamy grassland soils are fused into a N-climate surface (Fig. 12.15).

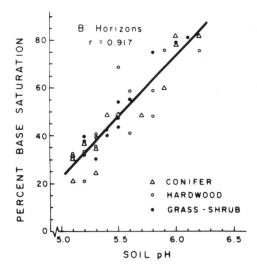

Fig. 14.2. High correlation of percentage base saturation and pH of B horizons of 39 profiles that differ mainly in state factor cl (1). Courtesy Soil Science Society America, Madison, WI

In the West Coast moisture transect the soil properties of the 97 sites fluctuate inordinately, from 0.18 to 10.61% for C, from 0.02 to 0.47% for N, and from 0 to 45 me/100 g for acidity (aci), all populations having high variances. The deterministic approach of clorpt assigns the fluctuations to the variations of the state factors in the transect belt, a postulate that is now put to test.

Within the sampling strip the state factors maP, maT, p, and r are barely related to each other, the highest correlation coefficient r being only 0.25 for dependency of slope on parent rock. Essentially, the four factors are independent, or orthogonal, and in stepwise conventional regression their coefficients are fairly stable.

For the linear model

$$s = a + k_1 \, maP + k_2 \, maT + k_3 p + k_4 \, sl + k_5 \, Fl + k_6 \, Lat$$

the factor contributions as R^2 are listed in Table 14.1. For all four s-properties maP is the most decisive, because most variable, and it explains 80.9% (R^2) of the variance of pac. Parent material is important for N and aci whereas maT and slope (sl) are relatively ineffective, especially for pac. Flora (Fl) influences pac and N adjusts to latitude (Lat).

Granting that soil organic matter is much more dynamic than clays, we may choose the latter as a stable physical and chemical matrix for the former and enter it in the above regression equation. It augments R and R^2 as seen in the two bottom lines in Table 14.1.

A high 95.6% of the pac variability is explained. For N the accord is lower, 68.6%, which is still substantial. Nitrogen may respond to additional factors not taken into account, or the linear model may need modification.

The computer's verdict of tangible linkages of soil properties to the state factors pertains to today's environment. Either the pedologically effective climate has been stable for a long time, or past climates are highly correlated with modern ones, or the chosen soil properties have readjusted themselves to today's precipitation.

Table 14.1. Variability of Soil Properties "Explained" by Variability of State Factors, Expressed as R^2 in Percent[a]

	Soil properties			
	N	C	aci	pac
Climate				
maP	22.8	45.2	45.5	80.9
maT	5.4	9.5	3.6	0.3
Parent rock	16.3	6.0	17.4	2.0
Slope	1.2	1.3	3.5	0.3
Flora	0.2	3.9	4.4	8.9
Latitude	5.9	3.1	1.9	0.4
Sum of R^2	51.8	69.0	76.3	92.8
Total R^2	68.6	75.5	80.4	95.6
Total R	0.823	0.870	0.897	0.978

[a] Jenny et al. (2).

B. Overview of Book

To summarize the book, a view of soil, its genesis, and ecological implications are outlined in the following pages.

Attitudes

For centuries land in Europe has been scarce. Owners of farmland possessed status and
were unlikely to starve. As science emerged its leaders studied and taught soil management (e.g., Sir Humphrey Davy, Justus von Liebig, Jean Baptiste Boussingault).

Meanwhile, in the New World nature had been building up soil fertility and the immigrants encountered an abundance of rich lands. No soil husbandry was needed.
When land gave out it was expendable and more could be had out West.

A wave of soil appreciation was sparked in the New Deal years of the early 1930s.
Erosion was destroying plowed soil of the Great Plains and dust was blown to the Atlantic Seaboard. H. H. Bennett proposed control of wind and water erosion that would
require the help of thousands of people. The idea coincided with President Roosevelt's
efforts to combat unemployment and the Soil Conservation Service was born. Soil became a national concern and soil conservation was taught in schools. In spite of suc-
cesses the movement is subsiding.

As this century is ending, people are becoming aware of resource limitations and
they will want to know more about soils and their responses to environmental change.

Soil Defined, or Not

An exact definition of soil is not given in this book. Popularly, soil is the stratum below the vegetation and above the hard rock, but questions quickly come to mind.
Many soils are bare of plants, temporarily or permanently; or, they may be at the bottom of a pond growing cattails. Soil may be shallow or deep, but how deep? Soil may
be stony, but surveyors exclude the larger stones. Most analyses pertain to fine earth
only. Some pretend that soil in a flower pot is not soil but soil material. It is embarras-
sing not to be able to agree on what soil is. In this the pedologists are not alone. Biolo-
gists cannot agree on a definition of life and philosophers on philosophy.

Comprehending Soil

Soil is a complex body and each step of inquiry reveals greater complexity than imagined. This book views soils in the context of land ecosystems and takes a holistic
stance in recognizing soil as a structured body made up of both biotic and abiotic components.

Soil genesis is examined from two complementary approaches distinguished by
pedologists as soil processes and state factor sequences. The first is based on reductionism (3), the method that reduces complex notions to simpler ones, such as resolving
humus to a set of specific molecules and oxidation to electron transfer. The conceptual tools are those of physics, chemistry, and biology and *Part A, Soil Processes,* deals
with atoms, molecules, colloids, enzymes, organisms, and the forces, potentials, and

reactions between them. The second approach, *Part B, State Factor Sequences,* is phenomenological and seeks to determine how soils in nature vary in space and time. Soil sampling is conducted according to the "clorpt" formula that lines up sites as climatic, organismic, topographic, parent material, and age sequences.

Brief Survey of Processes

Without *water,* soils would resemble the bare, coarse rubble on the moon. On Earth, soil is the water reservoir for land plants and the rates of filling it up and depleting it are crucial for crop growth and watershed management. Water is held in pores under negative pressure or suction and the plant-atmosphere combination must expend work to retrieve it by the roots. Soil texture and structure set limits to water held in available form and to its movement. The *water-balance method* ties soil water and evapotranspiration to climatic variables.

Soil particles (cobbles, gravel, sand, silt, clay) are crystalline or amorphous *assemblages of ions,* with Si^{4+}, Al^{3+}, O^{2-}, OH^- serving as frameworks and Na^+, K^+, Mg^{2+}, Ca^{2+} and many others as accessory companions. Rock crystals react with water and weather to minute *colloidal clay particles* that are either aluminosilicates (Si, Al, O, OH, K) such as montmorillonite and kaolinite or oxides such as gibbsite ($Al(OH)_3$) or goethite ($FeO(OH)_2$). On clay surfaces the ions Na^+, K^+, Ca^{2+}, Mg^{2+} are chemically reactive and *replaceable* by H^+ cations of the soil solution which derive from metabolic activity of roots and microbes and which are accompanied by such anions as HCO_3^-, NO_3^-, and $H_2PO_4^-$. Together the dissolved ions make up the *soil solution* which is a *supplier of nutrients to plants.* Particles that are not very soluble may interact with roots directly. The ions taken up by plants are returned to soil by leaf fall and by death and are then *recycled.*

Like clay, *humus* is a cardinal soil feature. The origin resides in the vert space where light rays impinge on chlorophyll molecules of green leaves and, jointly with carbon dioxide (CO_2) and water (H_2O), create sugars, carbohydrates, and many other organic compounds. When leaves fall to the ground they are macerated by insects, spiders, mites, and worms. Bacteria and fungi excrete enzymes that reduce the macromolecules to smaller units which then recombine to brown and black *humus substances* of colloidal size. They are studded with chemically active carboxyl (–COOH) and phenolic (–COH) groups. These humic and fulvic acids participate in ion exchange reactions and they solubilize Fe^{3+} and Al^{3+} ions. In some soils annual production of humus and its loss by oxidation are approximately balanced, humus then being at "steady state."

This broad spectrum of organic reactions owes part of its existence to a few bacterial species that possess the power of *fixing inert nitrogen gas* (N_2) from the air and converting it to ammonia and cellular amino acids. Nitrifying soil microbes oxidize ammonia to nitrate which is a macronutrient of the plant kingdom.

Additions of acids and neutral salts to clay and humus suspensions in pores or ponds initiate *flocculation* and the particles settle out as a gel. Conversely, alkali added to a gel will often *disperse* it to a suspension. Clay and humus particles themselves may flocculate and disperse each other. All these interactions shape the character of a soil, whether impermeable or open structured, stable or unstable. Thus, upon wetting a dry, montmorillonite clay soil the colloidal particles hydrate, repel one another, and cause *soil swelling* that may damage roots, move stones and fence posts, and induce slipping and slumping on slopes. Dispersed clay particles migrate from A to B horizons and in the extreme generate an impenetrable *clay pan layer*. When drying out it cracks into massive vertical columns.

In excessively wet soils iron and manganese are reduced, $Fe^{3+} \rightarrow Fe^{2+}$, $Mn^{3+} \rightarrow Mn^{2+}$, and become mobile. The ions congregate to brownish *mottles* and concretions that alternate with bleached light-gray matrix domains. The process is *gleysation*. In acid forest and brush lands humus complexes Fe^{3+} and Al^{3+} and in downward transfer creates whitish A2 horizons and Fe, Al-rich B horizons that may convert to iron hardpans. The process is *podzolization*. On old land surfaces in the humid tropics long-time leaching depletes the soil of bases and silica. Iron and aluminum oxides are left behind and the residues may harden to a laterite crust meters in thickness. The process is *laterization*. In soils of arid regions carbonate ($CaCO_3$) dissolved by rain water is carried downward to the depth of water penetration. During the dry season evapotranspiration dissipates the water, leaving the carbonates behind as seams and nodules of the *carbonate horizon of pedocals*.

Brief Survey of Soil Sequences (Pedogenic Functions)

Climate (cl), organisms (ϕ), topography (r), parent material (p), age or time (t), and any additional factors (...) are brought together in the clorpt formula as $s=f(cl,\phi,r,p,t,...)$ where s is any soil property. In sequences one of the factors is made to vary prominently whereas the remaining ones vary but little.

Segments of landscapes that possess similar values of cl, ϕ, r, p, ... but differ in dated ages, t, are lined up as *chronosequences*. They furnish reaction rates for long-time clay and humus productions, soil leaching, and horizon genesis. Accruement rates of total nitrogen in soil build-up are much less than predictions based on laboratory and short-time field tests. In the humid region soil genesis on level land surfaces leads eventually to unproductive ecosystems with endemic species.

Soils derived from different parent materials (p) are compared as *lithosequences*, provided all sites possess the same combinations of cl, ϕ, r, t, ... Granite and adjacent serpentinite ecosystems provide striking contrasts.

In *toposequences* topography (r) is the variable under similar constellations of cl, ϕ, p, t, ... Soils and vegetation are viewed along a slope or at different aspects. Downhill movement of soil constituents, including runoff-erosion and sliding, are dominant processes. Depth to water table is very important in the formation of salty and alkaline soils.

The century-old recognition of climate (cl) as a soil-forming factor is properly assessed as *climosequences* at comparable r, p, and t (or at steady state) and the sites must be accessible to a common influx of species (ϕ). Low temperatures tend to favor accumulations of humus, high ones enhance rates of weathering. With increasing precipitation (at a given temperature) humus rises, carbonates descend to greater soil depths, and leaching accelerates.

The role of organic species, the *biotic factor* (ϕ), at similar values of cl, r, p, t, ... is drastic in the early phases of ecosystem genesis. Particularly the nitrogen-fixing plants, by themselves or in cohabitation, acquire a head start in creating the soil humus capital. In the semihumid region of North America the invasion of forests into grasslands initiates humus oxidation, clay migration, and $CaCO_3$ translocation.

Since humans (ϕh) need minerals, nitrogen, and carbon compounds they must "live off the land." Often they treat the soil unwisely. In describing soil degradation in the Mediterranean region Naveh and Dan (4) successfully arrange the impact of man in the framework of state factors.

Ecological Perspectives

Since ecology champions the interplay of life and environment it would seem that the coupling of vegetation and soil should draw plant ecologists and pedologists together, but beyond agronomy and to some extent forestry collaboration has not been overwhelming.

Not surprisingly, soil microbiologists and soil zoologists are deeply involved in soil studies. Microbe-induced emissions of nitrous oxide (N_2O) from soil to the ozone layer in the stratosphere and health hazards of nitrate pollution evoke public interest. To physiologists litter and egesta are waste products, to ecologists they are vital cogs in ecosystem dynamics.

The ancient botanical riddle of *transpiration* is being broadened to the *soil-plant-atmosphere system,* and the soil physicist's concept of water potential is finding new opportunities. To what extent the ascent of soil water in plants serves as a vehicle of nutrient transport from roots to leaves, and what fraction of ions moves from old cells to new, is basic to growth and nutrient cycling.

Concern with *nutrition of plants in soils* focuses on the *root-soil boundary zone* which turns out to be vastly more complex than once believed. The bridges from cell wall to clay particles by way of mucigel and mycorrhiza hyphae intrigue both microbiologists and soil chemists. *Migration of nutrient ions to root surfaces* by water percolation, mass flow, diffusion in solution and on clay surfaces, including contact and chelation still has surprises in store. These mechanisms must be solved if we want to understand iron chlorosis on calcareous soils or the serpentine syndrome.

The International Biological Program supported the measuring of *production of biomass* which is a precursor to humus. Turnover rates of root mass will help pedologists to comprehend carbon and nitrogen depth functions and fluxes. The long-standing observation of high *soil nitrogen reservoirs* (compared to plant biomass) induced foresters and a school of ecologists (Ellenberg's) to elevate the element to a potent site factor for plants. The realization that *soil carbon storage* may exceed the amount of carbon in vegetation is aiding in assessments of ecosystems as sources and sinks of CO_2 in the global carbon problem.

Except for Whittaker's (5) gradient analysis most *vegetation sequences* stress time successions and climatic functions of species and biomass. Control of r and p and explicit involvement of soils would broaden and strengthen the conclusions. State factor analysis might be a tool for disentangling the causations of coexistence of natural grasslands and forests. Promising work lies ahead in clarifying the idea of plant climax and its imaginary twin the soil climax or mature soil. The possible transformations of a climatic climax to an edaphic one, as graphed at the end of Chapter 9, challenges ecologic and pedologic thought.

Climatic change is cited as a forceful agent in the creation of new species. The notable concentration of rare species, subspecies, and ecotypes on stress-soils advocates a role of soil genesis in organic evolution.

C. Review of Chapter

Pedogenic order in a landscape is unraveled by *stratified* random sampling along vectors of state factors.

The premise that soils in a given landscape are the result of particular constellations of cl, ϕ, r, p, t, ... is tested for the Pacific Coast transect. The individual factor roles that had been sorted out are reassembled by multivariate synthesis. The combination explains to a high degree the variability of organic matter (N, C) and acidity (aci, pac).

A brief overview of the entire book is included.

References

1. Blosser, D. L., and H. Jenny. 1971. *Soil Sci. Soc. Am. Proc.* 35: 1017-1018.
2. Jenny, H., A. E. Salem, and J. R. Wallis. 1968. In *Soil Organic Matter and Soil Fertility*, P. Salviucci, ed., pp. 5-37. Pont. Acad. Sci., Scripta var. 32, Vatican.
3. Lacey, A. R. 1976. *A Dictionary of Philosophy*. Routledge and Kegan Paul, London.
4. Naveh, Z., and J. Dan. 1973. In *Mediterranian Type Ecosystems*, F. di Castri and H. A. Mooney, eds., pp. 373-390. Springer-Verlag, New York.
5. Whittaker, R. H. 1967. *Biol. Rev.* 42: 207-264.

Appendix I. Names of Vascular Species Cited

Abies concolor (Gord. & Glend.) Lindl.	White fir
Abies grandis (Dougl.) Lindl.	Grand fir
Abies magnifica A. Murr.	Red fir
Acer pseudoplatanus L.	Sycamore
Acer saccharum Marsh.	Sugar maple
Agathis australis Hort. ex Lindl.	Kauri pine
Allenrolfea occidentalis (Wats.) Kuntze	Iodine bush
Alnus crispa (Ait.) Pursh	Alder
Alnus inokumai Murai a. Kusaka	Alder
Ammophila arenaria (L.) Link.	Beach or marram grass
Andropogon furcatus Muhlenb. =	
Andropogon gerardi Vitm.	Big bluestem
Andropogon scoparius Michx. =	
Schizachyrium scoparium (Michx.)	Little bluestem
Arctostaphylos myrtifolia (Parry)	Ione manzanita
Arctostaphylos nummularia Gray	Ft. Bragg manzanita
Arctostaphylos uva-ursi (L.) Spreng.	Bearberry
Artemisia tridentata Nutt.	Sage brush
Atriplex confertifolia (Torr. & Frém.) Wats	Shadscale
Atriplex cordulata Jeps.	Gray saltbush
Berberis nervosa Pursh.	Oregon grape
Betula alleghaniensis Britt.	Yellow birch
Betula verrucosa J.F. Ehrh. = *B. pendula* Roth	Birch
Brassica nigra (L.) Koch.	Mustard
Bromus rigidus Roth.	Brome grass
Calluna vulgaris (L.) Hull	Heather
Calcocedrus decurrens (Torr.) Florin =	
Libocedrus decurrens Torr.	Incense cedar
Carex maxima Scop. = *Carex pendula* Huds.	
Corylus avellana L.	Hazel
Cryptomeria japonica (L.f.) D. Don	Cedar, Sugi
Cupressus pygmaea (Lemmon) Sarg.	Mendocino cypress
Cynodon dactylon (L.) Pers.	Bermuda grass

Deschampsia flexuosa (L.) Trin.	Common hair grass
Erica carnea L.	Heather
Erica tetralix L.	Heather
Eriogonum apricum J.T. Howell	Wild buckwheat
Eriophorum vaginatum L.	Cotton grass
Eucalpytus globulus Labill.	Eucalyptus
Fagus grandifolia J.F. Ehrh.	Beech
Fagus sylvatica L.	Beech
Festuca arizonica Vasey	Fescue
Festuca scabrella Torr.	Rough fescue
Fraxinus excelsior L.	Ash
Gaultheria shallon Pursh.	Salal
Gaylussacia baccata (Wangenh.) C. Koch.	Huckleberry
Juniperus nana Willd. = *J. communis* L. ssp.	
nana Syme in Sowerby	Juniper
Lactuca sativa L. var. *Romaine*	Romaine lettuce
Larix occidentalis Nutt.	Western larch
Ledum glandulosum Nutt.	Labrador tea
Leptospermum ericoides A. Rich	Manuka shrub
Liriodendron tulipifera L.	Tulip poplar
Lolium italicum A. Braun = *L. multiflorum* Lam.	Rye grass
Medicago sativa L.	Alfalfa
Mercurialis perennis L.	Mercury
Muhlenbergia montana (Nutt.) Hitchc.	Muhly
Nardus stricta L.	Mat grass
Phacelia californica Cham.	Phacelia (herb)
Picea abies (L.) Karst.	Norway spruce
Picea engelmannii Parry ex Engelm.	Engelman spruce
Picea sitchensis (Bong.) Carr.	Sitka spruce
Pinus banksiana Lamb.	Jack pine
Pinus contorta Dougl. ex Loud.	Beach pine
Pinus contorta (Parl.) Vasey var. *bolanderi*	Bolander pine
Pinus jeffreyi Grev. & Balf. in A. Murr.	Jeffrey pine
Pinus longaeva D.K. Bailey	Bristlecone pine
Pinus muricata D. Don.	Bishop pine
Pinus ponderosa Dougl. ex P. & C. Lawson	Ponderosa pine
Pinus sabiniana Dougl.	Digger pine
Pinus strobus L.	White pine
Pinus sylvestris L.	Scots pine
Pinus taeda L.	Loblolly pine
Poa pratensis L.	Kentucky bluegrass
Populus balsamifera L.	Balsam poplar
Populus tremula L.	Aspen
Populus tremuloides Michx.	Aspen
Populus trichocarpa T. & G.	Poplar

Pseudotsuga menziesii (Mirb.) Franco.	Douglas fir
Pteridium aquilinum (L.) Kuhn	Bracken fern
Purshia tridentata (Pursh) DC.	Antelope bush
Quercus agrifolia Neé	Coast live oak
Quercus alba L.	White oak
Quercus douglasii H. & A.	Blue oak
Quercus ilex L.	Holly oak
Quercus kelloggii Newb.	Black oak (western)
Quercus petraea L. ex Liebl.	Durmast oak
Quercus robur L.	Truffle oak
Quercus velutina Lam.	Black oak (eastern)
Quercus wislizenii A.DC.	Interior live oak
Rhododendron ferrugineum L.	Alpenrose
Rhododendron macrophyllum D. Don	Rose-bay
Sarcobatus baileyi (Cov.) Jeps.	Greaswood
Salicornia subterminalis Parish	Pickleweed
Salix barclayi Anders.	Willow
Sequoia sempervirens (D. Don) Endl.	Coast redwood
Sequoiadendron giganteum (Lindl.) Buchh.	Sierra redwood
Sesleria coerulea (L.) Ard.	
Sorbus aucuparia L.	Mountain ash
Tilia americana L.	Basswood
Trifolium repens L.	White clover
Tsuga heterophylla (Raf.) Sarg.	Western hemlock
Vaccinium ovatum Pursh	Huckleberry
Vaccinium parvifolium Sm. in Rees	Huckleberry
Vaccinium uliginosum L.	Black bilberry
Vaccinium vitis-idaea L.	Redberry
Zea mays L.	Corn, Maize

[1] Arranged courtesy Alice Q. Howard, Herbarium, University of California, Berkeley, California.

Index